AF342656

Petroleum Geoscience of the West Africa Margin

The Geological Society of London
Books Editorial Committee

Chief Editor
RICK LAW (USA)

Society Books Editors
JIM GRIFFITHS (UK)
DAVE HODGSON (UK)
PHIL LEAT (UK)
NICK RICHARDSON (UK)
DANIELA SCHMIDT (UK)
RANDELL STEPHENSON (UK)
ROB STRACHAN (UK)
MARK WHITEMAN (UK)

Society Books Advisors
GHULAM BHAT (India)
MARIE-FRANÇOISE BRUNET (France)
ANNE-CHRISTINE DA SILVA (Belgium)
JASPER KNIGHT (South Africa)
MARIO PARISE (Italy)
SATISH-KUMAR (Japan)
VIRGINIA TOY (New Zealand)
MARCO VECOLI (Saudi Arabia)

Geological Society books refereeing procedures

The Society makes every effort to ensure that the scientific and production quality of its books matches that of its journals. Since 1997, all book proposals have been refereed by specialist reviewers as well as by the Society's Books Editorial Committee. If the referees identify weaknesses in the proposal, these must be addressed before the proposal is accepted.

Once the book is accepted, the Society Book Editors ensure that the volume editors follow strict guidelines on refereeing and quality control. We insist that individual papers can only be accepted after satisfactory review by two independent referees. The questions on the review forms are similar to those for *Journal of the Geological Society*. The referees' forms and comments must be available to the Society's Book Editors on request.

Although many of the books result from meetings, the editors are expected to commission papers that were not presented at the meeting to ensure that the book provides a balanced coverage of the subject. Being accepted for presentation at the meeting does not guarantee inclusion in the book.

More information about submitting a proposal and producing a book for the Society can be found on its website: www.geolsoc.org.uk.

GEOLOGICAL SOCIETY SPECIAL PUBLICATION NO. 438

Petroleum Geoscience of the West Africa Margin

EDITED BY

T. SABATO CERALDI
BP Exploration, UK

R. A. HODGKINSON
Dana Petroleum, UK

and

G. BACKE
BP Exploration (Epsilon) Ltd, Oman

2017

Published by
The Geological Society
London

THE GEOLOGICAL SOCIETY

The Geological Society of London (GSL) was founded in 1807. It is the oldest national geological society in the world and the largest in Europe. It was incorporated under Royal Charter in 1825 and is Registered Charity 210161.

The Society is the UK national learned and professional society for geology with a worldwide Fellowship (FGS) of over 10 000. The Society has the power to confer Chartered status on suitably qualified Fellows, and about 2000 of the Fellowship carry the title (CGeol). Chartered Geologists may also obtain the equivalent European title, European Geologist (EurGeol). One fifth of the Society's fellowship resides outside the UK. To find out more about the Society, log on to www.geolsoc.org.uk.

The Geological Society Publishing House (Bath, UK) produces the Society's international journals and books, and acts as European distributor for selected publications of the American Association of Petroleum Geologists (AAPG), the Indonesian Petroleum Association (IPA), the Geological Society of America (GSA), the Society for Sedimentary Geology (SEPM) and the Geologists' Association (GA). Joint marketing agreements ensure that GSL Fellows may purchase these societies' publications at a discount. The Society's online bookshop (accessible from www.geolsoc. org.uk) offers secure book purchasing with your credit or debit card.

To find out about joining the Society and benefiting from substantial discounts on publications of GSL and other societies worldwide, consult www.geolsoc.org.uk, or contact the Fellowship Department at: The Geological Society, Burlington House, Piccadilly, London W1J 0BG: Tel. +44 (0)20 7434 9944; Fax +44 (0)20 7439 8975; E-mail: enquiries@geolsoc.org.uk.

For information about the Society's meetings, consult *Events* on www.geolsoc.org.uk. To find out more about the Society's Corporate Affiliates Scheme, write to enquiries@geolsoc.org.uk.

Published by The Geological Society from:
The Geological Society Publishing House, Unit 7, Brassmill Enterprise Centre, Brassmill Lane, Bath BA1 3JN, UK

The Lyell Collection: www.lyellcollection.org
Online bookshop: www.geolsoc.org.uk/bookshop
Orders: Tel. +44 (0)1225 445046, Fax +44 (0)1225 442836

The publishers make no representation, express or implied, with regard to the accuracy of the information contained in this book and cannot accept any legal responsibility for any errors or omissions that may be made.

British Library Cataloguing in Publication Data

A catalogue record for this book is available from the British Library.
ISBN 978-1-78620-243-7
ISSN 0305-8719

Distributors
For details of international agents and distributors see:
www.geolsoc.org.uk/agentsdistributors

Typeset by Nova Techset Private Limited, Bengaluru & Chennai, India.
Printed and bound by CPI Group (UK) Ltd, Croydon CR0 4YY

Contents

Acknowledgements

The Geological Society thanks the following companies for their generous sponsorship of the Petroleum Group and the conference that led to this volume. The Petroleum Group is thanked for its contribution towards the colour printing in this volume.

Petroleum Group corporate sponsors at the time of the conference:

Sponsors of Petroleum Geoscience of the West Africa Margin conference:

The petroleum geology of the West Africa margin: an introduction

TERESA SABATO CERALDI[1]*, RICHARD HODGKINSON[2] & GUILLAUME BACKÉ[3]

[1]*BP Sunbury, Chertsey Rd, Sunbury on Thames TW16 7LN, UK*

[2]*Dana Petroleum, Kings Close, 62 Huntly Street, Aberdeen AB10 1RS, UK*

[3]*BP Oman, Bait Al Salam, STS Building – Building 75B – 2nd Floor, Dohat Al Adab Street, Way Number 42136, PO Box 1649 Madinat Qaboos, PC 130 (Al Azaiba) Muscat, Sultanate of Oman*

**Corresponding author (e-mail: teresa.ceraldi@uk.bp.com)*

Abstract: The continental margin of West Africa formed as result of the south-to-north rifting of Gondwana and the progressive separation of the South American and African continents. This margin has enjoyed a rich and varied exploration history and, in the 70 years or so since the first significant exploration began in the onshore area, the margin has emerged as a significant hydrocarbon-producing region. The amalgamation of hydrocarbon exploration approaches and imaginative ideas, leveraged with modern technologies, is yielding significant scientific and economic successes.

The main objective of this Special Publication is to provide an overview of the advances in our understanding of the crustal structure, tectonic evolution and Mesozoic to Cenozoic stratigraphy of the West Africa margin both onshore and offshore, with a particular focus on the petroleum geology. The papers contained in this Special Publication represent a selection from the 37 abstracts presented at the original conference in March 2014, which covered the entirety of the margin from South Africa to Morocco as well as stratigraphy from the crystalline basement to the most recent strata. The original abstracts from the conference are available through the Geological Society of London website at: http://www.geolsoc.org.uk/pgresources

The West Africa continental margin is defined as the coastal area extending from Morocco in the north to the Cape of Good Hope in South Africa in the south (Fig. 1). Twenty-three countries share the coastline along the West Africa margin, and although some of these countries have vast continental interiors (e.g. Democratic Republic of Congo), the coastline and territorial waters of the countries are actively explored for hydrocarbons and have been found to possess large economic resources, capturing academic and industrial interest. Presently, nine of these 23 countries in Africa are oil exporters: Nigeria, Angola, Equatorial Guinea, Congo (Brazzaville), Gabon, Cameroon, Ivory Coast, Democratic Republic of Congo (DRC) and Mauritania (US Energy Information Administration (EIA), http://www.eia.gov/).

The West Africa margin is one component of continental Africa, which formed as a direct result of the rifting of Gondwana and the progressive separation of the South American and African continents.

The South Atlantic can be divided into four segments (Moulin *et al.* 2005) from south to north: the 'Falkland Segment' (not covered in this volume), the 'South Segment', the 'Central Segment' and the 'Equatorial Segment' (Fig. 1), all of which are characterized by different tectonic, crustal and volcanic features. The South Segment, bounded to the south by the Agulhas–Falkland fracture zone and to the north by the Florianopolis Fracture Zone and the Walvis Ridge, is characterized by significant volcanic features identified in surface and subsurface data, such as extensive seaward-dipping reflectors. Evaporitic material, including the precipitation of a large volume of halite, filled the Central Segment during the Aptian, which is separated from the South Segment by the Florianopolis Fracture Zone and bounded to the north by the Romanche Fracture Zone. The Equatorial segment, separated from the Central Segment by the Romanche Fracture Zone and limited to the north at about the 10°N parallel near the Vema Fracture Zone, corresponds to a broad zone of lateral shearing in a transform margin setting (Torsvik *et al.* 2009).

These various tectonic domains along the West Africa margin are discussed the papers compiled in this volume.

From: SABATO CERALDI, T., HODGKINSON, R. A. & BACKE, G. (eds) 2017. *Petroleum Geoscience of the West Africa Margin*. Geological Society, London, Special Publications, **438**, 1–6.
First published online August 12, 2016, http://doi.org/10.1144/SP438.11
© 2017 The Author(s). Published by The Geological Society of London.
Publishing disclaimer: www.geolsoc.org.uk/pub_ethics

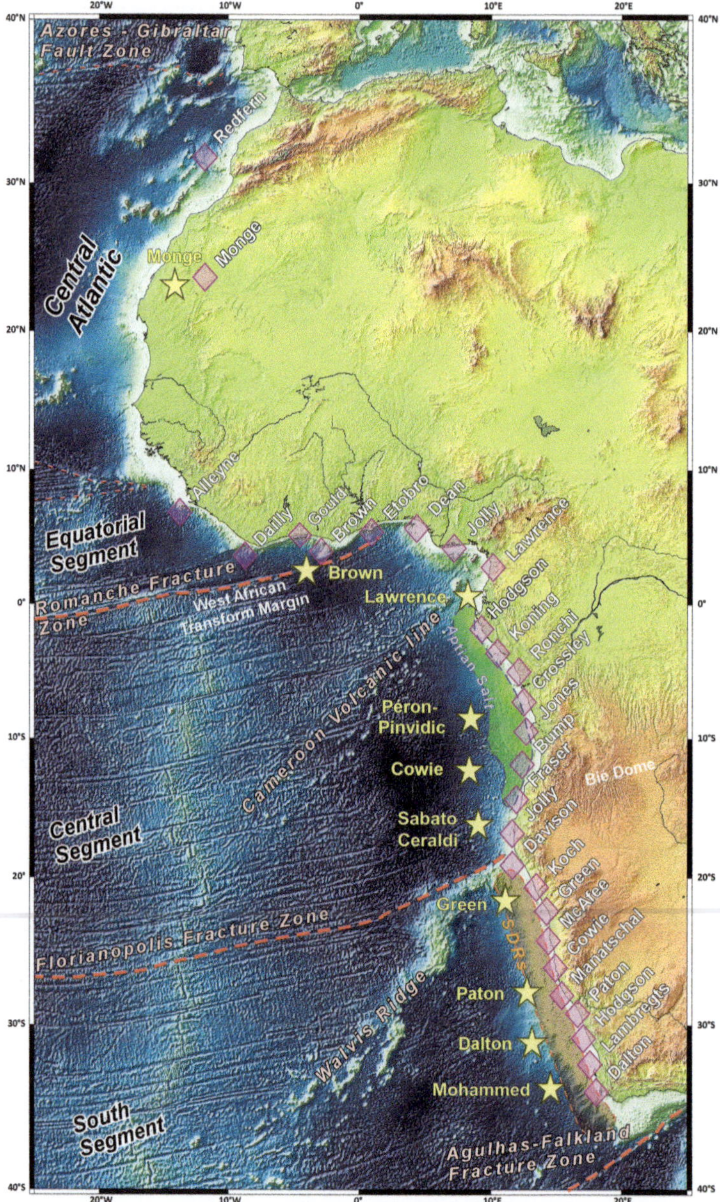

Fig. 1. Location map for the papers presented in this volume. The different tectonic domains and their structural boundaries along the margin are also indicated (Moulin *et al.* 2005; Torsvik *et al.* 2009). The green shading along the Angolan section of the Central Segment corresponds to the main area of salt deposition in the Aptian. The orange shading in the South Segment, along the Namibian and South African coast corresponds to the area where seismic reflection data display clear evidence for the presence of seaward-dipping reflectors (SDRs). Stars indicate the papers included in this volume, and purple diamonds show the published abstracts related to the original conference, available online from the Geological Society of London website (http://www.geolsoc.org.uk/pgresources; (Note: only the surname of the first author is used for each symbol).

The most widely accepted model for the opening of the Atlantic Ocean corresponds to the classic rift-to-drift transition. Rifting starts initially in the south during the Late Jurassic and propagates northwards, in effect 'unzipping' the palaeocontinent of Gondwana. Therefore, rifting in the Equatorial segment of the margin only initiated in the Late Cretaceous (Rabinowitz & LaBrecque 1979; Emery & Uchupi 1984).

In spite of the general agreement around this empirical model, the early stages of continental rifting, as well as the pre-drift positions of the major tectonic blocks forming the margin, remain a matter of debate. In the various models proposed in the literature, the locations and magnitudes of the gaps and overlaps between the reconstructed continental blocks differ greatly (Karner & Driscoll 1999; Karner *et al.* 2003; Aslanian *et al.* 2009; Moulin *et al.* 2010).

The quest for an integrated global model is economically as important for Africa and the world as it is scientifically. The locations and volumes of possible hydrocarbon reserves are underpinned by the tectonostratigraphic evolution of the margin. According to independent volumetric estimates, the West Africa margin holds the majority of the liquid hydrocarbon in sub-Saharan Africa (US EIA).

The synergy between academic studies and hydrocarbon exploration of the West Africa margin has led to the establishment of one of the most complete and varied datasets available in Earth sciences, and has brought together a large pool of experts in many fields. As a result, the wealth of data currently available to interpreters and modellers alike has considerably changed our understanding of the geometry, evolution, structure and depositional patterns of the margin and related sedimentary basins at all scales (Fig. 1). The interpretation of satellite-derived gravity data, for example, is enabling the development of basin- or margin-scale models that permit the characterization of the general structure and development of the margin (see **Péron-Pinvidic** *et al.* **2015**; **Cowie** *et al.* **2016**; **Paton** *et al.* **2016**). In the meantime, the interpretation of high-resolution 2D and 3D seismic data provides a detailed view of the stratigraphy, structures and volcanic bodies comprising these basins (see **Brown** *et al.* **2015**; **Dalton** *et al.* **2016**; **Lawrence** *et al.* **2016**; **Paton** *et al.* **2016**; **Sabato Ceraldi & Green 2016**). Finally, the wealth of drilling-related and outcrop data for the onshore region allow the characterization of the timing and amplitude of tectonic changes along the West Africa margin (see **Green & Machado 2015**; **Martín-Monge** *et al.* **2016**).

The timing of the conference was also coincident with the beginning of the exploration of the pre-salt stratigraphy in Angola. The industry focus on this new promising play had a strong bearing on the content of this volume.

This Special Publication is also characterized by an equal share of papers submitted by authors from both academic and industry backgrounds. This is certainly one of the key achievements of the conference, which enabled a high level of participation of colleagues from both academia and industry.

Between the conference in 2014 and the time of publication of the present volume, the petroleum industry has endured a period of drastic changes with the drive-down in the oil price by over 75% (from *c.* US$105 per barrel to *c.* US$35 per barrel in February 2016). These changes, principally implemented to install efficiencies into the industry to cope with a low-price environment, had a significant impact on the submission, review and final publication of the ten selected manuscripts. The editors would especially like to acknowledge the authors and the reviewers for their time and efforts. Without their support in these tough times for the industry, this volume would simply never have been published.

Summary of the selected papers

The original rationale for the Geological Society of London conference on the Petroleum Geoscience of the West Africa Margin in April 2014 was to exhibit some of the recent geoscientific work that had been undertaken along the margin. Previous publications from the Geological Society that focused on the West Africa margin, such as Cameron *et al.* (1999) and Arthur *et al.* (2003), established the general tectonic, stratigraphic and petroleum systems fabric of the margin. In the years following the latest Special Publication on the topic, significant progress and step-changes in the understanding of the margin plainly justified the organization of a new meeting. The conference highlighted the advances in the application of remote sensing technologies and 3D seismic data, which have allowed geoscientists to further interrogate the structural definition of the West Africa margin. Developing an understanding of the timing and characterization of the tectonostratigraphic variations both along strike, and between the conjugate components of the Atlantic margins, has proven to be a major step forwards since the millennium, with the Brazilian pre-salt discoveries being used as direct analogues for the Angolan sub-salt prospectivity. It was also shown that the geoscientific understanding of the margin is increasingly drawing on changes in climate as a contributing factor to prospectivity. Although perhaps a product of the post-millennium *Zeitgeist*, climate effects are directly interpreted to contribute not only in term of variations in the sediment flux along the margin through time, but also to the composition and diagenetic characteristics upon burial.

The selection of papers included in this volume is a product of the conference, drawing on these developments over the last 15 years. The papers show that as a geoscientific community, we are progressing with our overall understanding of the geological aspects of the margin and in many cases – through the synthesis of various data – are able to decipher and unlock the hydrocarbon potential of the West Africa margin.

The papers are organized as a north–south transect through the different tectonic domains along the margin. The following section presents the highlights of each manuscript in this order.

The paper by **Martín-Monge *et al.* (2016)** is the only one that focuses on an intracratonic basin, describing the hydrocarbon potential of the Taoudeni Basin (Mauritania) – the largest sedimentary basin in Africa. The basin is also a largely unexplored region; however, a well drilled in 1974 (Abolag-1) tested both a gas and a liquid hydrocarbon phase. This paper evaluates the source–reservoir pair potential of the basin and provides a link between the occurrence of petroleum and the potential sources via geochemical analysis. The play potential is also described for some of the oldest carbonate rocks on Earth that have a complex thermal history and large uncertainties on timing of generation, trap formation and preservation.

Brown *et al.* (2015) examine an unusual deepwater lithofacies from the West Africa transform margin. Based on well and seismic data, the paper describes aspects of the petrophysical analysis of a quartz-prone clastic interval with good porosity, which differs from drilling results that recorded a *c.* 200 m 'claystone' interval. The combination of the physical observation and wireline interpretation for the single exploration well was subsequently recognized as a regional lithofacies type based on data from a number of wells across several sub-basins along the margin. The consistency in the lithofacies, irrespective of the differing basins and differing sediment source provinces, created a challenge in formulating a model for the deposition of this lithofacies into a deep-water palaeoenvironment. Through the application of a variety of modern laboratory-based techniques used to characterize the rock, and research into palaeoclimates and modern day analogues, the authors interpret the lithofacies to be derived from far-field aeolian processes. Far-field events, such as this arid dusting of the West Africa transform margin during the Campanian, show that sedimentation in this region is not just a localized process, and that consideration must be given to the hinterland setting and processes acting thereupon during the exploration phase.

Lawrence *et al.* (2016) present evidence from the interpretation of seismic data for compressional deformation of intraplate oceanic crust in the eastern Gulf of Guinea, a region occupied by the Douala-Rio Muni Basin and the Cameroon Volcanic Line.

The Cameroon Volcanic Line is usually attributed to a mantle ('hotspot') origin but the interpretation given here indicates that strike-slip and/or compressional tectonics and the reactivation of the oceanic crustal fabric have played important roles in its evolution. A regional tectonic model that associates the deformation in the eastern Gulf of Guinea with regional 'Alpine far-field' compressive stress is also presented. The authors suggest that the evolution of the Cameroon Volcanic Line and the continent–ocean transition link has imparted a unique tectonostratigraphic history to the Douala-Rio Muni Basin, which has influenced its petroleum endowment – particularly in the different trapping mechanisms found.

Péron-Pinvidic *et al.* (2015) summarize observations from the South Atlantic Angola–Gabon rifted margin based on the interpretation of depth-migrated seismic reflection profiles. The mapping and characterization of these domains at the margin scale permit the along-strike structural and stratigraphic variability of the margin to be illustrated. The authors also identified key sections that can be considered as type examples of upper-plate and lower-plate margins and discuss some of the characteristics of these settings.

Cowie *et al.* (2016) offer a model for the accumulation of salt during the Aptian, a time of South Atlantic rifting. The methodology includes a quantitative analysis of deep seismic reflection and gravity anomaly data together with reverse post-break-up subsidence modelling. Inversion of gravity data is used to infer the Moho depth and crustal thickness, and residual depth anomalies (a technique usually used for the interpretation of oceanic crust) are applied to identify departures from oceanic bathymetry on hyperextended continental crust. In their paper, they propose that the Aptian salt was deposited approximately 0.2 (proximal) and 0.6 km (distal) below global sea-level and that the inner proximal salt subsided by post-rift (post-tectonic) thermal subsidence alone, whereas the outer distal salt formed during the synrift phase, prior to the break-up, resulting in additional tectonic subsidence.

Sabato Ceraldi & Green (2016) present a paper on the pre-salt lacustrine carbonates in both Brazil and Angola. They describe the pre-salt tectonostratigraphy with three significant events: the first is the development of synrift lakes (Valanginian–Hauterivian) during continental break-up and the formation of synrift graben. The formation of several deep basins is interpreted as leading to overfilling of freshwater lakes. The second event (Sag 1) in the Barremian–Aptian caused widespread subsidence through syn-kinematic stretching of the continental crust and/or continuous rifting. Lacustrine sediments comprising coarse coquina grainstones in fresh-to-brackish, balanced-filled interconnected lakes. The third event (Sag 2) in the Aptian is interpreted to have taken place during continuous subsidence and led to the isolation of the lakes. The base levels fell below sea-level, resulting in underfilled, hypersaline lakes with widespread microbialite growth. On the basis of integrated seismic data, well data, biostratigraphy, stable isotope analysis and tectonic models they develop a regional predictive model for reservoirs in Sag 2 and 3.

Green & Machado (2015) used apatite fission-track data from Cretaceous sediments and Precambrian crystalline basement from outcrop in the Namibe basin, Angola to describe the exhumation history of the region. They recognize Late Carboniferous–Early Permian and Jurassic exhumation events that preceded Early Cretaceous rifting. The most recent event occurred during the Cenozoic and resulted in uplift and erosion between 35 and 20 Ma. This timing is consistent with published analyses of river profiles that suggest that uplift of the margin began at *c.* 30 Ma. They recognize similar timing in uplift all along the West Africa margin and interpret this as the result of a continent-scale response to stresses related to tectonic plate movement.

In the paper by **Mohammed *et al.* (2016)** the pre-rift basement geometries have been characterized and then compared with the geometries of Cretaceous rift basin structures and the subsequent architectural elements of the post-rift margin. Half-graben depocentres migrate westwards through time within the continental synrift phase, at the same time as hard-linked arrays of basin-bounding faults become established, with lengths of approximately 100 km. The authors use the Southern Namibian margin as an example to demonstrate that the rift evolution and subsequent passive margin architecture are the products of lithospheric-thinning processes and crustal heterogeneity. They recognize that crustal heterogeneities play a role in influencing the dimensions and styles of fault evolution and in establishing the position of cross-cutting structures, and go on to highlight the role of pre-rift granitic bodies in influencing the location of deformation.

Dalton *et al.* (2016) present a large catalogue of gravity-driven structures across the Orange Basin in Namibia and South Africa. The structures documented here illustrate the role of multiple detachments levels, their activation timings and the related timings of fault activation within the gravity-driven system. These observations led Dalton *et al.* to propose an evolutionary model for the development of these gravity-driven systems that allows the role of the deposition of sediments in controlling the geometry of the structures to be established once placed within a well-defined stratigraphic framework.

The paper by **Paton *et al.* (2016)** addresses two previously unanswered questions: the location and continuation of the regionally extensive Gondwanian orogenic belt in the Orange Basin, offshore South Africa; and the reason that the east–west-trending Argentinian Colorado rift is perpendicular to the conjugate Orange Basin and the trend of the Atlantic Ocean spreading. This paper presents a new structural model for the southern South Atlantic by identifying the South African fold belt in the offshore domain for the first time. The trend of the fold belt changes from north–south to east–west offshore, correlating directly with the restored Colorado Basin. The authors suggest that the Colorado and Orange rifts form a tripartite system with the Namibian Gariep belt. All three rift branches were active during the rifting as Gondwana broke up, but during the Atlantic rift phase the Colorado Basin became an aulacogen while the other two branches developed in the present-day location of the South Atlantic.

The editors would like to acknowledge the contribution of the authors who devoted time and effort in the drafting and revision of the papers and the figures of this Special Publication. The editors would also like to extend their acknowledgement to the co-convenors of the conference, Matt Warner and Hannah Suttil, and the numerous named and anonymous reviewers who have contributed time and expertise to provide thorough and constructive reviews, which have considerably improved the quality of the present Special Publication: Anthony Bourne, Rob Crossley, Brian Cullen, Green Darryl, Tim Goodwin, Christian Heine, Richard Hodgkinson, Simon Holford, Sarah Houldsworth, Stephen Jones, Juliette Lamarche, Jean Malan, Webster Mohriak, Chris Morley, Alexis Anastas Nexis, Francisco Pangaro, Marta Perez-Gussinye, Nicholas Richardson, Donna Shillington, Taury Smith, Christina von Nicolai and Robert (Woody) Wilson.

References

ARTHUR, T., MACGREGOR, D.S. & CAMERON, N.R. (eds) 2003. *Petroleum Geology of Africa: New Themes and Developing Technologies.* Geological Society, London, Special Publications, **207**, http://sp.lyellcol lection.org/content/207/1

ASLANIAN, D., MOULIN, M. *ET AL.* 2009. Brazilian and African passive margins of the Central Segment of the South Atlantic Ocean: kinematic constraints. *Tectonophysics*, **468**, 98–112.

BROWN, A., BIRKHEAD, S., MCLEAN, D., TOWLE, P., WHITE, H. & WU, YAFEI. 2015. The Campanian quartz 'claystone' conundrum of the African Transform Margin. *In*: SABATO CERALDI, T., HODGKINSON, R.A. & BACKE, G. (eds) *Petroleum Geoscience of the West Africa Margin.* Geological Society, London, Special Publications, **438**. First published online December 23, 2015, http://doi.org/10.1144/SP438.3

CAMERON, N.R., BATE, R.H. & CLURE, V.S. (eds) 1999. *The Oil and Gas Habitats of the South Atlantic.*

Geological Society, London, Special Publications, **153**, http://sp.lyellcollection.org/content/153/1

COWIE, L., ANGELO, R.M., KUSZNIR, N., MANATSCHAL, G. & HORN, B. 2016. Structure of the ocean–continent transition, location of the continent–ocean boundary and magmatic type of the northern Angolan margin from integrated quantitative analysis of deep seismic reflection and gravity anomaly data. *In*: SABATO CERALDI, T., HODGKINSON, R.A. & BACKE, G. (eds) *Petroleum Geoscience of the West Africa Margin.* Geological Society, London, Special Publications, **438**. First published online January 25, 2016, http://doi.org/10.1144/SP438.6

DALTON, T.J.S., PATON, D.A. & NEEDHAM, D.T. 2016. Influence of mechanical stratigraphy on multi-layer gravity collapse structures: insights from the Orange Basin, South Africa. *In*: SABATO CERALDI, T., HODGKINSON, R.A. & BACKE, G. (eds) *Petroleum Geoscience of the West Africa Margin.* Geological Society, London, Special Publications, **438**. First published online January 19, 2016, http://doi.org/10.1144/SP438.4

EMERY, K.O. & UCHUPI, E. 1984. *The Geology of the Atlantic Ocean.* Springer, New York.

GREEN, P.F. & MACHADO, V. 2015. Pre-rift and synrift exhumation, post-rift subsidence and exhumation of the onshore Namibe Margin of Angola revealed from apatite fission track analysis. *In*: SABATO CERALDI, T., HODGKINSON, R.A. & BACKE, G. (eds) *Petroleum Geoscience of the West Africa Margin.* Geological Society, London, Special Publications, **438**. First published online December 23, 2015, http://doi.org/10.1144/SP438.2

KARNER, G.D. & DRISCOLL, N.W. 1999. Tectonic and stratigraphic development of the West African and eastern Brazilian margins; insights from quantitative basin modelling. *In*: CAMERON, N.R., BATE, R.H. & CLURE, V.S. (eds) *The Oil and Gas Habitats of the South Atlantic.* Geological Society, London, Special Publications, **153**, 11–40, http://doi.org/10.1144/GSL.SP.1999.153.01.02

KARNER, G.D., DRISCOLL, N.W. & BARKER, D.H.N. 2003. Synrift regional subsidence across the West African continental margin: the role of lower plate ductile extension. *In*: ARTHUR, T.J., MACGREGOR, D.S. & CAMERON, N.R. (eds) *Petroleum Geology of Africa: New Themes and Developing Technologies.* Geological Society, London, Special Publications, **207**, 105–129, http://doi.org/10.1144/GSL.SP.2003.207.6

LAWRENCE, S.R., BEACH, A., JACKSON, O. & JACKSON, A. 2016. Deformation of oceanic crust in the eastern Gulf of Guinea: role in the evolution of the Cameroon Volcanic Line and influence on the petroleum endowment of the Douala-Rio Muni Basin. *In*: SABATO CERALDI, T., HODGKINSON, R.A. & BACKE, G. (eds) *Petroleum Geoscience of the West Africa Margin.* Geological Society, London, Special Publications, **438**. First published online February 18, 2016, http://doi.org/10.1144/SP438.7

MARTÍN-MONGE, A., BAUDINO, R. *ET AL.* 2016. An unusual Proterozoic petroleum play in Western Africa: the Atar Group carbonates (Taoudeni Basin, Mauritania). *In*: SABATO CERALDI, T., HODGKINSON, R.A. & BACKE, G. (eds) *Petroleum Geoscience of the West Africa Margin.* Geological Society, London, Special Publications, **438**. First published online January 7, 2016, http://doi.org/10.1144/SP438.5

MOULIN, M., ASLANIAN, D. *ET AL.* 2005. Geological constraints on the evolution of the Angolan margin based on reflection and refraction seismic data (ZaïAngo project). *Geophysical Journal International*, **162**, 793–810.

MOULIN, M., ASLANIAN, D. & UNTERNEHR, P. 2010. A new starting point for the south and equatorial Atlantic Ocean. *Earth-Science Reviews*, **98**, 1–37.

MOHAMMED, M., PATON, D., COLLIER, R.E.L., HODGSON, N. & NEGONGA, M. 2016. Interaction of crustal heterogeneity and lithospheric processes in determining passive margin architecture on the southern Namibian margin. *In*: SABATO CERALDI, T., HODGKINSON, R.A. & BACKE, G. (eds) *Petroleum Geoscience of the West Africa Margin.* Geological Society, London, Special Publications, **438**. First published online March 16, 2016, http://doi.org/10.1144/SP438.9

PATON, D.A., MORTIMER, E.J., HODGSON, N. & VAN DER SPUY, D. 2016. The missing piece of the South Atlantic jigsaw: when continental break-up ignores crustal heterogeneity. *In*: SABATO CERALDI, T., HODGKINSON, R.A. & BACKE, G. (eds) *Petroleum Geoscience of the West Africa Margin.* Geological Society, London, Special Publications, **438**. First published online March 9, 2016, http://doi.org/10.1144/SP438.8

PÉRON-PINVIDIC, G., MANATSCHAL, G., MASINI, E., SUTRA, E., FLAMENT, J.M., HAUPERT, I. & UNTERNEHR, P. 2015. Unravelling the along-strike variability of the Angola–Gabon rifted margin: a mapping approach. *In*: SABATO CERALDI, T., HODGKINSON, R.A. & BACKE, G. (eds) *Petroleum Geoscience of the West Africa Margin.* Geological Society, London, Special Publications, **438**. First published online December 2, 2015, http://doi.org/10.1144/SP438.1

RABINOWITZ, P.D. & LABRECQUE, J. 1979. The Mesozoic South Atlantic Ocean and evolution of its continental margins. *Journal of Geophysical Research*, **84**, 5973–6002.

SABATO CERALDI, T. & GREEN, D. 2016. Evolution of the South Atlantic lacustrine deposits in response to Early Cretaceous rifting, subsidence and lake hydrology. *In*: SABATO CERALDI, T., HODGKINSON, R.A. & BACKE, G. (eds) *Petroleum Geoscience of the West Africa Margin.* Geological Society, London, Special Publications, **438**. First published online June 21, 2016, http://doi.org/10.1144/SP438.10

TORSVIK, T.H., ROUSSE, S., LABAILS, C. & SMETHURST, M.A. 2009. A new scheme for the opening of the South Atlantic Ocean and the dissection of an Aptian salt basin. *Geophysical Journal International*, **177**, 1315–1333, http://doi.org/10.1111/j.1365-246X.2009.04137.x

Deformation of oceanic crust in the eastern Gulf of Guinea: role in the evolution of the Cameroon Volcanic Line and influence on the petroleum endowment of the Douala-Rio Muni Basin

STEVE R. LAWRENCE[1]*, ALASTAIR BEACH[2], OWAIN JACKSON[3] & ANDREW JACKSON[1]

[1]*ERCL, Dragon Court, 15 Station Road, Henley-on-Thames RG9 1AT, UK*

[2]*Exploration Outcomes, 1 Huntly Gardens, Glasgow G12 9AS, UK*

[3]*Glencore E & P, 50 Berkeley Street, London W1J 8HD, UK*

**Corresponding author (e-mail: steve.lawrence@ercl.com)*

Abstract: Evidence is presented from the interpretation of seismic data for the compressional deformation of intra-plate oceanic crust in the eastern Gulf of Guinea, a region occupied by the Douala-Rio Muni Basin and the Cameroon Volcanic Line. The deformation has taken the form of the reactivation of syn-kinematic structures in the fabric of the oceanic crust. The origin of the Cameroon Volcanic Line is usually attributed to a mantle hotspot, but interpretation of the seismic data indicates that strike-slip and/or compressional tectonics and reactivation of the fabric of the oceanic crust have all played an important role in its evolution. A regional tectonic model is presented that associates deformation in the eastern Gulf of Guinea with regional alpine-related far-field compressive stress. In this way, a causal link is invoked between deformation of the oceanic crust and strike-slip movement on the Central African Shear Zone, establishing a continent–ocean tectonic link. The evolution of the Cameroon Volcanic Line and the continent–ocean tectonic link has imparted a unique tectonostratigraphic history to the Douala-Rio Muni Basin and has influenced its petroleum endowment. The structural and combination trapping potential has been greatly enhanced in this environment and the structural architecture of the basin has influenced the depositional patterns of the deep water reservoir sands.

Several examples of intra-plate deformation of the oceanic crust under compressive stress have been documented. The most widely studied example is in the central Indian Ocean, where the deformation has the form of long-wavelength folding ('buckling') and the reactivation of fracture zones and the ridge-parallel fault fabric (Bull 1990; Bull & Scrutton 1992; Beekman *et al.* 1996). Deformation of the oceanic crust has also been reported in the Equatorial and South Atlantic (Fig. 1). Post-Eocene compressional reactivation of fracture zone faults has been described by Jones (2003) in the eastern Equatorial Atlantic and Briggs *et al.* (2009) reported deformation of the oceanic crust characterized by the compressional reactivation of intra-crustal faults in the central Gulf of Guinea (western Niger Delta). We present evidence for the intra-plate deformation of oceanic crust in the eastern Gulf of Guinea (about 300 km east of the study area of Briggs *et al.* 2009) (Fig. 1) where this deformation can be shown to have played an important part in the tectonostratigraphic evolution and petroleum endowment of the Douala-Rio Muni Basin.

Regional setting and structure

The study area is located in the eastern Gulf of Guinea on the West African margin of the northern South Atlantic Ocean (Fig. 1). The structural fabric of oceanic crust in the South Atlantic is shown on the satellite-derived gravity map in Figure 1 (reproduced from Sandwell *et al.* 2014). Several major fracture zones, characterized by linear gravity lows, can be traced from the Mid-Atlantic Ridge eastwards across the South Atlantic towards the continental margin of West Africa (Fig. 1). As the West African continental margin is approached, the fabric of the oceanic crust becomes obscured by the widespread occurrence of discrete quasi-circular gravity highs, interpreted to be the product of intra-plate volcanism. The St Helena Seamount Chain forms a prominent NE–SW-trending volcanic line extending over 2700 km from the volcanic island of St Helena to the coast of Cameroon and bisecting the eastern Gulf of Guinea study area (Fig. 1). The northeastern segment of this volcanic chain, extending from the island of Pagalu in the SW up to Mount Cameroon and then to onshore

From: SABATO CERALDI, T., HODGKINSON, R. A. & BACKE, G. (eds) 2017. *Petroleum Geoscience of the West Africa Margin*. Geological Society, London, Special Publications, **438**, 7–26. First published online February 18, 2016, http://doi.org/10.1144/SP438.7

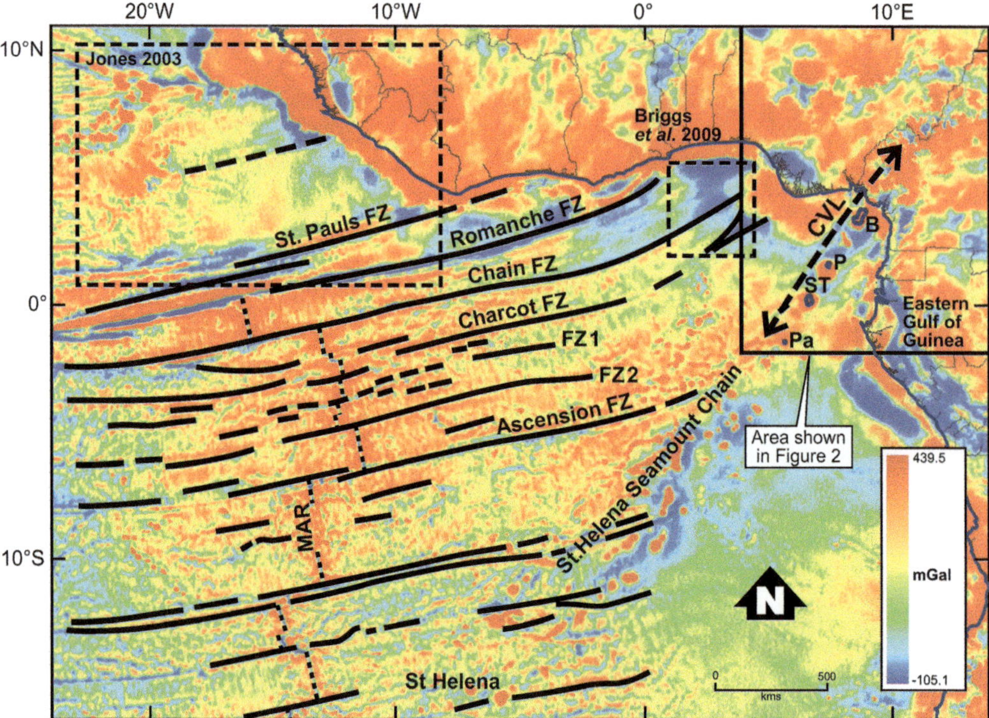

Fig. 1. Satellite gravity image of the Equatorial and South Atlantic Ocean (source Sandwell *et al.* 2014) showing the regional study area in the eastern Gulf of Guinea, the Cameroon Volcanic Line and the fabric of the oceanic crust in the South Atlantic with named fracture zones. Unnamed fracture zones are designated FZ1 and FZ2. The areas of deformation of oceanic crust studied by Jones (2003) and Briggs *et al.* (2009) are also shown. B, Bioko; CVL, Cameroon Volcanic Line; FZ, fracture zone; MAR, Mid-Atlantic Ridge; Pa, Pagalu; P, Príncipe; ST, São Tomé.

Cameroon, is usually referred to as the Cameroon Volcanic Line, or just the Cameroon Line (Njome & de Wit 2014).

The regional study area is occupied by the Douala-Rio Muni Basin, forming part of the West African passive margin basin system lying between the Niger Delta and the North Gabon Basin (Fig. 2). The Rio Muni-Douala Basin developed as an oblique margin with stretching and rifting of continental crust prior to separation, which was tectonostratigraphically placed in the late Aptian/early Albian by Exploration Consultants Ltd (2001), Lawrence *et al.* (2002), Turner *et al.* (2003), Moulin *et al.* (2010) and Seton *et al.* (2012). More recent studies by Heine *et al.* (2013) and Gaina *et al.* (2013) estimated the first formation of oceanic crust from about 120 Ma (mid–late Aptian). Early break-up along the oblique margin is thought to have been influenced by pull-apart tectonics linked to transtensional movement on the Central African Shear Zone (Fairhead 1988; Binks & Fairhead 1992; Genik 1993; Maurin & Guiraud 1993), thus providing an early established link between continental tectonics and the development of an ocean basin (Fairhead & Binks 1991).

The Douala-Rio Muni Basin is bisected by the oceanic segment of the Cameroon Volcanic Line, formed by a *c.* 700 km-long NNE–SSW-trending linear belt of volcanic islands and seamounts extending from Mount Cameroon through the islands of Bioko, Príncipe and São Tomé to Pagalu (Fig. 2). A seamount to the SW of Bioko, lying in the territory of Equatorial Guinea, has historically been named the De Santarem–Escobar Bank, but is here abbreviated to the Santesc Seamount. The Cameroon Volcanic Line extends onshore onto continental crust for another 1000 km as far as the Kapsiki Plateau (in the definition of Njome & de Wit 2014). The volcanic ages along the Cameroon Volcanic Line do not show a systematic time progression and are all considerably younger than the underlying oceanic crust (Fig. 2). The oldest recorded volcanic ages are 30.6 ± 2.1 Ma for the offshore segment on the island of Príncipe (Dunlop & Fitton 1979) and

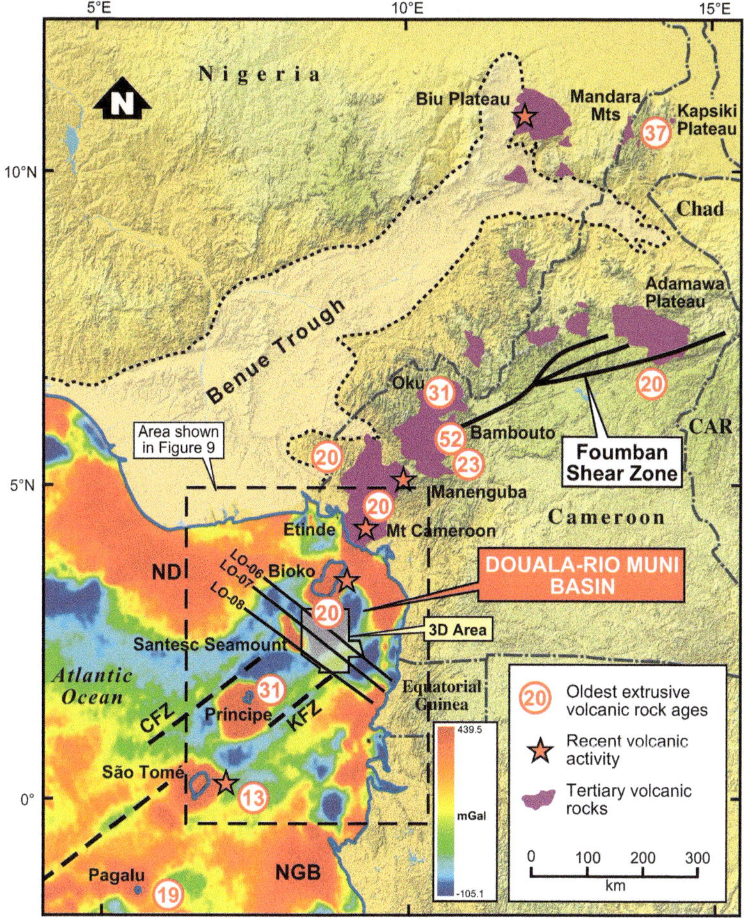

Fig. 2. Regional setting of the Douala-Rio Muni Basin and Cameroon Volcanic Line. Offshore satellite gravity image taken from Sandwell *et al.* (2014). Onshore topography taken from USGS (2014) – Shuttle Radar Topography Mission NASA/JPL/NIMA and onshore colour shading taken from GEBCO (2014). Volcanic record shown by oldest volcanic (extrusive) dates (Ma), mostly from Dunlop & Fitton (1979), Fitton & Dunlop (1985) and Fitton (1987), but also Mount Bambouto (Lee *et al.* 1994; Moundi *et al.* 2007) and Mount Oku (Njilah 1991). Location of regional 2D (LO-) seismic profiles and 3D seismic study area are shown. CFZ, Cameroon Fracture Zone; KFZ, Kribi Fracture Zone; ND, Niger Delta; NGB, North Gabon Basin.

51.79 ± 1.21 Ma for the onshore segment at Mount Bambouto (Moundi *et al.* 2007).

The structure of the Douala-Rio Muni–North Gabon sector of the West African margin and the Cameroon Volcanic Line was investigated previously by Meyers & Rosendahl (1991), Meyers *et al.* (1996, 1998) and Rosendahl & Groschel-Becker (2000) using deep-imaging seismic data acquired in 1989 as part of the West African PROBE Study. The structure of the oceanic crust interpreted from these data is a series of NE–SW-oriented fracture zones extending across the Cameroon Volcanic Line and intersecting the continental margin (Meyers *et al.* 1998). Two of these structures were identified

as the Cameroon Fracture Zone and the Kribi Fracture Zone (Fig. 2).

Meyers & Rosendahl (1991) and Meyers *et al.* (1998) also investigated the nature of the features comprising the Cameroon Volcanic Line using West African PROBE Study seismic data. They showed that the volcanic centres of the Cameroon Volcanic Line are not made up of thick piles of intra-plate volcanic material. This supported the view of Hedberg (1969), who proposed from gravity modelling that the islands of the Cameroon Volcanic Line are underlain by thick sedimentary sections lying on oceanic basement. Meyers & Rosendahl (1991) and Meyers *et al.* (1998) concluded that the topographic expression of the islands and seamounts was produced by crustal uplift rather than by the building of shield volcanoes with thick crustal roots. They dated the uplift/buckling as Miocene based on correlation of the uplift unconformity with neighbouring well controls.

Meyers *et al.* (1998) interpreted the formation of the islands and seamounts of the Cameroon Volcanic Line in terms of the buckling/uplift of oceanic lithosphere and in this way provided the first geophysical evidence for the deformation of the oceanic crust and a tectonic component to the volcanic islands of the Cameroon Volcanic Line. This interpretation was supported by the following geological evidence:

(1) Cretaceous sediments (dated as Turonian) cropping out on the island of Bioko (De Villalta & Assens 1967);
(2) Upper Eocene–Oligocene sediments encountered within 250 m of the surface (i.e. near sealevel) on São Tomé Island in the Ubabudo-1 borehole drilled in 1990 (ANP-STP 2012*c*). Note that the lithified sandstones of unknown age occurring at the surface on the island of São Tomé (i.e. above sea-level) (Hedberg 1969) may be age-equivalent to those encountered in Ubabudo-1.

Interpretation of seismic data

We used the interpretation of offshore 2D and 3D seismic data to further investigate the nature of deformation of the oceanic crust in the eastern Gulf of Guinea regional study area (Fig. 2).

Interpretation of regional 2D seismic data

Deep-imaging long-offset 2D seismic data (prefix LO-) were acquired by GeSeis Technologies Ltd in 2005. The dataset consists of 2518 km of 8 km offset, 107-fold 2D seismic data recorded to 14 s, forming a regional grid over the whole offshore territory of Equatorial Guinea. Interpreted versions of lines LO-6, LO-7 and LO-8 are shown in Figure 3 (line locations in Fig. 2).

Lines LO-7 and LO-8 traverse the Santesc Seamount of the Cameroon Volcanic Line from the Niger Delta thrust front in the NW to the Rio Muni margin in the SE (Fig. 3a, b). The Kribi Fracture Zone (first named by Meyers *et al.* 1996, 1998) is characterized by a ridge-and-furrow morphology where it separates so-called proto-oceanic crust from oceanic crust along the Rio Muni margin (Fig. 3a, c). The term 'proto-oceanic' was introduced for the thicker crust on the Rio Muni side by Meyers *et al.* (1996, 1998) and later used by Lawrence *et al.* (2002) and Wilson *et al.* (2003). The top

oceanic crust and the reflection Moho can be discerned on both sides of the Kribi Fracture Zone, showing a crustal thickness of 1.6–1.75 s twoway travel time for oceanic crust and 2.25–3.5 s two-way travel time for proto-oceanic crust. This translates to thicknesses of 4.8–5.25 and 6.75–10.5 km, respectively, and compares with values of 5–8.3 km recorded to the west by Briggs *et al.* (2009) (both assuming a crustal velocity of 6000 m s^{-1}). For a discussion of the nature and composition of proto-oceanic crust, see Meyers *et al.* (1996), Rosendahl & Groschel-Becker (2000) and Wilson *et al.* (2003). No significant posttransform reactivation can be discerned along the Kribi Fracture Zone.

The Malabo Hinge Zone is best observed on LO-6 and LO-7 as a crustal inflection corresponding to an inclined intra-crustal surface and a monoclinal structure in the overlying sedimentary section (Fig. 3a, c). Deformation affects the sedimentary section up to the Early Tertiary, thus identifying a phase of deformation here designated as D_1. The Malabo Hinge Zone is analogous to the structures parallel to the fracture zone described by Briggs *et al.* (2009) as intra-crustal faults and interpreted as thrusts detaching at the Moho.

LO-7 and LO-8 show the interpretation of the Santesc Seamount as a fold–fault structure involving the basement and deforming the overlying sedimentary section up to a Base Miocene unconformity (Fig. 3a, b). This structure, here called the Seamount Inversion Structure, formed during early Miocene deformation, here designated deformation phase D_2. Lines LO-7 and LO-8 show the continuity of the sedimentary section across the Santesc Seamount and, although intrusive rocks and vent/cone features confirm the volcanic influences of the seamount, the seismic data demonstrate the absence of a significant volcanic root or intrusive body (Fig. 3a, b). This interpretation of the Santesc Seamount and the Seamount Inversion Structure therefore supports that of Meyers & Rosendahl (1991) and Meyers *et al.* (1998) in their recognition of deformation/uplift of oceanic crust underlying the volcanic islands and seamounts of the Cameroon Volcanic Line. An earlier formed basement-involved structure within the Seamount Inversion Structure highlighted on LO-8 (Fig. 3b) is here called the Block V Structure. This is interpreted as a product of a late Cretaceous (Santonian) deformation phase here designated D_0.

LO-6 crosses the Cameroon Volcanic Line between the volcanic island of Bioko and the Santesc Seamount (Fig. 3c). A structure here called the Monoclinal Fold Structure, with the steep limb on the SE side, can be observed affecting the sedimentary section. The monocline is interpreted as an asymmetrical fold formed by movement on a

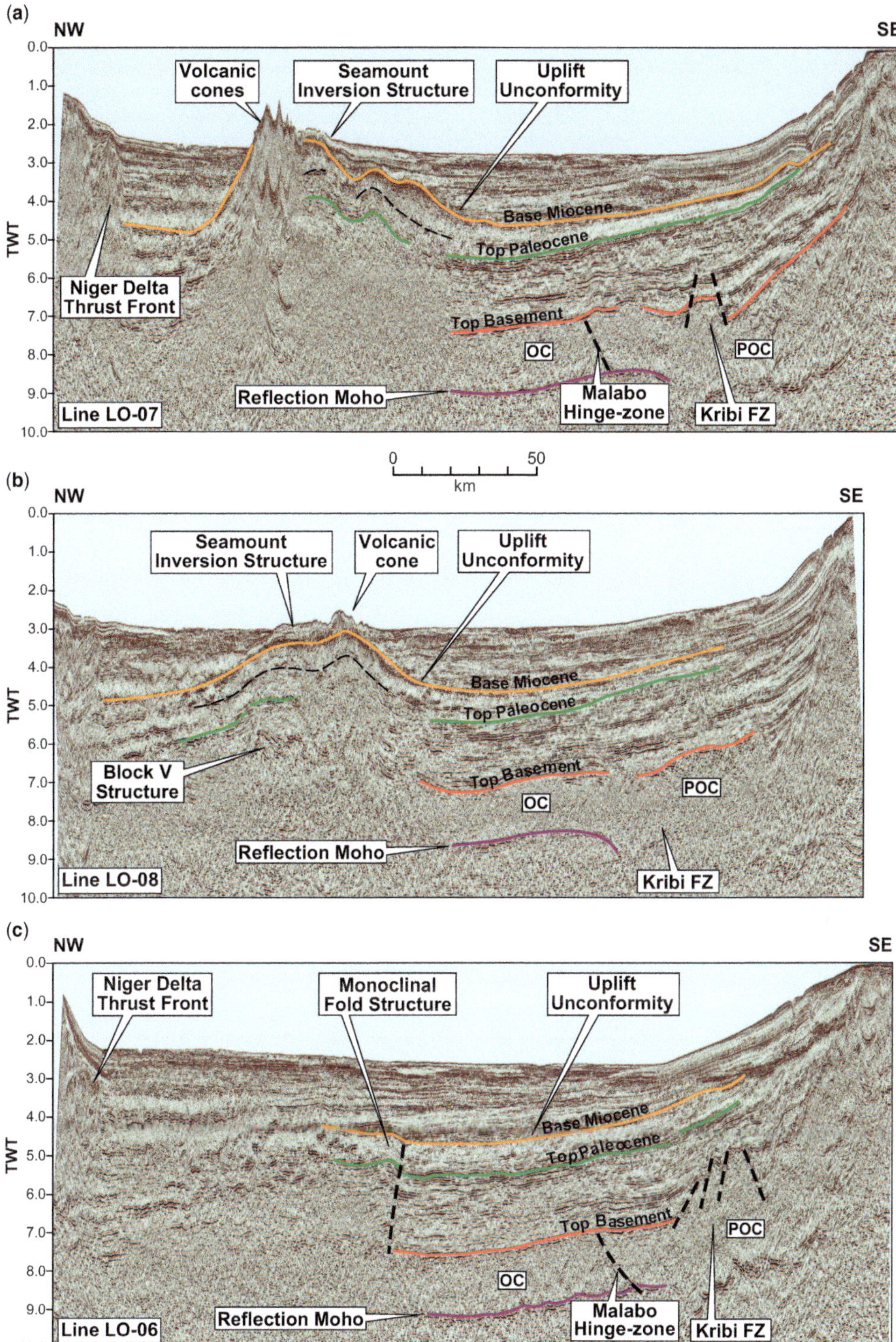

Fig. 3. Seismic profiles (**a**) LO-7, (**b**) LO-8 and (**c**) LO-6 across the Douala-Rio Muni Basin and Cameroon Volcanic Line. Location of profiles shown on Figure 2. FZ, fracture zone; OC, oceanic crust; POC, proto-oceanic crust.

deep crustal fault and has the same timing as the D_2 deformation associated with the Seamount Inversion Structure – that is, early Miocene.

LO-7 and LO-8 also show thickening of the Early Tertiary section (between Top Paleocene and Base Miocene) into the Santesc Seamount (Fig. 3a, b). One explanation is that this is an expression of flexural moat development in response to volcanic loading similar to that interpreted around the Canary Islands (Collier & Watts 2001) and the Cape Verde Islands (Ali *et al.* 2003). Significantly, the thickening is not observed on LO-6, which crosses the Cameroon Volcanic Line between the 'volcanic' Santesc Seamount and Bioko Island (Fig. 3c). If this interpretation is correct, the timing of moat subsidence would broadly correspond to the first recognized offshore volcanic activity (i.e. Oligocene on Príncipe Island) and just predate the D_2 (early Miocene) phase of deformation and uplift. It should be noted, however, that Meyers *et al.* (1998) did not recognize flexural moats associated with the islands of the Cameroon Volcanic Line and attributed their non-development to the lack of sufficient volcanic load.

Interpretation of 3D seismic data

The 3D seismic data were acquired between 2010 and 2014 in five separate surveys over an aggregate area of 7750 km^2, forming a 3D study area of >9000 km^2 on the eastern side of the Cameroon Volcanic Line between Bioko Island and the Santesc Seamount (Fig. 2). The data were acquired by BGP under the auspices of Geoex International using a streamer length of 6 km.

The interpretation of the 3D data has provided a closer view of the deformation of oceanic crust along the eastern flank of the Cameroon Volcanic Line between Bioko Island and the Santesc Seamount. The structural interpretation of the 3D study area is based on the mapping of several seismic horizons regionally correlated and stratigraphically identified by ties to proprietary well data. Two-way travel time structure maps are presented for the Top Basement (Fig. 4), Top Paleocene (Fig. 5) and Base Miocene (Fig. 6). Examples of interpreted arbitrary 3D seismic sections are displayed in Figure 7. The structure of the 3D study area is summarized in Figure 8, which shows the inherited (syn-kinematic) structural fabric of oceanic crust and the deformation of the overlying sedimentary section, here referred to as a reactivation structure. The structural features of the 3D study area are further described in the following:

Syn-kinematic structure:

- Spreading ridge structure: NW–SE linear faulted ridges and troughs (Figs 4 & 7b) are interpreted as structures formed at the spreading ridge. Faults have a 3–5 km spacing and throw westwards with displacements of up to 200 ms two-way travel time (estimated at *c.* 400 m). Note that there is no observable curvature of the spreading ridge fabric against the Kribi Fracture Zone or the Malabo Hinge Zone as might be expected for slow-spreading axes such as the Mid-Atlantic Ridge.
- Structure parallel to the fracture zone:
  - The Kribi Fracture Zone is seen crossing the extreme SE corner of the 3D study area (Figs 4 & 7a).
  - The Malabo Hinge Zone crosses NE–SW through the southern part of the 3D study area. It is confirmed as a fracture zone parallel, crustal hinge line formed by an inclined intra-crustal surface associated with displacement of *c.* 90 ms two-way travel time (estimated at *c.* 180 m) at Top Basement level (Figs 4 & 7a). This is analogous to the intra-crustal faults of Briggs *et al.* (2009), which also have a fracture zone orientation.
  - A basement fault structure crosses NE–SW through the northern part of the 3D study area. It has expression as a deep crustal fault forming a structure with an elevation of up to 500 ms two-way travel time (estimated at *c.* 1000 m) at Top Basement level (Figs 4 & 7c). A NW–SE-trending offset of the basement fault trend is observed on Figure 4.

Reactivation structure:

- Rotational/compressional deformation along the Malabo Hinge Zone. A monoclinal fold has formed in the overlying sedimentary section up to the Early Tertiary (Figs 5 & 7a). This is attributed to deformational phase D_1. This deformation is analogous to the compressional deformation reported by Briggs *et al.* (2009).
- Folding along NW–SE arcuate axial trends between the Kribi Fracture Zone and the Malabo Hinge Zone and across the Malabo Hinge Zone (Fig. 5). This folding is also attributed to deformational phase D_1.
- Compressional deformation crosses the northern part of the 3D study area where it is expressed as the prominent NE–SW-oriented Monoclinal Fold Structure first observed on regional 2D seismic data and formed over the basement fault structure (Figs 5, 6 & 7c). The timing of the deformation is given by the Base Miocene unconformity and an early Miocene growth sequence as D_2. A NW–SE-trending offset of

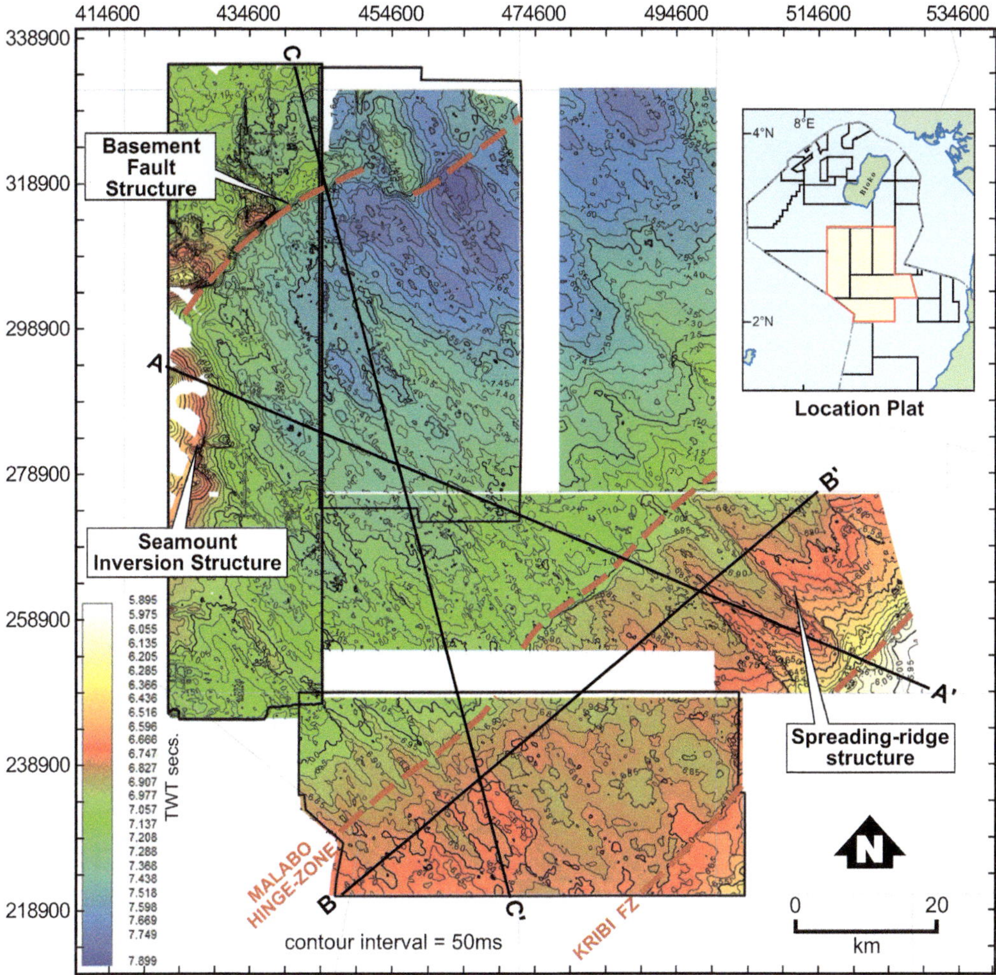

Fig. 4. 3D seismic study area: Top Basement two-way travel time structure map showing syn-kinematic structure of oceanic crust. Location of seismic sections in Figure 7 are shown. FZ, fracture zone.

the monoclinal fold trend is observed on Figures 5 and 6.

- Folding forming the eastern flank of the Seamount Inversion Structure along the western edge of the 3D study area (Figs 5, 6 & 7a). This can also be seen to have formed during deformational phase D$_2$.
- Subtle reactivation of spreading ridge normal faults can be seen (Fig. 7c).

Regional tectonics

A regional structural framework for the eastern Gulf of Guinea is presented in Figure 9. This is derived from the seismic interpretation described in this paper and builds on the previous interpretations of Meyers & Rosendahl (1991), Meyers *et al.* (1996) and Meyers *et al.* (1998). This structural framework is rationalized in the following sections in terms of a regional tectonic history incorporating the compressional deformation of oceanic crust and its effect on the evolution of the Cameroon Volcanic Line.

Deformation of oceanic crust

The intra-plate deformation described in the eastern Gulf of Guinea study area is interpreted entirely in terms of the reactivation of the syn-kinematic crustal fabric:

The Malabo Hinge Zone: the Malabo Hinge Zone structure is analogous to the thrust structures interpreted by Briggs *et al.* (2009) as reactivated intra-crustal faults detaching at the Moho. The

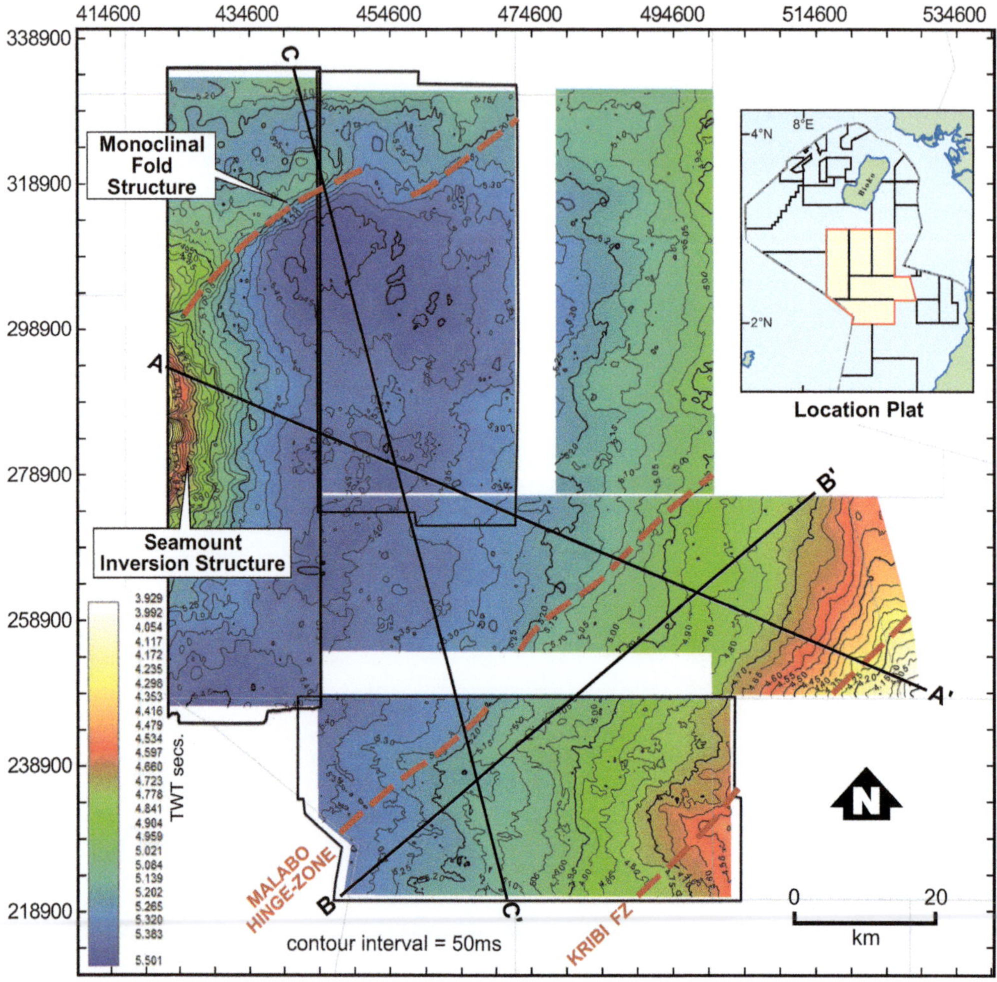

Fig. 5. 3D seismic study area: Top Paleocene two-way travel time structure map showing reactivation structure along the Malabo Hinge Zone and the Monoclinal Fold Structure. Location of seismic sections in Figure 7 are shown. FZ, fracture zone.

origin of the fracture zone parallel (non-transform) intra-crustal structure is unknown, but reactivation along the Malabo Hinge Zone is interpreted *sensu* Briggs *et al.* (2009) as compressional movement along an intra-crustal surface.

Cameroon Fracture Zone: The West African PROBE Study and industrial seismic data in São Tomé–Príncipe territory to the west of the Cameroon Volcanic Line (e.g. ANP-STP 2010, 2012*a*) show that the Cameroon and Fernando Poo fracture zones are expressed as a series of basement ridges with NW-facing scarps displaying elevations of up to 1.5 s two-way travel time. This implies a series of original left-handed transforms (i.e. the offset across the transform

is to the left) with the younger crust on the southern side (on the African plate) and dextral strike-slip movement on the original transform. The interpretation presented here invokes two expressions of reactivation along the projected Cameroon Fracture Zone. To the east of the Cameroon Volcanic Line, basement fault displacement and monoclinal folding in the overlying sedimentary section is interpreted as compressional (or transpressional) reactivation of a fracture zone fault. The other expression is the folding/uplift forming the Seamount Inversion Structure and interpreted as transpressional inversion along a 'jog' between separate fault segments of the Cameroon Fracture Zone. Together these structures form a zone of tectonic reactivation of the

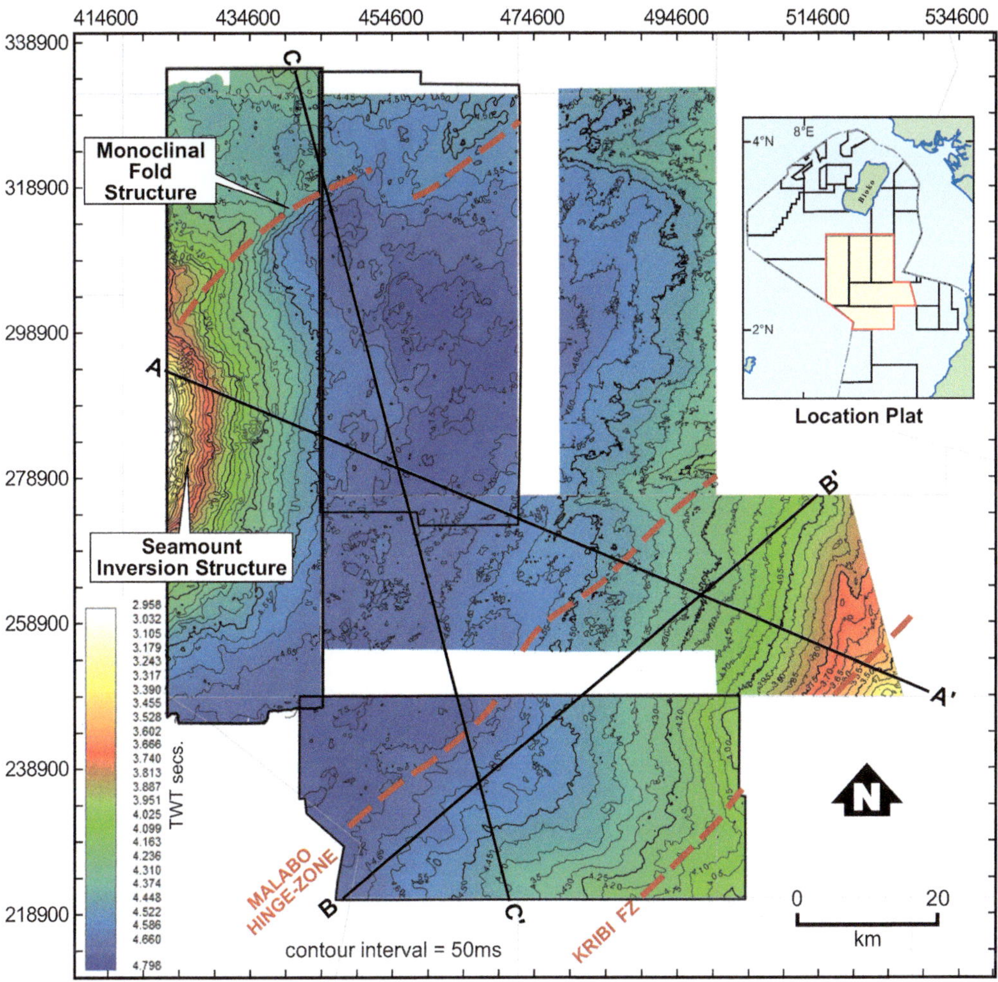

Fig. 6. 3D seismic study area: Base Miocene two-way travel time structure map showing reactivation structure along the Monoclinal Fold Structure and the Seamount Inversion Structure. Location of seismic sections in Figure 7 are shown. FZ, fracture zone.

Cameroon Fracture Zone, here called the Bioko Deformation Zone. On Figure 9, the Bioko Deformation Zone has been projected northeastwards on the basis of deformation observed on proprietary seismic data between the 3D study area and onshore Cameroon. This projection purposely implies a link between the Bioko Deformation Zone and basement shear zones interpreted as second-order splays of the Foumban Shear Zone (of the Central African Shear Zone) (Fig. 9).

The Cameroon Volcanic Line

There have been many published attempts to explain the formation of the Cameroon Volcanic Line.

Njome & de Wit (2014) provided a useful historical review of previous and prevailing theories. It is generally accepted that there is a mantle source for the volcanic rocks of the Cameroon Volcanic Line (Gallacher & Bastow 2012; De Plaen *et al.* 2014; Elsheikh *et al.* 2014; Adams *et al.* 2015) and the tectonic environment for the formation of the Cameroon Volcanic Line is usually assumed to have been extensional (Njome & de Wit 2014). However, the association of tectonic activity other than purely extensional along the Cameroon Volcanic Line has also been reported by several workers. Meyers *et al.* (1998) explained the buckling/uplift of oceanic crust below the volcanic centres as a product of ductile creep from the re-heating of the lithosphere. Others have invoked strike-slip tectonics associated

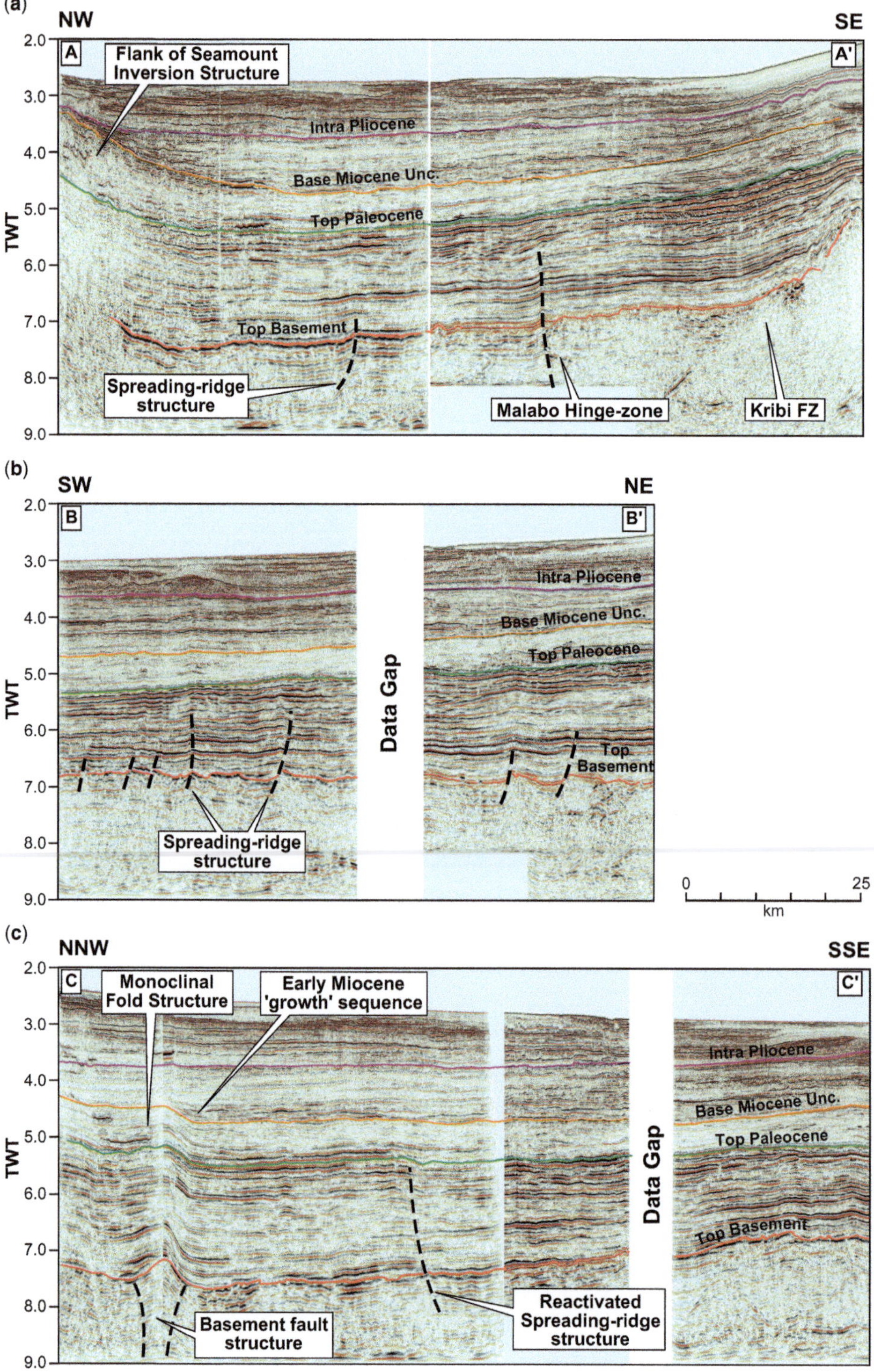

Fig. 7. 3D full-stack seismic sections (**a**) A–A′, (**b**) B–B′ and (**c**) C–C′. Location of lines shown on Figures 4, 5 and 6. FZ, fracture zone.

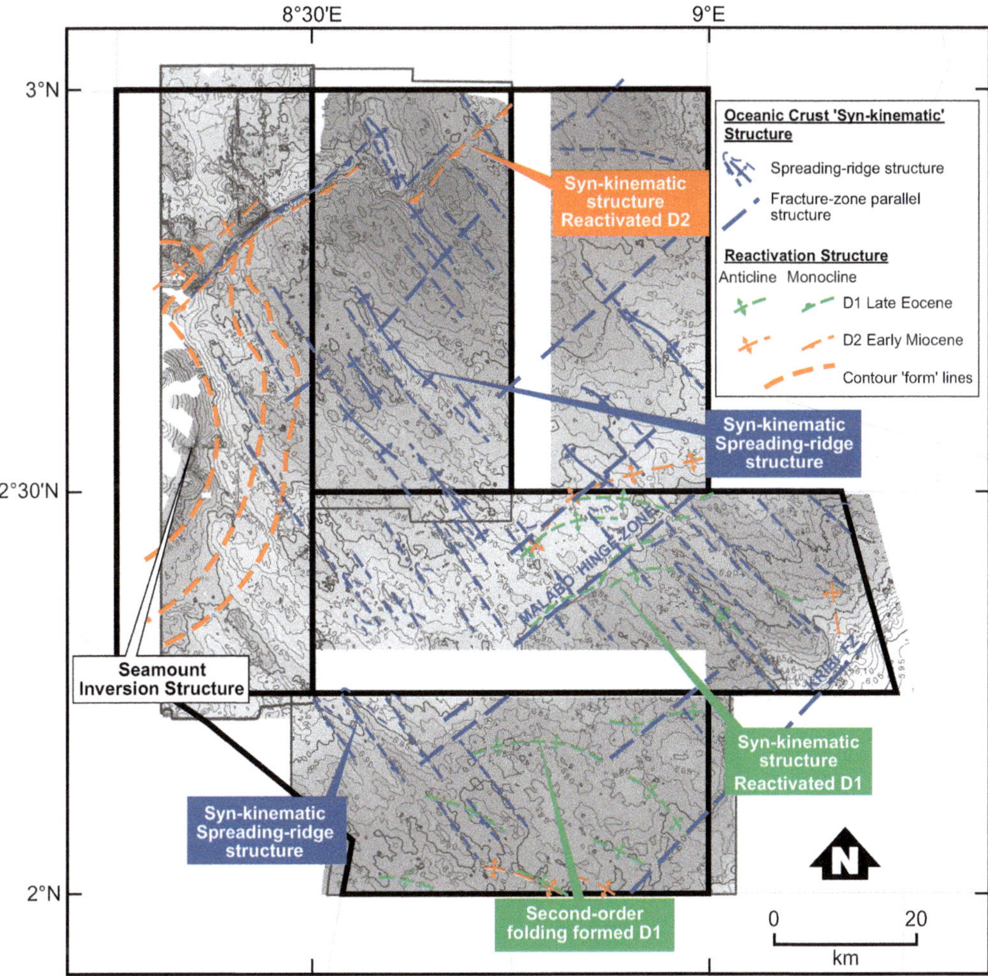

Fig. 8. Structural elements of 3D seismic study area showing syn-kinematic spreading zone and fracture zone structure and reactivation structure. Grey-scale background is Top Basement two-way travel time structure. Regional location of 3D study area is shown on Figure 2.

with the Cameroon Volcanic Line (Moreau *et al.* 1987; Deruelle *et al.* 1991). Moreau *et al.* (1987) mapped en echelon fault and lineament orientations along the landward portion of the Cameroon Volcanic Line and proposed that they were formed by 030° NE-oriented sinistral shear. Deruelle *et al.* (1987) mapped similar trending faults on Mount Cameroon and concluded that it formed above tension gashes produced by sinistral shear along the Foumban Shear Zone.

Tectonic model

We present a model for the compressional deformation of oceanic crust in the eastern Gulf of Guinea. In other examples of the compressional deformation of oceanic crust in the central Indian Ocean (Bull 1990; Bull & Scrutton 1992) and the Equatorial/South Atlantic (Jones 2003; Briggs *et al.* 2009), the structural fabric of the crust has played an important role depending on the regional stress orientation. In the central Indian Ocean the stress orientation is such that spreading ridge faults have been compressionally reactivated and fracture zone faults have been subject to strike-slip movement (Bull 1990; Bull & Scrutton 1992). In addition, long-wavelength buckle folds have formed with axes perpendicular to the regional stress (Beekman *et al.* 1996). In the Equatorial Atlantic major and minor fracture zones have reactivated in response to regional stresses (Jones 2003). In the South Atlantic, intra-crustal faults (with fracture

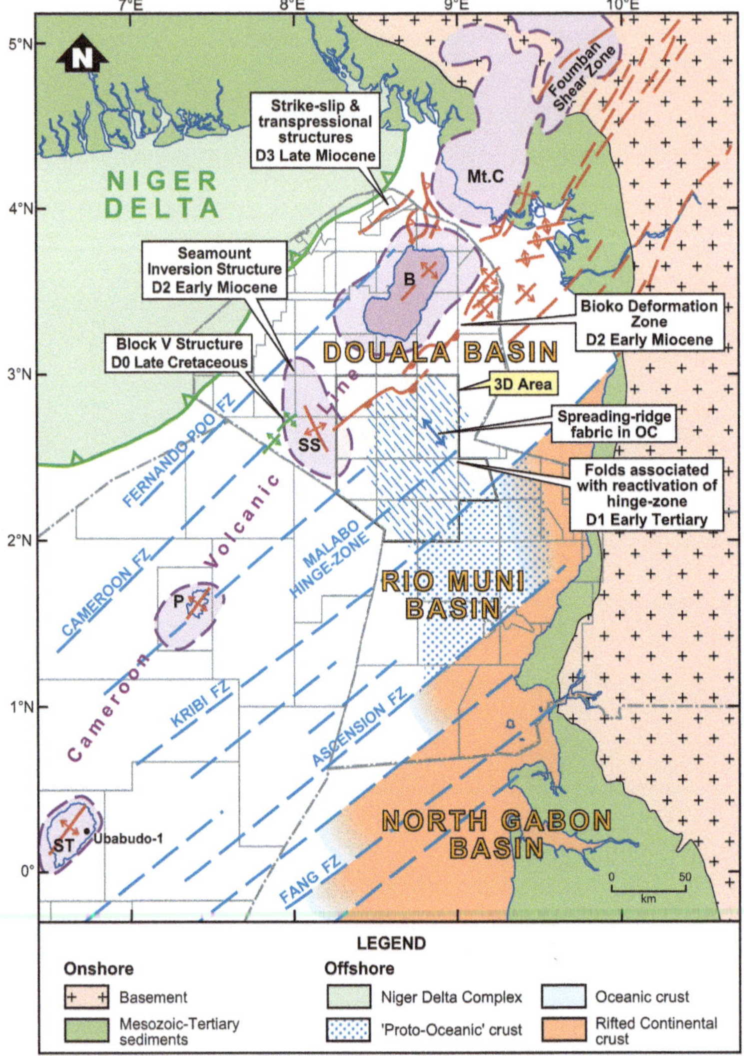

Fig. 9. Regional structural framework of eastern Gulf of Guinea. Structural fabric of oceanic crust is expressed as spreading ridge structure and fracture zones, including the reactivated Cameroon Fracture Zone here called the Bioko Deformation Zone. B, Bioko; FZ, fracture zone; Mt. C, Mt Cameroon; OC, oceanic crust; P, Príncipe; SS, Santesc Seamount; ST, São Tomé.

zone orientations) have reactivated as thrust faults in response to NW–SE-oriented regional stress (Briggs *et al.* 2009).

As in these other examples, it is proposed that deformation of the oceanic crust in the eastern Gulf of Guinea is a product of regional stress inducing deformation influenced or partly controlled by mechanically weak zones formed by the inherited syn-kinematic fabric of the oceanic crust. In addition to the syn-kinematic structural fabric (the spreading ridge structure and fracture zone faults), another fabric element is introduced here. This is

the line of weakness invoked by Fairhead & Wilson (2005) to explain the formation of the Walvis Ridge and was also used by them to explain the Cameroon Volcanic Line. In their model, the release of plate stress emanating from the ridge axis and propagating into the African (Atlantic) plate forms a long-lived crustal/lithospheric weak zone. They postulated that shear and wrench displacement along these zones can occur either over short or longer time periods depending on the duration of the stress imbalance. To explain the formation of the Cameroon Volcanic Line, Fairhead & Binks (1991) and

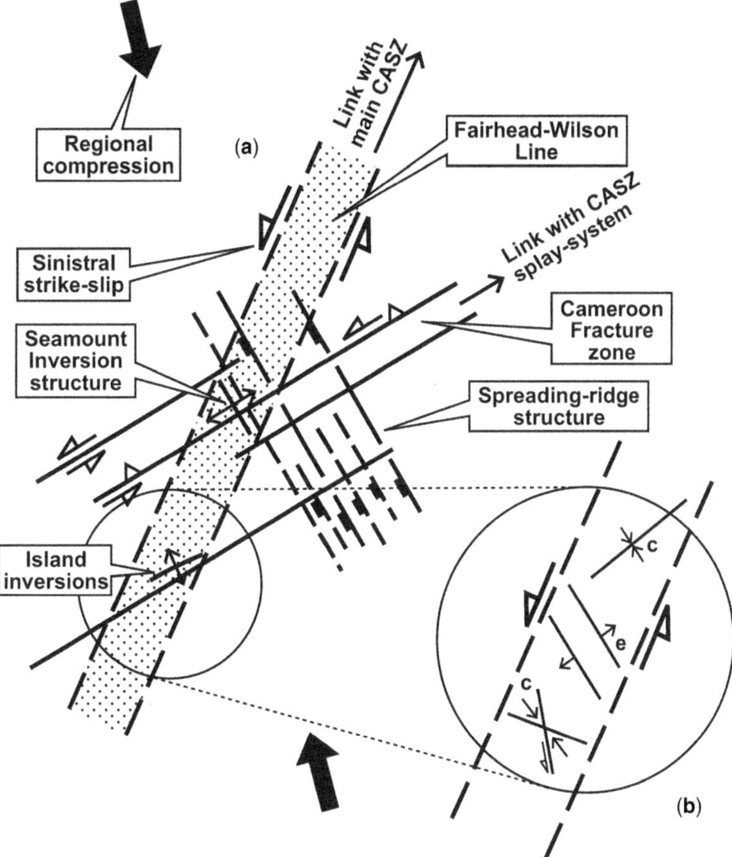

Fig. 10. (**a**) Model for deformation of the oceanic crust in the eastern Gulf of Guinea under regional NW–SE or NNW–SSE-oriented compression. (**b**) Orientation patterns of faults and fold axes associated with sinistral simple shear under transpression at fracture zone 'crossings'. c, compression; e, extension. CASZ, Central African Shear Zone.

Fairhead & Wilson (2005) invoked stress release via the Ascension Fracture Zone to reactivate a segment of the St Helena Seamount Chain. In the model presented here, this zone of crustal weakness is used to explain the alignment of the Cameroon Volcanic Line island/seamount inversion structures and we refer to it as a Fairhead–Wilson Line.

We attribute the deformation to regional NNW–SSE-directed compression (Fig. 10a). One manifestation of the deformation is folding and faulting involving transpressional reactivation of the NE–SW-oriented Cameroon Fracture Zone faults. Another is the induction of sinistral strike-slip movement (transpression) along the NNE–SSW-oriented Fairhead–Wilson Line (Fig. 10a). In this way, island/seamount inversion structures formed as a result of the complex interactions taking place where the fracture zones crossed the Fairhead–Wilson Line. Internal stress patterns induced along the Fairhead–Wilson Line and at fracture zone crossings (Fig. 10b) theoretically predict the transpressional island/seamount inversion structures and NE–SW extension across the spreading ridge structure, allowing volcanic breakthrough. The orientation of the Seamount Inversion Structure is attributed to deformation along a transpressional 'jog' between offset segments of the Cameroon Fracture Zone (Fig. 10a). Spreading ridge faults are oriented broadly parallel to the regional stress and are thereby unlikely to have seen significant reactivation.

Tectonic history

Compressional deformation is well documented on continental crust in the region. Transpressional reactivation or the inversion of faults along the Central African Shear Zone occurred during tectonic phases in the Santonian and late Eocene (Guiraud & Bosworth 1997; Reynolds & Jones 2004; Guiraud *et al.* 2005). Similarly, phases of deformation and tectonic uplift have been identified along the Rio Muni margin in the Santonian, late Eocene and mid- and late Miocene on the basis of seismic interpretation and apatite fission track analysis (Exploration Consultants Ltd 2001; Lawrence *et al.* 2002). The Abakaliki fold belt along the southern edge of the Benue Trough was formed by deformation in the Santonian (Guiraud & Bosworth 1997).

The interpretation of regional 2D and 3D seismic data outlined here, together with other proprietary seismic data in the area, indicates that four main phases of oceanic crust deformation/reactivation have taken place in the eastern Gulf of Guinea region:

- D_0 Santonian. The effects of a late Cretaceous deformation identified within the Seamount Inversion Structure, forming the Block V Structure, is attributed to this deformation phase (Fig. 3).This episode was used by Briggs *et al.* (2009) as a cause for the compressional deformation of oceanic crust in the central Gulf of Guinea region.
- D_1 Late Eocene. Observed as compressional reactivation along the Malabo Hinge Zone. Also recognized as thick-skinned deformation along the northeastern projection of the Bioko Deformation Zone to the south of Mount Cameroon (Fig. 3).
- D_2 Early Miocene. Observed as compressional reactivation along the Cameroon Fracture Zone forming the Seamount Inversion Structure and the Bioko Deformation Zone. Meyers & Rosendahl (1991) and Meyers *et al.* (1998) have shown that the inversion structures of Príncipe and São Tomé also formed at this time.
- D_3 Late Miocene. NW-verging, thick-skinned, probably strike-slip related, thrusting/reverse faulting observed along the western side of Bioko and Mount Cameroon (Fig. 9) (Lawrence *et al.* 2002).

The phases of compressional deformation of oceanic crust recognized in the late Cretaceous, late Eocene and early Miocene correlate with those observed on continental crust. The latter are generally attributed to Alpine-style far-field effects and correspond to the timing of changes in plate motion (Africa with respect to Eurasia) at 84 Ma (Santonian), 37 Ma (Pyrenean) and 22 Ma (Guiraud & Bosworth 1997).

A causal link is posited between deformation affecting oceanic crust and that affecting continental crust. It is proposed that the crust in both domains is responding to Alpine-style far-field stresses. In this way, a continuity of the reactivated Cameroon Fracture Zone system (the Bioko Deformation Zone) and the Fairhead–Wilson Line with the Foumban Shear Zone and its splays (part of the Central African Shear Zone) is described in terms of a continent–ocean tectonic link. This extends the regional tectonic model proposed by Fairhead & Binks (1991), which involved shear/wrench fault zones extending into continental Africa from the Gulf of Guinea and invoked a complex interaction between continental and oceanic tectonic processes.

This history of oceanic crust deformation and its association with the development of the Cameroon Volcanic Line prompts a comparison with another volcanic chain developed along the West African margin: the Canary Islands. This is also associated with the compressional deformation of oceanic crust (most prominently expressed on the island of Fuerteventura; Steiner *et al.* 1998), where the timing of deformation and uplift is reported to be pre-early Oligocene (Le Bas *et al.* 1986). Post-moat (i.e. early Miocene) uplift of Fuerteventura is also observed on offshore seismic data (Collier & Watts 2001). In this way, at least a temporal link can be invoked between the deformation of oceanic crust associated with the Canary Islands and the reactivation of major structures on the neighbouring continent forming the High Atlas (Anguita & Hernan 2000).

Petroleum endowment of the Douala Basin

The Douala-Rio Muni Basin is an evolving petroleum province with the discovery of several oil, gas and gas condensate fields in shallow and intermediate water depths on both sides of the Cameroon Volcanic Line (Fig. 11). The 3D seismic data described here form part of the activities now extending exploration out to the distal deep water parts of the basin (Fig. 11).

A summary of the petroleum geology (reservoirs and source rocks) established for the Douala-Rio Muni sector of the West African margin is presented in Figure 12. Petroleum systems are associated with source rock intervals recognized regionally as Lower Cretaceous (Aptian, Aptian–Albian), Upper Cretaceous (Albian–Turonian) and Palaeogene (see Exploration Consultants Ltd 2001) (Fig. 12). In the distal basin setting on oceanic crust, synrift and transitional source rocks of the continental Rio Muni margin can be discounted, so that the main source potential lies in the Upper Cretaceous and Palaeogene section. The Cretaceous source intervals potentially include the recognized Atlantic Cenomanian–Turonian oceanic anoxic event (OAE-2). Proof for the occurrence of Upper Cretaceous source rocks in the deep oceanic basin is provided by geochemical fingerprinting of oils recovered from seeps on the islands of São Tomé and Príncipe

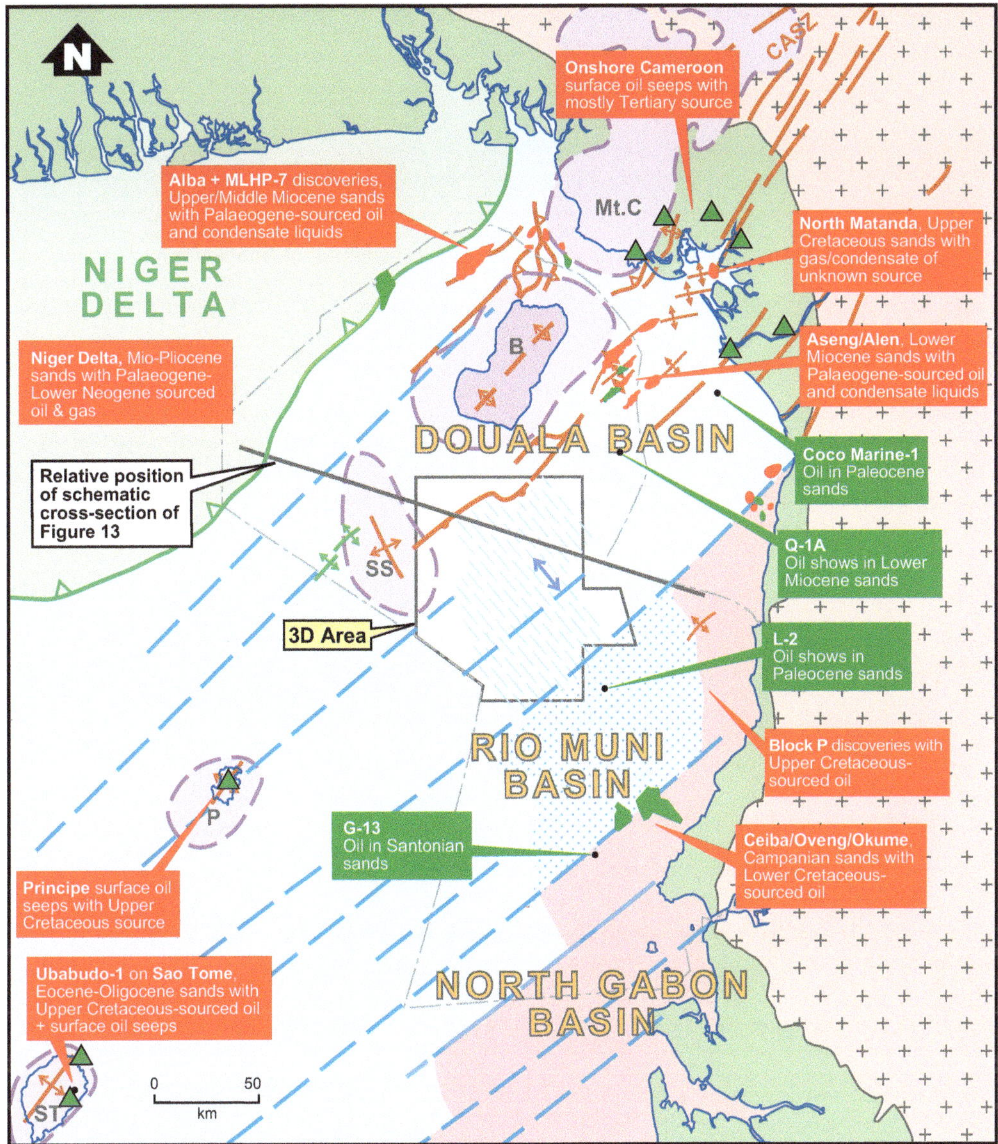

Fig. 11. Douala-Rio Muni Basin petroleum province. Gas and gas condensate fields, red; oil fields, green; surface oil seeps, green triangles. The relative position of schematic cross section in Figure 13 is shown. B, Bioko; CASZ, Central African Shear Zone; Mt. C, Mt Cameroon; P, Príncipe; SS, Santesc Seamount; ST, São Tomé.

(Exploration Consultants Ltd 2001; ANP-STP 2012*b*). Furthermore, because the São Tomé sample was recovered from oil seeping into the cellar of the Ubabudo-1 well and oil shows were reported in the cores from the Palaeogene sandstone interval (ANP-STP 2012*c*), this is likely to be reservoired oil. The Palaeogene source interval has been identified from the geochemistry of oils and condensate liquids from the producing fields of Alen-Aseng and Alba as well as oil seeps onshore Cameroon (unpublished internal Glencore report).

Proved or potential reservoirs recognized in fields or discoveries establish the important plays occurring in the eastern Gulf of Guinea (Figs 11 & 12):

• Cretaceous sands in Ceiba, Okume-Oveng, Venus (Block P) (Campanian), Jupiter (Block P)

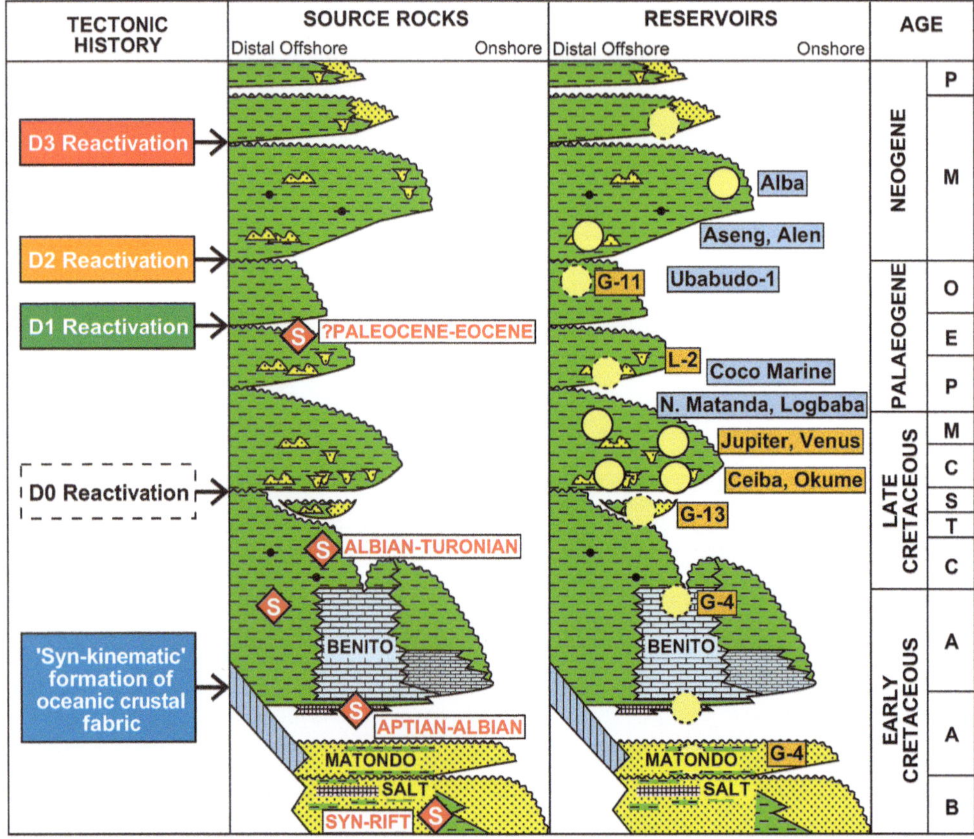

Fig. 12. Stratigraphic column for Douala-Rio Muni Basin System showing elements of petroleum geology related to tectonic history. Reservoirs attributed to fields (solid circles) and discoveries (dashed circles) in Douala Basin (blue) and Rio Muni Basin (orange). Source rock intervals depicted as red diamonds.

(Maastrichtian) and G-13 (Santonian) of Rio Muni; and Logbaba and North Matanda (Campanian–Maastrichtian) of the proximal Douala Basin.

- Palaeogene sands in Coco Marine (Paleocene) of the Douala Basin; Paleocene sands of L-2 of Rio Muni, Paleocene–Eocene–Oligocene sands in the extreme distal setting represented by Ubabudo-1 (drilled on São Tomé Island; ANP-STP 2012*c*).
- Lower Miocene sands in Aseng-Alen and Q-1A of the Douala Basin.
- Middle–Upper Miocene sands in the Alba and Etinde (MHLP-7) discoveries (Isongo Sands).

With the exception of the Ubabudo-1 sands, all are interpreted as variably confined (upper–middle fan) turbidites. The Paleocene–Eocene–Oligocene sands of Ubabudo-1 are interpreted as more distal lower fan sands (ANP-STP 2012*b*, *c*). They are non-volcanogenic and largely non-lithic in nature and therefore have an African continental provenance. Exploration studies have shown that Cretaceous and Palaeogene depositional systems have a principal transport direction westwards or northwestwards away from the Rio Muni margin, but that Lower Miocene depositional systems have a principle transport direction southwestwards along the axis of the Douala Basin. Although not the subject of this paper, the 3D seismic data described here confirm that these recognized Cretaceous and Tertiary reservoir fairways extend out to the distal parts of the Douala and Rio Muni basins.

The petroleum geology described here will be familiar to academic and industry researchers of Atlantic margin basins. The uniqueness of the Douala-Rio Muni Basin, however, is related to its tectonostratigraphic evolution and its association with the deformation of oceanic crust and the development of the Cameroon Volcanic Line. The episodic deformation of oceanic crust and the overlying sedimentary section has provided a critical

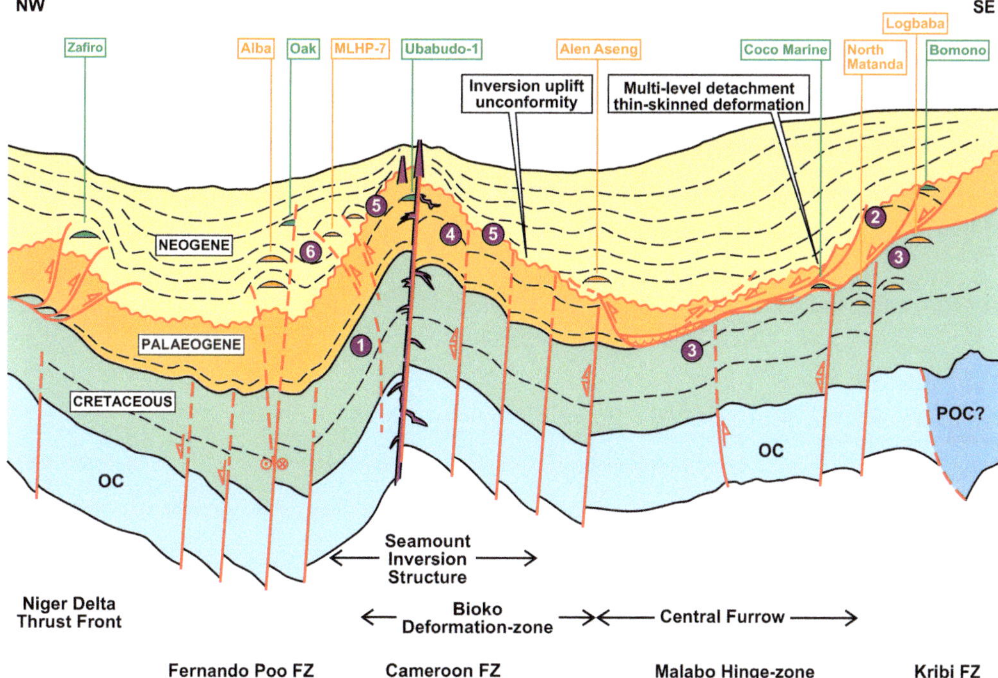

Fig. 13. Vertically exaggerated schematic structural cross-section of Douala-Rio Muni Basin and Cameroon Volcanic Line (relative position shown on Fig. 11) showing structural setting of hydrocarbon habitats (fields and discoveries; orange denotes wet gas condensate; green denotes oil). Purple circles denote deformation phases: 1, Santonian fracture zone reactivation (D_0); 2, Paleocene thin-skinned deformation; 3, Late Eocene reactivation of Malabo Hinge Zone (D_1); 4, Oligocene moat development; 5, Early Miocene folding and fracture zone reactivation (D_2); 6, Late Miocene strike-slip tectonics and thrusting (D_3). FZ, fracture zone; OC, oceanic crust; POC, proto-oceanic crust.

control on hydrocarbon habitats in the region. This control has been either direct by the formation of structural or stratigraphic–structural combination traps, or indirect by the control of reservoir sand distribution. The tectonic events and structures responsible for controlling hydrocarbon habitats are illustrated schematically in Figure 13.

These structural settings have not been fully explored in the deep water, distal parts of the Douala-Rio Muni Basin. Potential D_0- and D_2-related structures and D_2 inversion structures associated with the Bioko Deformation Zone are poorly explored. In addition, the Ubabudo-1 well on São Tomé Island provides a glimpse of the potential for large D_2 structures, such as the Seamount Inversion Structure, to contain oil in continentally derived reservoir sands. Although the occurrence of igneous intrusive rocks is an obvious risk in this setting, the interpretation has shown that volcanic and intrusive activity is localized and that a significant part of the sedimentary section is likely to be free of their deleterious effects on reservoir quality.

Late Cretaceous–Tertiary burial combined with a declining heat flow from cooling oceanic crust provides a progressive and uninterrupted maturation history for vertically separated source rock intervals in the Upper Cretaceous and Tertiary rocks. Both source rock intervals have the potential to have become oil-mature in different parts of the basin. The key control on which of the recognized source rock intervals becomes oil-mature favourably late is the amount of Mio-Pliocene burial, which is greater moving northeastwards in the basin. Because of the continuous maturation and hydrocarbon generation history of one or both source rocks and the episodic nature of the deformation, timing is not a major risk element in the Douala-Rio Muni Basin.

Conclusions

Interpretation of regional 2D and sub-regional 3D seismic data has highlighted the compressional deformation of oceanic crust in the eastern Gulf of

Guinea. This deformation has reactivated the syn-kinematic crustal structure, especially the Cameroon Fracture Zone, in several key tectonic episodes.

This deformation played a role in the evolution of the Cameroon Volcanic Line, where volcanic islands and seamounts are attributed to the inversion of fracture zone faults at intersections with a zone of crustal weakness called a Fairhead–Wilson Line. A mechanical link is invoked between this deformation of oceanic crust and that on continental crust in the form of transpressional movement on the Central African Shear Zone, described as a continent–ocean tectonic link.

The deformational history has influenced the tectonostratigraphic history and thus the petroleum endowment of the Douala Basin. The structure has provided a critical control on hydrocarbon habitats, either directly by the formation of structural or combination traps, or indirectly by controlling the deposition of reservoir sands. Late Cretaceous–Tertiary burial, combined with a declining heat flow from cooling oceanic crust, provides a progressive and uninterrupted maturation history for vertically separated source rock intervals in the Upper Cretaceous and Tertiary strata.

The authors thank Geoex International for allowing the use of 3D seismic data and the Ministry of Mines, Industry and Energy, Government of Equatorial Guinea for allowing the interpretation and presentation of regional 2D seismic data. C. Heine and W. Mohriak are thanked for their helpful reviews of the manuscript.

References

ADAMS, A.N., WIENS, D.A., NYBLADE, A.A., EULER, G.C., SHORE, P.J. & TIBI, R. 2015. Lithospheric instability and the source of the Cameroon Volcanic Line: evidence from Rayleigh wave phase velocity tomography. *Journal of Geophysical Research*, **120**, 1708–1727.

ALI, M.Y., WATTS, A.B. & HILL, I. 2003. A seismic reflection profile study of lithospheric flexure in the vicinity of the Cape Verde Islands. *Journal of Geophysical Research*, **108**, EPM-4-19.

ANGUITA, F. & HERNAN, F. 2000. The Canary Islands origin: a unifying model. *Journal of Volcanology and Geothermal Research*, **103**, 1–26.

AGENCIA NACIONAL DO PETROLEO DE SÃO TOMÉ E PRÍNCIPE (ANP-STP) 2010. Introduction to the Hydrocarbon Prospectivity of São Tomé and Príncipe, www.stp-eez.com

AGENCIA NACIONAL DO PETROLEO DE SÃO TOMÉ E PRÍNCIPE (ANP-STP) 2012a. 1st (EEZ) licensing round – Gulf of Guinea Regional Geology, www.stp-eez.com

AGENCIA NACIONAL DO PETROLEO DE SÃO TOMÉ E PRÍNCIPE (ANP-STP) 2012b. 1st (EEZ) licensing round – Petroelum Geology, www.stp-eez.com

AGENCIA NACIONAL DO PETROLEO DE SÃO TOMÉ E PRÍNCIPE (ANP-STP) 2012c. 1st (EEZ) licensing round – Ubabudo-1, www.stp-eez.com

BEEKMAN, F., BULL, J.M., CLOETINGH, S. & SCRUTTON, R.A. 1996. Crustal fault reactivation facilitating lithospheric folding/buckling in the central Indian Ocean. *In*: BUCHANAN, P.G. & NIEUWLAND, D.A. (eds) *Modern Developments in Structural Interpretation, Validation and Modelling*. Geological Society, London, Special Publications, **99**, 251–263, http://doi.org/10.1144/GSL.SP.1996.099.01.19

BINKS, R.M. & FAIRHEAD, J.D. 1992. A plate tectonic setting for the Mesozoic rifts of West and Central Africa. *Tectonophysics*, **213**, 141–151.

BRIGGS, S.E., CARTWRIGHT, J. & DAVIES, R.J. 2009. Crustal structure of the deepwater west Niger Delta passive margin from the interpretation of seismic reflection data. *Marine and Petroleum Geology*, **26**, 936–950.

BULL, J.M. 1990. Structural style of intra-plate deformation, Central Indian Ocean Basin: evidence for the role of fracture zones. *Tectonophysics*, **184**, 213–228.

BULL, J.M. & SCRUTTON, R.A. 1992. Seismic reflection images of intraplate deformation, central Indian Ocean, and their tectonic significance. *Journal of the Geological Society, London*, **149**, 955–966, http://doi.org/10.1144/gsjgs.149.6.0955

COLLIER, J.S. & WATTS, A.B. 2001. Lithospheric response to volcanic loading by the Canary Islands: constraints from seismic reflection data in their flexural moat. *Geophysical Journal International*, **147**, 660–676.

DE PLAEN, R.S.M., BASTOW, I.D., CHAMBERS, E.L., KEIR, D., GALLACHER, R.J. & KEANE, J. 2014. The development of magmatism along the Cameroon Volcanic Line: evidence from seismicity and seismic anisotropy. *Journal of Geophysical Research, Solid Earth*, **119**, 4233–4252.

DE VILLALTA, J.F. & ASSENS, J. 1967. Hallazgo de un ammonites Cretacico en la isla volcanica de Fernando Poo (Guinea ecuatorial espanola). *Acta Geologica Hispanica*, **11**, 117–118.

DERUELLE, B., N'NI, J. & KAMBOU, R. 1987. Mount Cameroon: an active volcano of the Cameroon line. *Journal of African Earth Sciences*, **6**, 197–214.

DERUELLE, B., MOREAU, C. ET AL. 1991. The Cameroon line: a review. *In*: KAMPUNZI, A.B. & LUBALA, R.T. (eds) *Magmatism in Extensional Structure Settings: the Phanerozoic African Plate*. Springer, Berlin, 275–327.

DUNLOP, H.M. & FITTON, J.G. 1979. A K-Ar and Sr-isotopic study of the volcanic rocks of the island of Principe, West Africa – evidence for mantle heterogeneity beneath the Gulf of Guinea. *Contributions to Mineralogy and Petrology*, **71**, 125–131.

EXPLORATION CONSULTANTS LTD 2001. *An Integrated Study of Structural and Stratigraphic Development and Source Rock Maturation History of the Rio Muni Basin, Equatorial Guinea*. Unpublished non-exclusive report.

ELSHEIKH, A.A., GAO, S.S. & LIU, K.H. 2014. Formation of the Cameroon Volcanic Line by lithospheric basal erosion: insight from mantle seismic anisotropy. *Journal of African Earth Sciences*, **100**, 96–108.

FAIRHEAD, J.D. 1988. Mesozoic plate reconstructions of the central South Atlantic Ocean: the role of the West and Central African rift system. *Tectonophysics*, **155**, 181–191.

FAIRHEAD, J.D. & BINKS, R.M. 1991. Differential opening of the Central and South Atlantic Oceans and the opening of the West African rift system. *Tectonophysics*, **187**, 191–203.

FAIRHEAD, J.D. & WILSON, M. 2005. Plate tectonic processes in the South Atlantic Ocean: do we need deep mantle plumes? *In*: FOULGER, G.R., NATLAND, J.H., PRESNALL, D.C. & ANDERSON, D.L. (eds) *Plates, Plumes and Paradigms*. Geological Society of America, Special Papers, **388**, 537–553.

FITTON, J.G. 1987. The Cameroon line, West Africa: a comparison between oceanic and continental alkaline volcanism. *In*: FITTON, J.G. & UPTON, B.G.J. (eds) *Alkaline Igneous Rocks*. Geological Society, London, Special Publications, **30**, 273–291, http://doi.org/10.1144/GSL.SP.1987.030.01.13

FITTON, J.G. & DUNLOP, H.M. 1985. The Cameroon line, West Africa, and its bearing on the origin of oceanic and continental alkali basalt. *Earth and Planetary Science Letters*, **72**, 23–38.

GAINA, C., TORSVIK, T.H., VAN HINSBERGEN, D.J.J., MEDVEDEV, S., WERNER, S.C. & LABAILS, C. 2013. The African Plate: a history of oceanic crust accretion and subduction since the Jurassic. *Tectonophysics*, **604**, 4–25.

GALLACHER, R.J. & BASTOW, I.D. 2012. The development of magmatism along the Cameroon Volcanic Line: evidence from teleseismic receiver functions. *Tectonics*, **31**, TC3018.

GEBCO 2014. Gridded bathymetry data, http://www.gebco.net/data_and_products/gridded_bathymetry_data/gebco_30_second_grid/ [last accessed 14 January 2015].

GENIK, G.J. 1993. Petroleum geology of Cretaceous–Tertiary rift basins in Niger, Chad, and Central African Republic. *AAPG Bulletin*, **77**, 1405–1434.

GUIRAUD, R. & BOSWORTH, W. 1997. Senonian basin inversion and rejuvenation of rifting in Africa and Arabia: synthesis and implications to plate-scale tectonics. *Tectonophysics*, **282**, 39–82.

GUIRAUD, R., BOSWORTH, W., THIERRY, J. & DELPLANQUE, A. 2005. Phanerozoic geological evolution of Northern and Central Africa. *Journal of African Earth Sciences*, **43**, 83–143.

HEDBERG, J.D. 1969. *A geological analysis of the Cameroon trend*. PhD thesis, Princeton University.

HEINE, C., ZOETHOUT, J. & MULLER, R.D. 2013. Kinematics of the South Atlantic rift. *Solid Earth*, **4**, 215–253.

JONES, E.J.W. 2003. Seismic evidence for pervasive deformation of oceanic sediments in the Eastern Equatorial Atlantic. *Geo-Marine Letters*, **23**, 102–109.

LAWRENCE, S.R., MUNDAY, S. & BRAY, R. 2002. Regional geology and geophysics of eastern Gulf of Guinea (Niger Delta to Rio Muni). *Leading Edge*, **21**, 1112–1117.

LE BAS, M.J., REX, D.C. & STILLMAN, C.J. 1986. The early magmatic chronology of Fuerteventura, Canary Islands. *Geological Magazine*, **123**, 287–298.

LEE, D.C., HALLIDAY, A.N., HALL, C.M. & FITTON, J.G. 1994. Similarities and differences in isotopic and chemical compositions between continental and oceanic basalts in the Cameroon Line. *In*: LANPHERE, M.A., DALRYMPLE, G.B. & TURRIN, B.D. (eds) *Abstracts of the 8th International Conference on Geochronology, Cosmochronology and Isotope Geology*, Berkeley, CA. US Geological Survey, Circular, 1107.

MAURIN, J.-C. & GUIRAUD, R. 1993. Basement control in the development of the early Cretaceous West and Central African rift system. *Tectonophysics*, **228**, 81–95.

MEYERS, J.B. & ROSENDAHL, B.R. 1991. Seismic reflection character of the Cameroon volcanic line: evidence for uplifted oceanic crust. *Geology*, **19**, 1072–1076.

MEYERS, J.B., ROSENDAHL, B.R., GROSCHEL-BECKER, H., AUSTIN, J.A. & RONA, P. 1996. Deep penetrating MCS imaging of the rift-to-drift transition, offshore Douala and North Gabon basins, West Africa. *Marine and Petroleum Geology*, **13**, 791–835.

MEYERS, J.B., ROSENDAHL, B.R., HARRISON, G.A. & DING, Z.-D. 1998. Deep-imaging seismic and gravity results from the offshore Cameroon Volcanic Line, and speculation of African hotlines. *Tectonophysics*, **284**, 31–63.

MOREAU, C., REGNOULT, J.-M., DERUELLE, B. & ROBINEAU, B. 1987. A new tectonic model for the Cameroon line, Central Africa. *Tectonophysics*, **142**, 317–334.

MOULIN, M., ASLANIAN, D. & UNTERNEHR, P. 2010. A new starting point for the South and Equatorial Atlantic Ocean. *Earth-Science Reviews*, **98**, 1–37.

MOUNDI, A., WANDJI, P. *ET AL.* 2007. Les basalts Eocene's a affinite transitionelle du plateau bamoun, temoins d'un mantellique enrichi sous la ligne volcanique du Cameroun. *Comptes Rendus Geosciences*, **339**, 396–406.

NJILAH, I.K. 1991. *Geochemistry and petrogenesis of Tertiary–Quaternary volcanic rocks from Oku-Ndu area, N.W. Cameroon*. PhD thesis, University of Leeds.

NJOME, M.S. & DE WIT, M.J. 2014. The Cameroon Line: analysis of an intraplate magmatic province transecting both oceanic and continental lithospheres: constraints, controversies and models. *Earth-Science Reviews*, **139**, 168–194.

REYNOLDS, D.J. & JONES, C.R. 2004. Tectonic evolution of the Doba and Doseo Basins, Chad: controls on trap formation and depositional setting of the Three Fields area, Chad. *AAPG Bulletin*, **88**, 15–16.

ROSENDAHL, B.R. & GROSCHEL-BECKER, H. 2000. Architecture of the continental margin in the Gulf of Guinea as revealed by reprocessed deep-imaging seismic data. *In*: MOHRIAK, W. & TAIWANI, M. (eds) *Atlantic Rifts and Continental Margins*. American Geophysical Union Geophysical Monograph Series, **115**, 85–103.

SANDWELL, D.T., MULLER, R.D., SMITH, W.H.F., GARCIA, E. & FRANCIS, R. 2014. New global marine gravity model from CryoSat-2 and Jason-1 reveals buried tectonic structure. *Science*, **346**, 65–67.

SETON, M., MULLER, R.D. *ET AL.* 2012. Global continental and ocean basin reconstructions since 200 Ma. *Earth-Science Reviews*, **113**, 212–270.

STEINER, C., HOBSON, A., FAVRE, P., STAMPFLI, G.M. & HERNANDEZ, J. 1998. Mesozoic sequence of Fuerteventura (Canary Islands): witness of early Jurassic sea-floor spreading in the central Atlantic. *Geological Society of America Bulletin*, **110**, 1304–1317.

TURNER, J.P., ROSENDAHL, B.K. & WILSON, P.G. 2003. Structure and evolution of an obliquely-sheared continental margin: Rio Muni, West Africa. *Tectonophysics*, **374**, 41–55.

USGS 2014. Shuttle Radar Topography Mission (SRTM), https://lta.cr.usgs.gov/SRTM [last accessed 14 January 2015].

WILSON, P.G., TURNER, J.P. & WESTBROOK, G.K. 2003. Structural architecture of the ocean–continent boundary at an oblique transform margin through deep-imaging seismic interpretation and gravity modelling: equatorial Guinea, West Africa. *Tectonophysics*, **374**, 19–40.

The Campanian quartz 'claystone' conundrum of the African Transform Margin

ALLEN BROWN*, SCOTT BIRKHEAD, DAVID MCLEAN,
PHILIP TOWLE, HOWARD WHITE & YAFEI WU

Anadarko Petroleum Corporation, The Woodlands, Texas 77380, USA

**Corresponding author (e-mail: allen.brown@anadarko.com)*

Abstract: An unusual deep-water lithofacies has been penetrated by numerous wells along the equatorial margin of West Africa by Anadarko Petroleum and other operators. An exploration well drilled in 2009 was initially thought to contain very thick, high-quality sands in the Lower Campanian section. Wireline log interpretations calculated the section to be a quartz-rich reservoir rock with 67–83% quartz based on X-ray diffraction analysis and effective porosity values as high as 25%. This petrophysical analysis appeared to differ from the well site geological description of 200 m of 'claystone'. The conflict between field observations and the interpreted wireline data led to the re-evaluation of earlier wells and interest in gathering data from this unusual lithofacies from subsequent wells. Field observations (from wellsite geologists) derived from this Lower Campanian section have been consistent in describing this lithofacies as 'claystone'. Sidewall cores from several of the deep-water wells that penetrated this lithofacies were also consistently described as claystones. These descriptions were from visual inspection and conclusions from laboratory analysis utilizing scanning electron microscopy, X-ray diffraction analysis, laser particle size analysis and thin section point counts as evaluation tools. The sidewall cores typically contain >90% silt- and clay-sized particles with extremely low measured permeability as a result of highly reduced pore throat diameters. With the presence of such small grain sizes, these samples have high measured seal capacities. It is important to recognize that this unusual lithofacies exists and can easily be mistaken for a high-quality, quartz-rich sand when, in fact, it is a very low permeability 'claystone' with a high percentage of quartz. Wireline data over this lithofacies alone can be very misleading and rock samples must be analysed to validate or invalidate the log interpretations. Misidentifying this lithofacies as an exploration target could have obvious financial ramifications. The challenge of this regional study was to formulate a model that could explain the presence of this deep-water lithofacies over an extensive area encompassing several basins with different sediment source provenances. Based on well sample analysis, palaeoclimatic research and the use of modern day analogue examples, the deposition of aeolian-derived clay-sized quartz particles is proposed as the most likely source of this lithofacies.

An unusual quartzose 'claystone' lithofacies has been observed in many of the 65+ deep-water wells (exploration, appraisal and development) drilled along the African Transform Margin between Sierra Leone and Benin, a distance of approximately 1800 km. The locations of these wells span three sedimentary basins with different provenance terranes providing sedimentary fill for deep-water, palaeodepositional environments characterized by mud-dominated turbidites, sand-dominated turbidites and mass transport deposits. More than 40 of these wells have penetrated a 90–275 m thick Middle to Lower Campanian section that has been described in the field as a medium to dark grey non-calcareous 'claystone', but which appears to be sand-rich based on the wireline log interpretation. Similar 'claystone' descriptions have been reported from the same stratigraphic interval based on laboratory analysis. Petrophysically, this sequence is commonly interpreted as a porous sandstone with gamma ray (GR) values averaging 30–40 GAPI units, but with some values as low as 15 GAPI. The neutron density responses are similar to those expected from relatively 'clean' sandstones with low volumes of clay (V_{clay}). A regional cross-section of the GR curves shows the presence of this correlative interval across the study area (Fig. 1). This paper reviews the properties of this quartzose 'claystone' lithofacies using wellsite, wireline and laboratory data and presents a model within which the scale of the lithofacies distribution can be suitably accommodated.

Regional context

The study area for this unusual deep-water facies encompasses the entire African Transform Margin and stretches from Sierra Leone in the west to Benin in the east. It consists of multiple basins,

From: SABATO CERALDI, T., HODGKINSON, R. A. & BACKE, G. (eds) 2017. *Petroleum Geoscience of the West Africa Margin*. Geological Society, London, Special Publications, **438**, 27–48.
First published online December 23, 2015, http://doi.org/10.1144/SP438.3

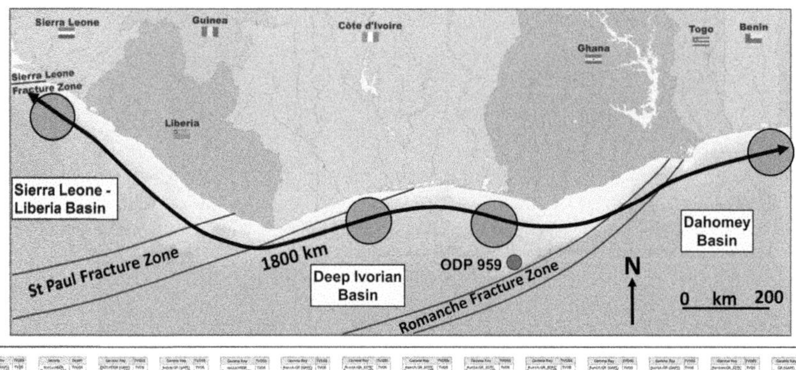

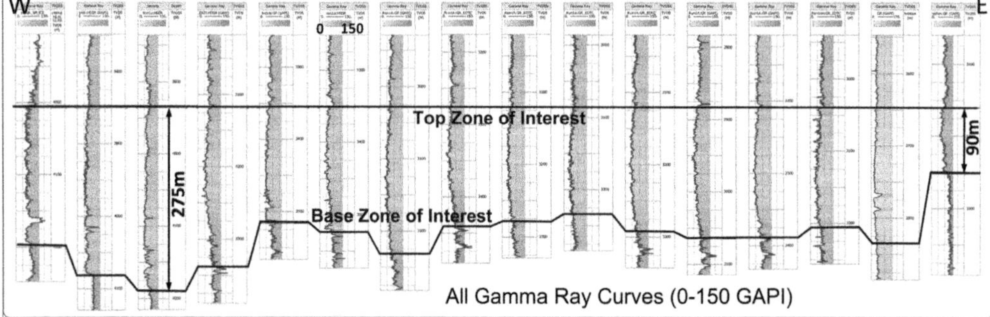

Fig. 1. Well gamma ray curves showing the regional extent of the 'Campanian quartz claystone' across three basins. The circles represent the areas with well data used in this study.

including the Sierra Leone–Liberia Basin, the Deep Ivorian Basin and the Dahomey Basin, separated by two major fracture zones, the St Paul Fracture Zone to the NW and the Romanche Fracture Zone to the SE.

The basins of the African Transform Margin share a common evolutionary history. Tectonic reconstructions indicate that this part of the central and southern Atlantic Ocean were the last to separate, with a continent–continent connection across the Romanche and St Paul fracture zones until Late Albian or Early Cenomanian times (Mascle & Blarez 1987; Mascle *et al.* 1988; Antobreh *et al.* 2009).

The sedimentary basins of the western African Transform Margin are floored by a synrift section. Early Cretaceous rifting propagated westwards and was controlled by dextral transtension on the incipient fracture zones, occurring over a 10–11 myr period commencing in the Late Barremian (Nemčok *et al.* 2012). The earliest onset of rifting is dated as Late Barremian based on the age of the basal transgressive sediments in the Benue Trough (Popoff 1988; De Matos 1992; Giraud & Maurin 1992); thick Aptian and Barremian age sediments have also been penetrated offshore Ghana (Akpati 1978; Kesse 1986). Rifting and the deposition of predominantly non-marine sediments together with interbedded intrusive volcanics in a 'synrift' sequence continued through to at least the Mid-Albian, with the rifting to drifting transition occurring in the Late Albian. An open marine connection through the Equatorial Atlantic by the late Albian has been established by results from ODP Leg 159 (Pletsch *et al.* 2001) and recent exploration drilling results along the western African Transform Margin confirm widespread marine sedimentation by this time.

The transition from rift to drift sedimentation is marked by a strong angular discordance, the Mid-Cretaceous Unconformity. Early post-rift sedimentation, from the late Albian to late Cenomanian, is characterized by an onlapping sequence of dominantly outer shelf to upper bathyal fine-grained clastic sediments deposited in a series of basins undergoing rapid thermal subsidence. These sediments were deposited in discrete, mineralogically distinct sub-basins along the Ivorian margin (Wozazek & Krawinkel 2002) and a similar setting throughout the African Equatorial Margin can be interpreted using regional seismic data. The overlying late Cenomanian to Eocene section is dominated by marine sediments deposited on a passive margin (Clifford 1986). Shallow-marine sedimentation occurred in the areas proximal to the present-day coastline, while further offshore the section is dominated by deep-marine clastic sediments consisting of fine-grained hemipelagic shales and claystones, together with upper slope to basin floor turbidites consisting of coarse-grained sandstones and transported fine-grained silts, shales and claystones. A marked regionally extensive erosional unconformity dated as Oligo-Miocene is observed throughout the study area, after which shallow- to deep-marine sedimentation was re-established. Generalized stratigraphic columns covering the study area are shown in Figure 2 (Brownfield & Charpentier 2006; Conn & Rodriguez 2011).

Early exploration along the Ivorian margin was concentrated in shallow water along the shelf and was focused on structural targets with reservoirs of mid- to late Albian age. The discovery of the Jubilee Field (Ghana) in 2007 stimulated deep-water drilling throughout the African Transform Margin, primarily targeting Turonian-aged stratigraphically defined traps. In addition to post-Jubilee discoveries in the Turonian, oil has been found in the Cenomanian, Coniacian, Campanian and Maastrichtian strata. The focus of this paper is the lower to middle Campanian sections found in the deep-water parts of the study area. The Campanian is characterized by the deposition of numerous stacked turbidite systems that were deposited in a deep-water mid- to upper slope environment. Regionally, the abundance of these Campanian turbidite systems has caused considerable seismic imaging issues for

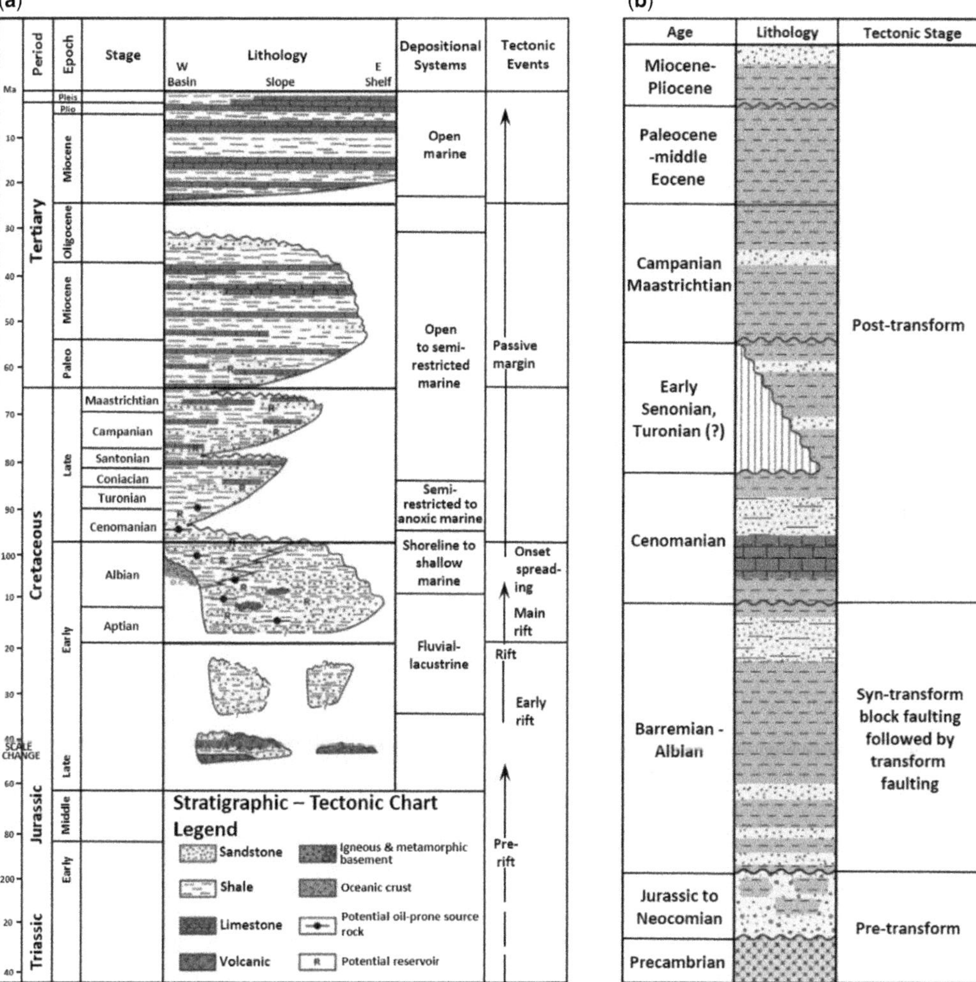

Fig. 2. Generalized stratigraphic columns for (a) the Liberia–Sierra Leone Basin (modified from Conn & Rodriguez 2011, data courtesy of TGS) and (b) the Ivorian Basin (modified from Brownfield & Charpentier 2006, data courtesy of USGS).

targets below the Campanian. These Campanian turbidite deposits contain sands with excellent reservoir quality, but they are only occasionally found to be hydrocarbon-bearing. This is primarily due to challenged migration pathways from the Ceno-Albian source rock. Accumulations have primarily been found within structures associated with faulting that assisted hydrocarbon migration from the deeper source. Located between these major Campanian sand-rich turbidites are time-equivalent deposits of a very unusual lithofacies. This lithofacies appears to be sand-rich based on log analysis, but further study indicates it to be a non-reservoir rock.

Observations and analytical data

Wireline log interpretation anomalies

In addition to the regional lower 'Campanian quartz claystone', several examples of thinner quartz-rich claystone lithofacies have been observed in Turonian, Santonian and Maastrichtian strata across the study area. Of particular interest is an example of 'apparent sands' found in the upper Turonian section and shown on the wireline log in Figure 3. Initial log analysis indicated that this 40 m zone consisted of about 65% net sand within an inter-bedded sand/claystone sequence. However, the wellsite geologist described this zone as medium to dark grey, non-calcareous claystone, which we questioned at the time. Five rotary sidewall cores (SWCs) were acquired to evaluate the reservoir quality and also to use in fluid extraction analysis to identify trace amounts of hydrocarbons. The results of the SWC analysis were surprising. The cleanest sands cored had average GR values of 30–42 GAPI. The average porosities of the SWCs ranged between 13 and 17%, but the measured permeability ranges were very low at 0.0002–0.05 mD. The average quartz content determined by X-ray diffraction (XRD) analysis is 73%, with a high of 80%. The average laser particle size analysis (LPSA) indicated 59% silt-sized and 31% clay-sized grains. Thin sections revealed a siliceous shale with silt- and clay-sized quartz. The work on these Turonian SWCs was the first actual laboratory analysis of this quartz-rich claystone facies that was thought to be reservoir sand.

Another quartz-rich claystone lithofacies had previously been recognized in the lower to mid-Campanian, where it was much thicker and more widespread than the Turonian example. This section was always identified as claystone by wellsite geologists and yet petrophysical analysis indicated a good quality reservoir sandstone. The reason for this apparent contradiction was never fully understood. However, the laboratory analysis from these thin upper Turonian sands seemed to validate what the wellsite geologists were describing as claystones. With this insightful information, a regional study of the lower to mid-'Campanian quartz claystone' was undertaken to determine whether the petrological characteristics were similar to the Turonian 'sands' studied previously.

A computer-processed interpretation (CPI) log shows the difference between this quartzose 'claystone' lithofacies and the contrasting responses of the shales above and below it in example 1 (Fig. 4). The shale intervals at the top and base of the illustrated well-log cross-section have common GR and neutron density responses. Shales commonly have high total porosities with virtually no effective porosity because few pore spaces are interconnected, which results in a low permeability. The cleanest GR section (3900–4100 m) showed a neutron density response trending towards a clean sand with a total porosity of 15–25%. The interpreted effective porosities displayed on the CPI log indicated values nearly as high as the total porosity. This suggested a high percentage of interconnected pore throats. As a result of the contrasting interpretations of claystone versus high-porosity sandstone, drill cuttings in this well and cuttings or SWCs in other wells were subjected to mercury injection capillary pressure (MICP) seal analysis in the laboratory. The results proved that the permeability in the tested samples was extremely low, of the order of 10^{-3} mD. These data, which appeared to contradict the CPI interpretation of the wireline data, proved problematic. We might conclude that the effective porosity was accurate, i.e. the pore throats were connected, but the pore spaces were so small, generally 1–2 μm in diameter, that the measured permeability was low. The small pore sizes are a result of the small grain sizes that dominate the rock framework: 0.001–0.062 mm (10–4 ϕ on the 1922 Wentworth grain size scale).

An early version of the CPI log analysis of example 1 indicated that 95% of the interval between 3900 and 4100 m was net sand with >10% porosity. Later editions of the CPI logs, which incorporated the laboratory analyses, indicated that the lithology in Figure 4 is a lithofacies composed of a high percentage of clay- and silt-sized quartz grains. We have coined this unusual lithofacies the 'Campanian quartz claystone conundrum' or just the 'Conundrum'.

Mercury injection capillary pressures seal analysis

Several avenues of investigation were undertaken to help understand this high-quartz 'claystone'. A very valuable tool was MICP seal analysis, which

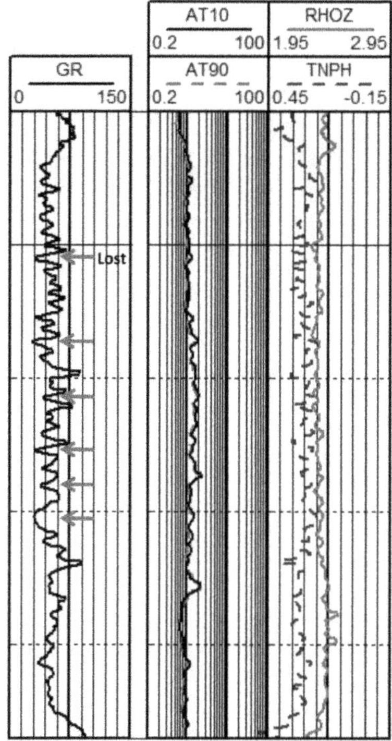

Fig. 3. Wireline log with locations of sidewall cores targeting clean thin sands for reservoir analysis. The top core was lost.

helped to validate either the wireline results or the sample descriptions from the field and laboratory. The premise was that clean, high-porosity and permeable sandstone would yield low capillary entry pressures (P_e) of <10 psia. However, the P_e values derived from the Conundrum samples between 4010 and 4100 m in example 1 (Fig. 4) were 735–1050 psia. This equates to a seal capacity for an oil column of 475–805 m at 7.5% mercury (Hg) saturation. For gas, the seal capacity ranges between 235 and 395 m for the same Hg saturation. A mercury saturation of 7.5% is frequently used to indicate the point at which the hydrocarbon phase in the seal rock is continuous enough to cause the seal to leak. These results support the field and laboratory observations for a quartz-rich claystone versus an initial wireline interpretation of high-porosity reservoir sandstone. The MICP tests were performed on the largest drill cuttings sampled from five 10 m intervals in the Conundrum interval as no SWC was acquired. These tests on cuttings were run a few years after the well was drilled as part of the emerging Lower Campanian Conundrum study. The permeability derived from the MICP analysis ranged between 0.006 and 0.02 mD. The

MICP pressure plot and wireline log with the Conundrum sample intervals tested highlighted are shown in Figure 5.

A different MICP plot for SWCs below the Conundrum section of example 1 is shown in Figure 6a. The original seal analysis in this section of the well was derived from five SWCs that were acquired in 'clean' shales (>75 GAPI) with locations shown to the right of the GR curve. The objective was to determine the potential seal capacity of the three top seals as well as two intra-formational seals. The seal capacities of these shales were extremely high, with an oil column height potential in excess of 3000 m (1450 m for gas) with 7.5% Hg saturation. One SWC from a thin sand at 4202.6 m was tested for permeability utilizing MICP analysis so that the results could be compared with the CMS-300 porosity–permeability for thin sands immediately above and below 4200 m. The first-entry pressure (P_e) for this sandstone was an extremely low 7 psia with a measured permeability of 36 mD. The zoomed-in wireline log in Figure 6b shows the thin sand that was tested along with the results of several claystone seals immediately above and below. The first Hg entry pressures for the Conundrum facies cuttings were 100–150 times greater than that of the thin sandstone tested at 4202.6 m. The permeability derived from MICP analyses of the Conundrum lithofacies was 1800–6000 times less than the permeability of the sandstone at 4202.6 m. This is a significant difference in permeability values between the Conundrum zone that was originally calculated as a good quality reservoir sand (15–25% porosity) and the vertically proximal sandstones that are true reservoirs with low P_e values and much higher measured permeabilities.

Analyses were performed to determine the whole-rock mineralogy and particle size distribution. The XRD results indicated that the Conundrum interval in example 1 was composed of 71–83% quartz, with most of the remaining fraction composed of clay minerals. LPSA showed that this 'quartz claystone' section consisted of 51% clay-sized particles and 45% silt-sized particles. Even though there was a high percentage of quartz grains, the small grain size was the contributing factor to such a low permeability. The CPI log correctly identified the presence of interconnected pore space; however, for the silt- and clay-sized grains, the pore throats were small and the resultant low permeability was readily demonstrated by the high P_e values of the MICP seal analysis. Scanning electron microscopy (SEM) images at 40 000× magnification revealed pore diameters of only 1–2 μm. A similar analytical approach was used to evaluate several wells from the greater study area. The results were similar to this example.

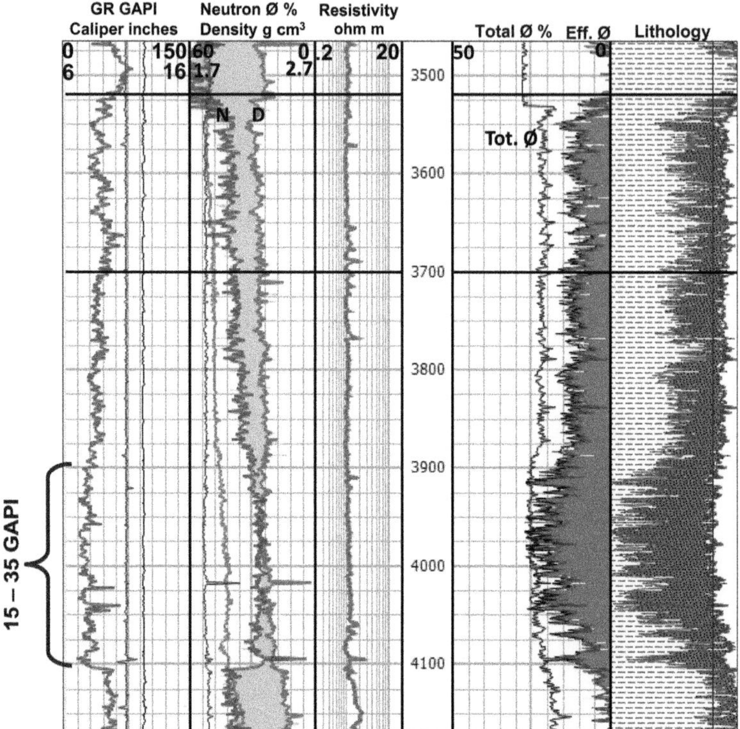

Fig. 4. Example 1 of the Conundrum section characterized by low gamma ray results, but described as a claystone. The wireline log indicates a high-quality reservoir sand between 3900 and 4100 m, but this tested as a competent seal .This zone contains 71–83% quartz based on X-ray diffraction analysis. Particle size analysis indicated 45% silt-sized and 51% clay-sized grains.

Example 2 in Figure 7 shows wireline and lithology logs from two wells located 6 km apart. The wellsite lithology log and the wireline GR–resistivity responses of the Conundrum were compared. The GR values ranged between 20 and 45 GAPI. Again, the same anomaly was observed of an 'apparent sand' interval (based on the GR log) that was described as 'claystone', but tested as a competent seal. MICP seal analysis was performed for four different sample intervals and the oil column height potential ranged between 370 and 510 m at 7.5% Hg saturation. For gas, the seal capacity ranged between 185 and 250 m for the same Hg saturation. The permeability was between 0.001 and 0.009 md with calculated porosities between 10.6 and 18.6%.

Seismic examples

Example 3 is a 42 km east–west seismic line (seismic line 1) through well 2 (Fig. 8), which shows that the Conundrum interval in well 2 is characterized by relatively low-amplitude, parallel reflectors interpreted as a quiet, deep-water depositional environment. This contrasts with the high-energy, stacked Campanian turbidite systems located east of well 2. There is a difference in the type of Campanian slope facies between well 2 and wells drilled to the east that penetrate these Campanian slope–channel complexes. The Conundrum interval is preserved in the subsurface in areas located between major, Lower Campanian, channel systems that are typically sand-rich. These sand-rich systems have effectively removed the Conundrum by erosion.

Example 4 is a NW–SE seismic line (seismic line 2) that illustrates the difference in seismic facies between well 2 (Conundrum) and well 3 (sandy Campanian system) (Fig. 9). The mapped lower reflector (yellow) near the base of the Conundrum in well 2 can be mapped to the base of the Conundrum in well 3. However, the mapped upper reflector (pink) near the top of the Conundrum in well 2 is clearly terminated by the Campanian channel system and is not present in well 3.

Well 3 is unique in that two different lithofacies have been interpreted within the Conundrum interval. Both wells have clean GR sections, but very

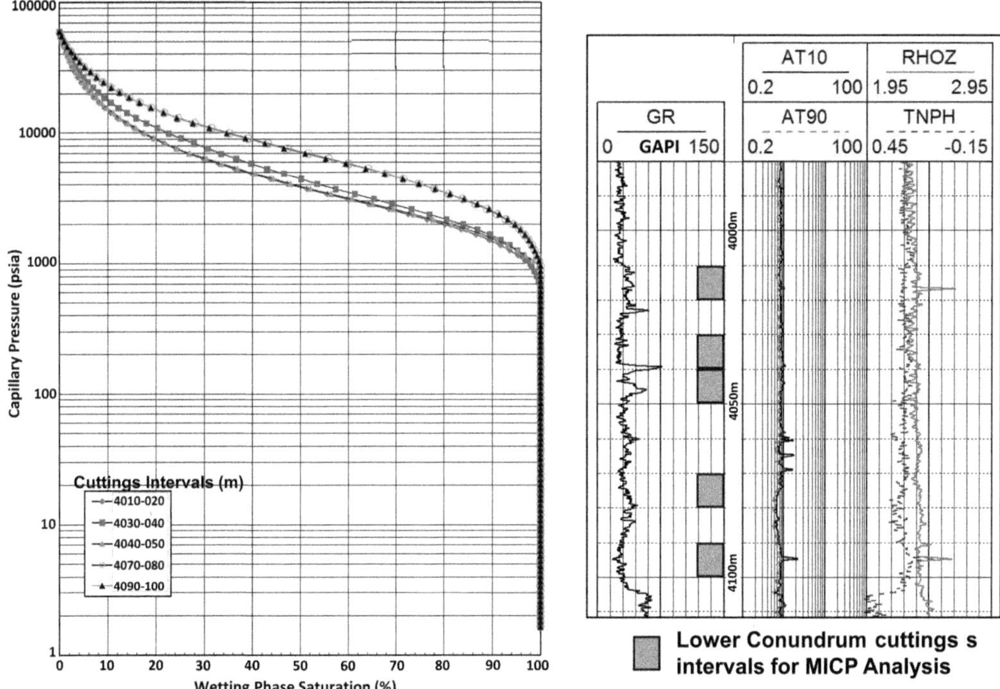

Fig. 5. Mercury injection capillary pressure (MICP) results showing relatively high entry pressures (P_e) from lower 'Conundrum' cuttings. These pressures equate to potential seal capacities for an oil column of 475–805 m at 7.5% Hg saturation. The seal potential for a gas column is 235–395 m at the same Hg saturation. Intervals tested are highlighted on the log.

different lithological descriptions. The cross-section in Figure 10 shows the corresponding wellsite lithology logs tied to the wireline log curves.

Both wells have similar GR and resistivity log responses, but these sections are different based on the observed lithologies. A thick Conundrum interval is preserved in well 2, but well 3 penetrated a much thinner Conundrum section at the bottom of the overall interval. Wells 2 and 3 have the characteristic dark grey to black, non-calcareous claystone in the Conundrum interval. However, well 3 has a 100 m section consisting of medium- to coarse-grained, moderately sorted sand associated with a slope–channel system. This coarser-grained system is slightly younger and has removed or replaced most of the Conundrum. A thin claystone interval is present above this coarse sand-rich system and is interpreted to be either an abandonment phase lithofacies or may represent an 'off-axis' penetration. There is a marked difference in the description of this claystone compared with the Conundrum. It is much lighter in colour and more calcareous. The youngest Campanian fan system immediately above this abandonment or off-axis

facies has eroded into the older systems, as seen on seismic line 2. The Lower Campanian lithofacies map in Figure 11 shows the locations of wells 1, 2 and 3 and seismic lines 1 and 2.

Petrographic analysis

Example 5 compares two intervals from the same wellbore that have SWCs with similar GR values of 45 GAPI (Fig. 12). However, very different lithologies were interpreted from these SWCs by visual inspection.

Two are cores of coarse-grained sandstone (1 and 2 on the left) and the other is a claystone core (SWC A). The depth difference between the two zones is several hundred metres. The open-hole GR log curve shows a 120–135 GAPI shale baseline against a 45 GAPI Conundrum facies value. LPSA from SWC A indicates the presence of 71% clay-sized and 29% silt-sized grains (Fig. 13).

An expanded lithology log below core A shows an additional 100 m of section (Fig. 14). Two additional core points, B and C, are marked on the logs. Cores A and C were studied using XRD, thin

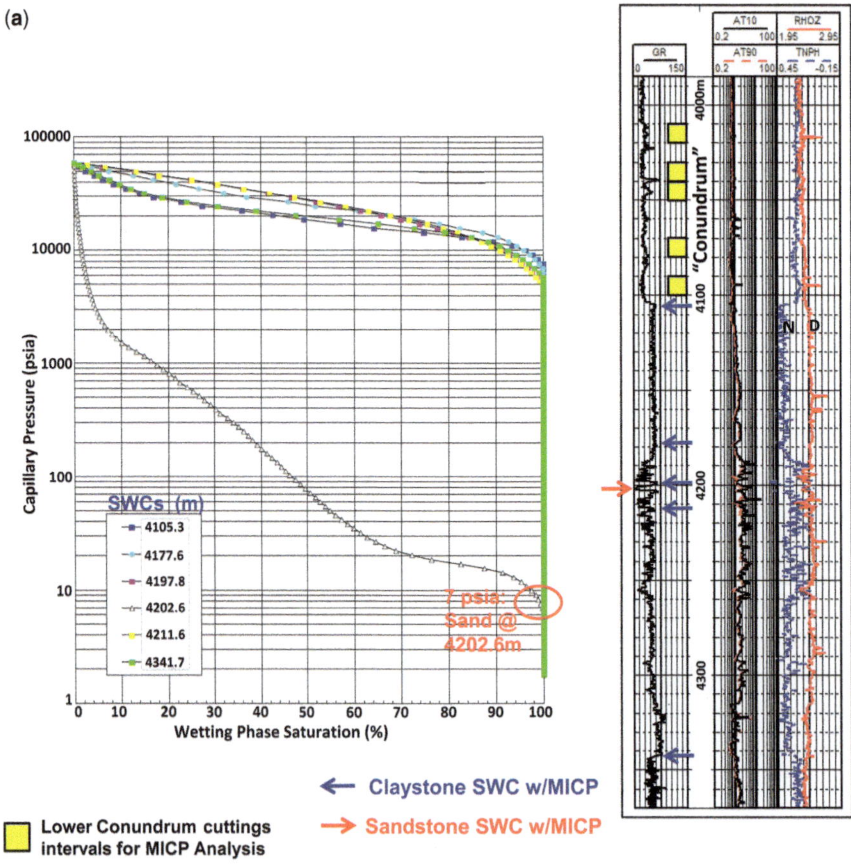

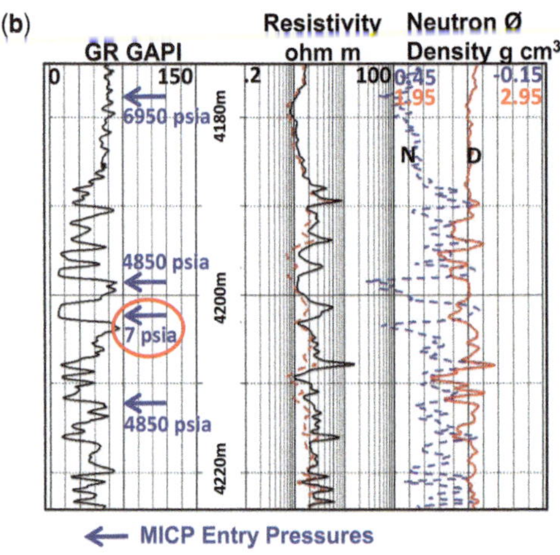

Fig. 6. (a) Mercury injection capillary pressure (MICP) pressure plot from sidewall cores (SWCs) with locations of cores on wireline log. Blue arrows indicate SWCs targeted specifically for top and inter-formational seal capacities and have first-entry pressures between 4850 and 7600 psia. This equates to potential seal capacities for an oil column of 3000–3900 m at 7.5% Hg saturation. Note first-entry pressure of 7 psia in a high-quartz sandstone at 4202.6 m versus Conundrum entry pressures of 735–1050 psia from previous figure. (b) Zoomed-in wireline log from (a) with SWC locations and MICP first-entry pressures. Tests from three claystone seals and one thin porous sand.

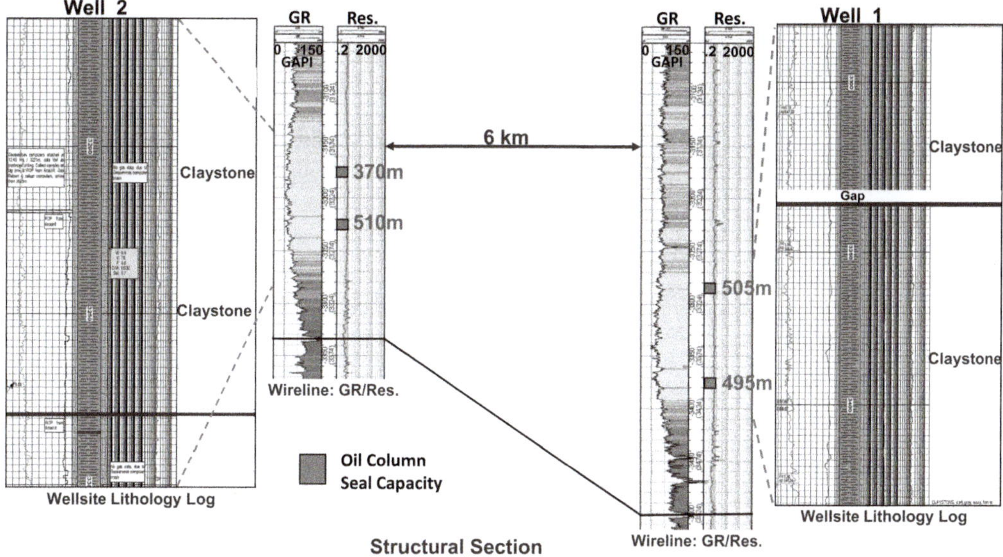

Fig. 7. Example 2 of low gamma ray (GR) zones described as claystones with good seal capacities up to 510 m potential oil column at 7.5% Hg saturation. Potential for a gas column is up to 250 m at the same Hg saturation.

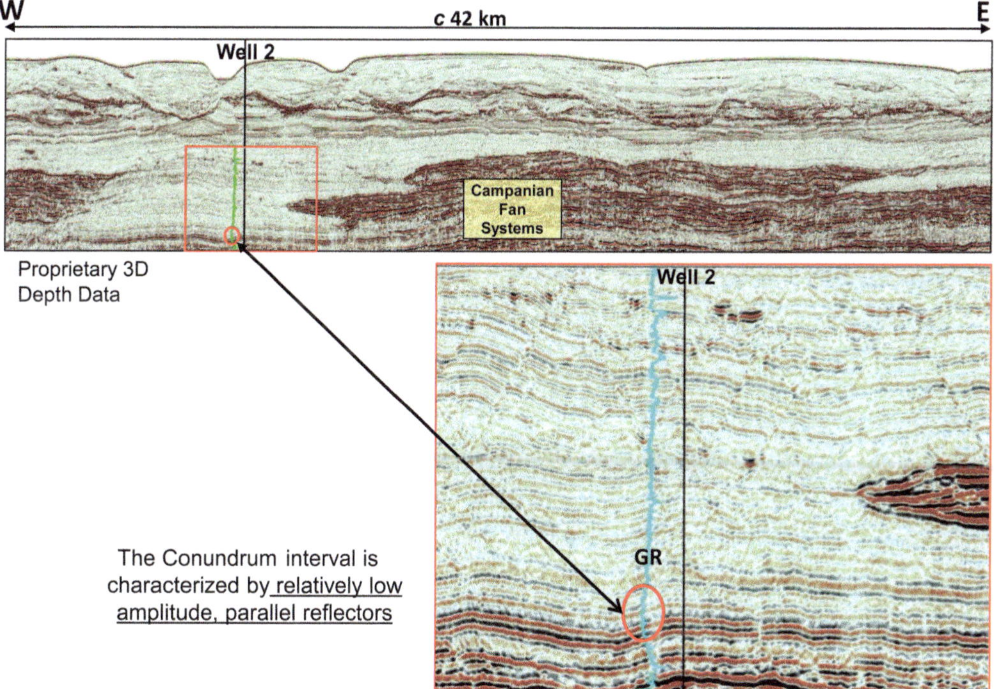

Fig. 8. Example 3 is the east–west seismic line 1 that crosses the Conundrum lithofacies as seen previously in well 2. Note that 8–10 km east and west of well 2 the Conundrum is not present due to the presence of sand-rich lower Campanian turbidite fan systems. Zoomed-in section shows parallel reflectors of the circled Conundrum interval.

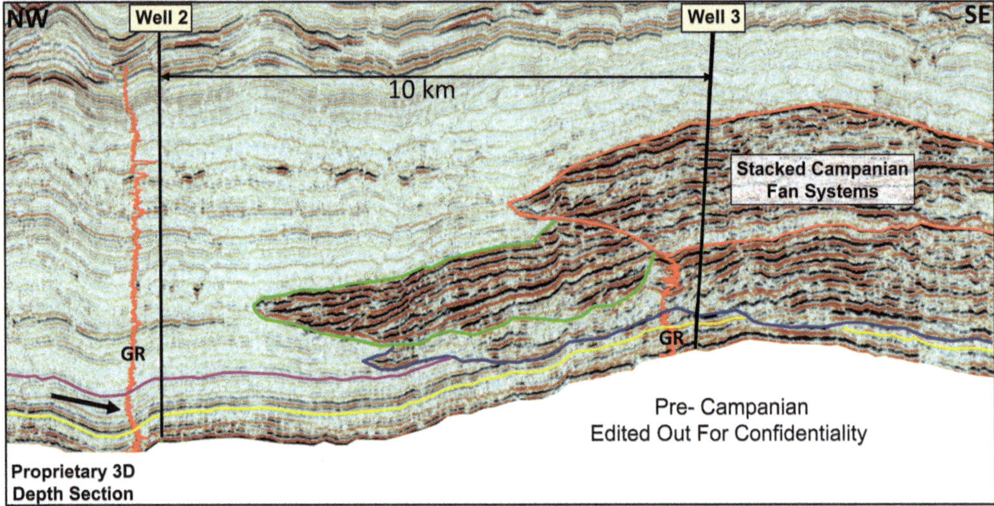

Fig. 9. Example 4 is the NW–SE seismic line 2 that ties well 2 (Conundrum) and well 3 (sand-rich in same interval). The arrow in well 2 points to the Conundrum interval between the two interpreted horizons.

sections and point counts. A thin section was made from core B along with point count analysis. Cores A and C were described as siliceous shales. Core B was described as 'clayey' siliceous shale. Photographs of cores B and C look almost identical and both were described as shales. However, core B

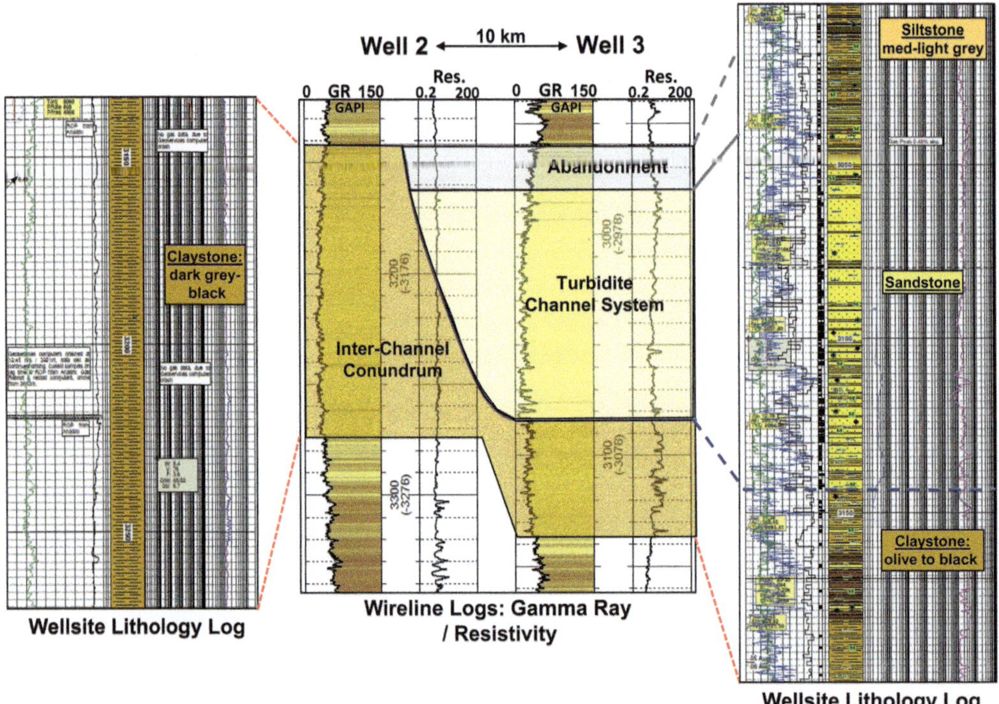

Fig. 10. Well logs and corresponding lithology logs demonstrate full preservation of the Conundrum lithofacies in well 2, but only partial preservation at the base of the section in well 3.

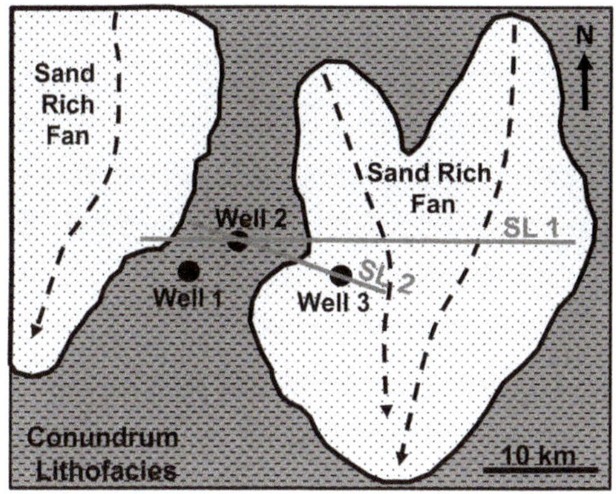

Fig. 11. Lower Campanian schematic facies map with well locations 1, 2 and 3 and seismic lines (SL) 1 and 2.

had a GR value of about 125 API compared with 50 GAPI for core C. It is interesting to note that although core A has the lowest GR value of 45 GAPI, it consists entirely of clay- and silt-sized particles with no very fine sand grains present based on the LPSA data.

Thin sections under plain polarized light from cores A, B and C are shown in Figure 15. Again, there is very little difference in appearance between the true argillaceous shale and this 'apparent' quartz-rich sand based on the low GR response.

Scanning electron microscopy

Rotary SWCs were acquired within the Conundrum lithofacies section shown on the lithology log in Figure 16. Three cores were taken specifically to study the fabric of the Conundrum lithofacies in greater detail using high-magnification SEM. Prior to the SEM study, XRD was performed on all three cores and MICP seal analysis was conducted on the upper two cores.

The MICP results indicated a range of 580–730 m capillary top seal capacity to contain an underlying oil column at 7.5% Hg saturation. Equivalent gas column ranges were 260–315 m at the same Hg saturation. The percentage of quartz determined by XRD analysis varied between 75 and 82% and the total clay content was 16–22%, with 2–3% pyrite making up the remainder of the mineral composition. A thin section from the sample recovered

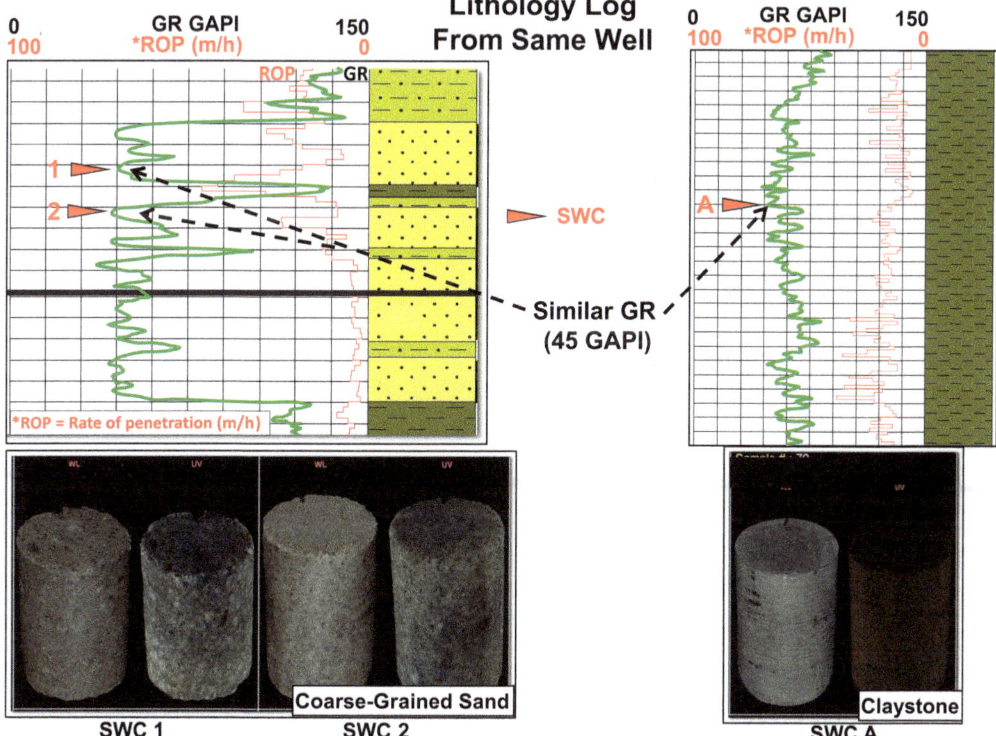

Fig. 12. Example 5 compares similar gamma ray (GR) values where sidewall cores (SWCs) were taken in the same hole section. The sand-rich section on the left is several hundred metres above the 'claystone' section on the right.

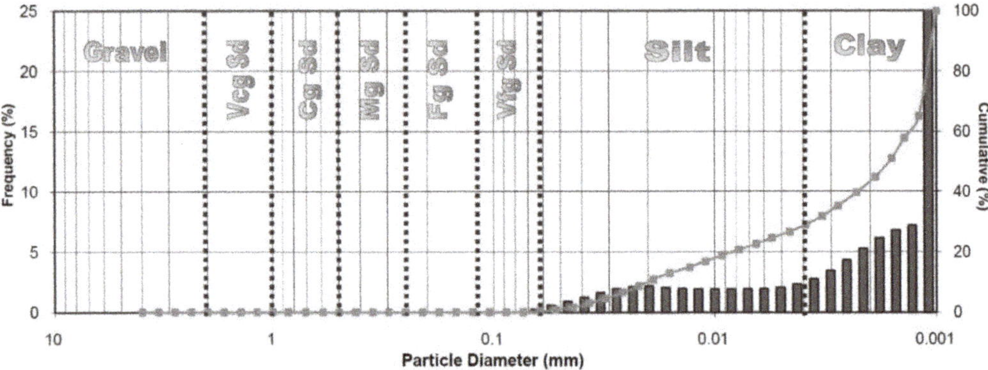

Fig. 13. Laser particle size analysis of sidewall core A showing a distribution of 29% silt-sized and 71% clay-sized particles.

at 4138 m appeared to be composed of claystone, but the XRD analysis indicated that quartz comprised 82% of the sample, the highest percentage observed for the three analysed cores. This SWC at 4138 m was chosen for SEM as a result of its high total quartz content. Low magnification SEM images (300× and 6000×) from the SWC at 4138 m did not resolve the internal architecture of the sample (Fig. 17). The SEM image at 300× magnification was non-descript with no visible porosity and appeared shaley.

The SEM image at 6000× magnification revealed detrital clays, trace amounts of quartz microcrystals and micropores. Magnifications of 40 000× and 60 000× (Fig. 18) revealed an abundance of euhedral quartz, which commonly suggests an authigenic origin. The proposal of an aeolian source (Towle *et al.* 2012; Brown *et al.* 2014) is

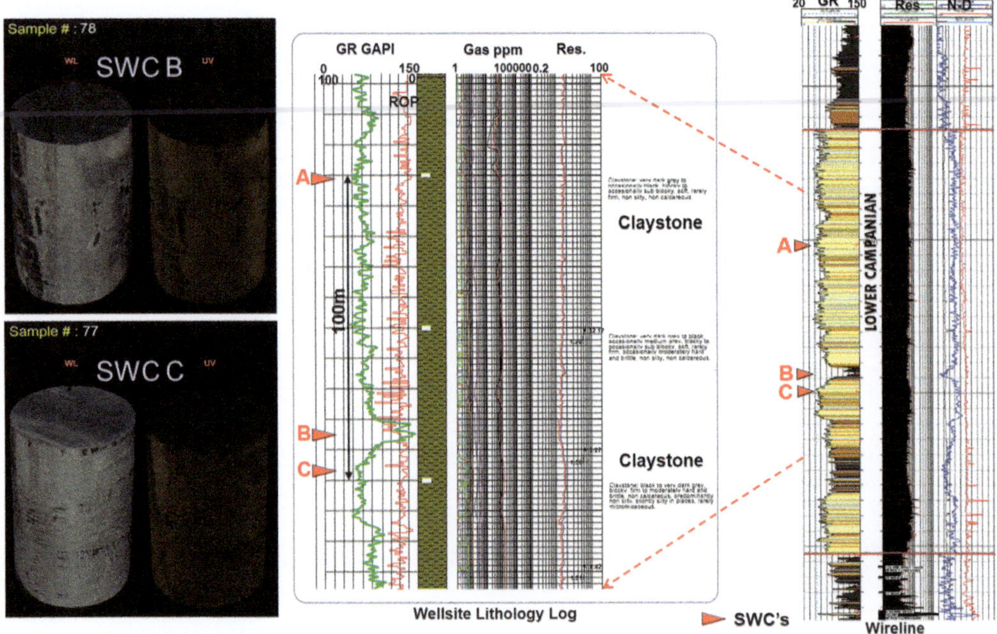

Fig. 14. Expanded lithology log (below core A, Fig. 10) with core locations, wireline data, core photos and laser particle size analysis results. Note that sidewall core (SWC) A is 90% clay- and silt-sized quartz based on laser particle size analysis.

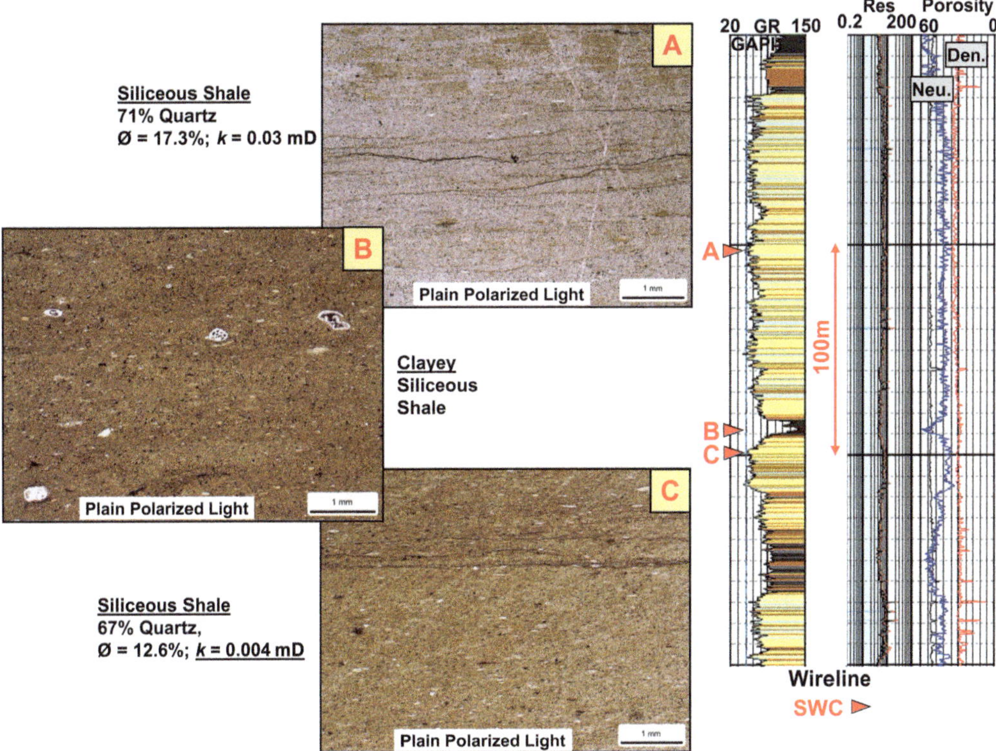

Fig. 15. Thin sections from rotary sidewall core (SWCs) along with wireline data (gamma ray (GR), resistivity, porosity). Visually there is little difference between thin sections B and C, but there was a sharp contrast in GR values where the cores were acquired.

therefore difficult to defend if the origin of the quartz is diagenetic. The absence of rounded to sub-rounded, frosted or abraded quartz grains (albeit very small) typical of an aeolian origin caused a shift in the study to determining the source of silica for the formation of the euhedral quartz crystals. To further investigate this, high-resolution field-emission SEM (FESEM) was used.

Field-emission scanning electron microscopy

An FESEM image has a resolution down to 1.5 nm, which is three to six times higher than a standard SEM image. A thin section from the same SWC (4138 m) was ion-milled (essentially sand-blasted at the atomic scale) to create a perfectly clean, flat surface. The sample was then viewed at various magnifications between 4550× and 128 600× using FESEM. The greatest magnification that could be interpreted without distortion was 120 000×. A low magnification of 4551× (Fig. 19) revealed

the presence of large (5–20 µm) monocrystalline quartz crystals as well as considerable microcrystalline quartz that vaguely appeared to be aggregated or 'clumped'. Micropores were clearly visible. A magnification of 15 000× showed that the 'aggregate' nature of the microcrystalline quartz was readily apparent and that the detrital clays were more visible.

The individual microcrystalline aggregates varied in size between 1 and 3 µm. Discrete aggregates of microcrystalline quartz within a fine detrital clay matrix were clearly visible at 120 000× magnification (Fig. 20). The individual quartz 'grains' or 'clumps' within these flocculated packages were usually <0.5 µm. Quartz grains this small are highly susceptible to dissolution as a result of their large surface area (Cecil 2004). These individual quartz grains may still be of aeolian origin, but are so small that the typical characteristic features are not apparent. These could be remnant quartz grains that did not undergo dissolution, while the majority were dissolved and were the source of silica for the euhedral quartz crystals observed in Figure 18

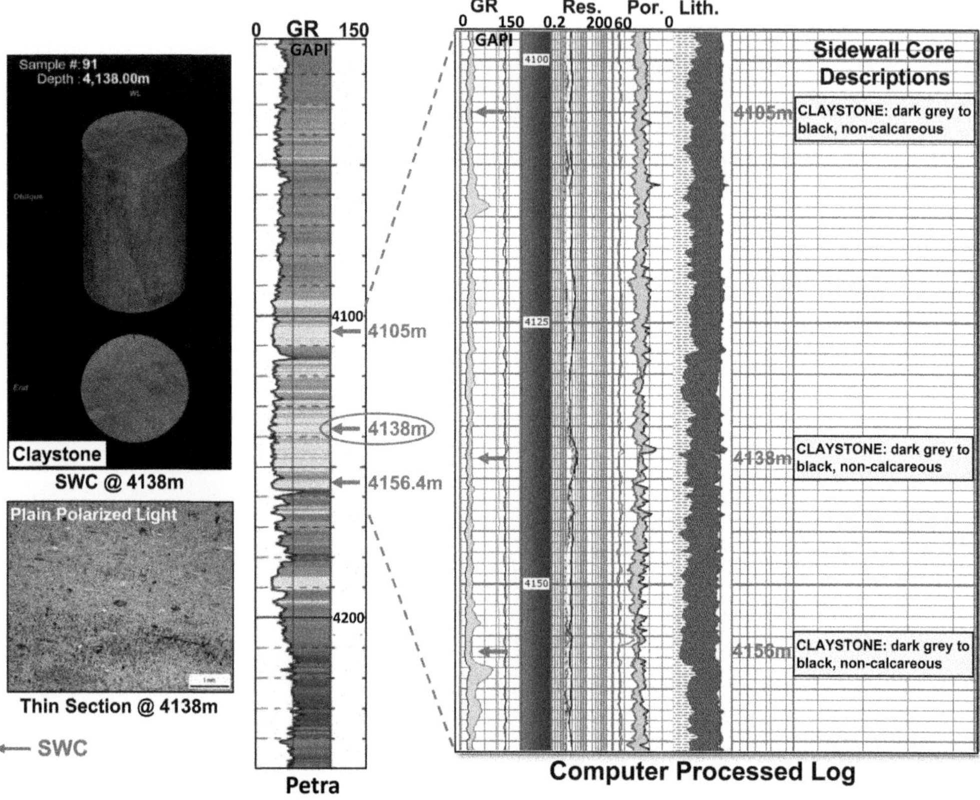

Fig. 16. Example 6 of low gamma ray (GR) zone described as claystone with sidewall core (SWC) locations, thin section and core photo at 4138 m used for scanning electron microscopy images.

(Brown *et al.* 2014). In addition, these small, remnant grains might have acted as the nuclei for re-precipitated silica (Cecil 2004).

An alternative interpretation may be that the origin of these <0.5 μm quartz grains may be related to the upwelling of nutrient-rich oceanic currents

Fig. 17. Scanning electron microscopy images from sidewall core at 4138 m (see Fig. 16).

Fig. 18. Higher magnification scanning electron microscopy images from sidewall core at 4138 m. Note that at 60 000× the pore throats are about 1 µm in diameter; aggregates of microcrystals are outlined in the lower left image.

and the resulting deposition of siliceous pelagic sediments. Any portion of the siliceous sediment that dissolved would have probably re-precipitated as authigenic quartz or chert. This might be explained by the coast-parallel prevailing winds interpreted for the Mid-Campanian, which would have directed deep bottom-water currents towards the edge of the continental shelf immediately inboard of the study area.

Depositional model of the origin of the Conundrum lithofacies

Original hypothesis

The origin of the Mid- to Lower Campanian Conundrum lithofacies was originally proposed by Towle *et al.* (2012) as aeolian. The source of the silt- and clay-sized quartz grains for this early model was

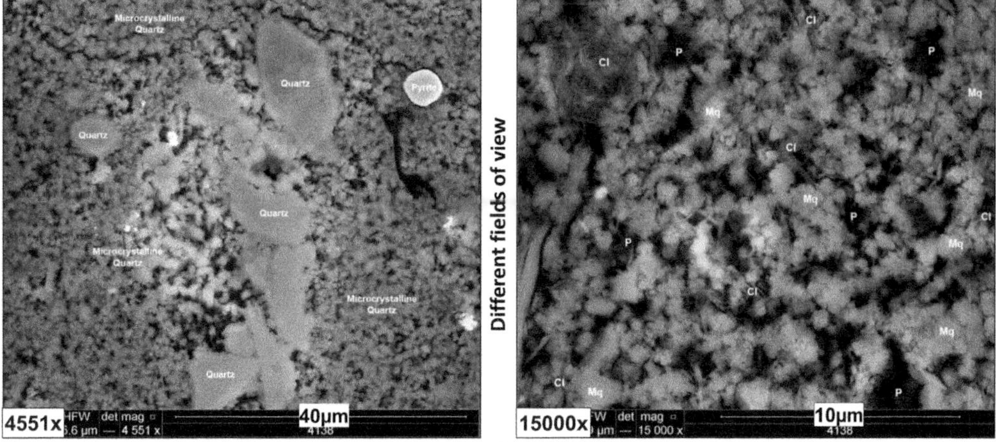

P = Micropore; Cl = Detrital Clay; Mq = Microcrystalline Quartz

Fig. 19. Field-emission scanning electron microscopy image from ion-milled section of sidewall core at 4138 m. Note discreet aggregates of microcrystalline quartz on the right at 15 000× magnification.

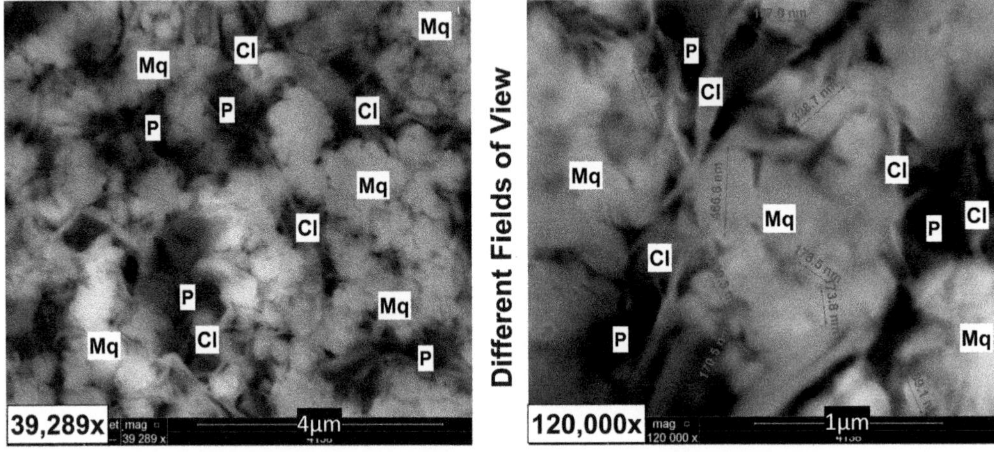

P = Micropore; Cl = Detrital Clay; Mq = Microcrystalline Quartz

Fig. 20. Field-emission scanning electron microscopy image from ion-milled section of sidewall core at 4138 m at higher magnifications.

postulated to have originated from the arid regions of central and southern Africa, as interpreted by Scotese (2014) (Fig. 21). The aeolian transportation of microscopic quartz grains (desert dust) from southern Africa could have resulted from seasonal prevailing winds originating from this region and transporting fine-grained material into the study area. The distance from the centre of the palaeodesert to the northern end of the study area is about 5000 km. As no Campanian palaeowind interpretations were available, a Turonian prevailing wind interpretation (Fig. 22) (Scotese 2014) was used as a proxy for conditions that may have existed in the lower to mid-Campanian. This assumption was made because thin intervals of the Conundrum facies have also been observed within the Upper Turonian and Santonian of three or four wellbores in the study area. Aeolian source material may have blanketed this large study area both offshore and onshore as the result of thousands of desert dust storms that occurred over several million years during lower Campanian time. There are several modern-day examples, such as the Saharan and Gobi deserts, where significant wind storms have carried very fine-grained material many thousands of kilometres.

Acquisition of a continuous Campanian core (ODP 959D)

The ODP Leg 159, Site 959 core site (Figs 1 & 23) is located 155 km from the southwestern coast of Ghana (Beckmann *et al.* 2005) in a water depth of 2090 m. The core site is situated landwards of the crest of the Cote d'Ivoire–Ghana marginal ridge, where a thicker Tertiary–Upper Cretaceous section is present (Fig. 24) (Beckmann *et al.* 2005). Four cores were acquired in 1995 from holes A–D. A Pleistocene to Late Albian section was recovered (Mascle *et al.* 1996). At the 959D site, coring began at 417 m below the seafloor and was stopped at 1159 m (441 m less than planned) as a result of poor rates of penetration in late Albian shales; a total core section of 742 m was acquired.

The pre-drill main objective at the ODP 959 site was to obtain cores to study the different stages of development of both the extensional basin and the bordering marginal ridge (Mascle *et al.* 1996). This was intended to span the sedimentary record of the Deep Ivorian Basin synrift and post-rift period as well as the syn-transform and subsequent vertical behaviour of the marginal ridge. Continuous coring was aimed at determining the ages and facies of sediments within the active transform margin (Mascle *et al.* 1996).

The entire cored section is shown on the right-hand side of Figure 25 (Mascle *et al.* 1996). The curves are from post-coring wireline logging. Of particular interest to this study was the 70 m section of the Campanian Conundrum section. The interval had an average GR of 25–45 GAPI, a porosity >30% and was described as a 'black claystone'. This is consistent with the same observations made from cuttings and SWC analysis in the many oil industry wellbores that have encountered the Conundrum lithofacies. On the left-hand side of Figure 25 (Luning & Kolonic 2003) is a zoomed-in

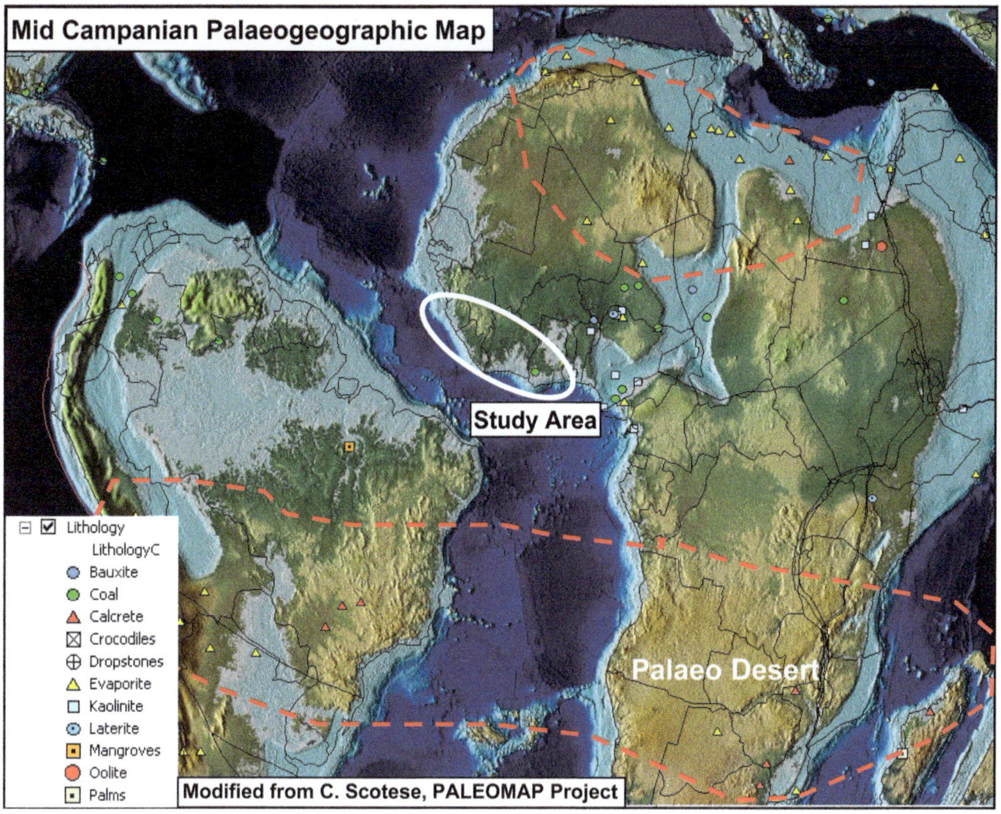

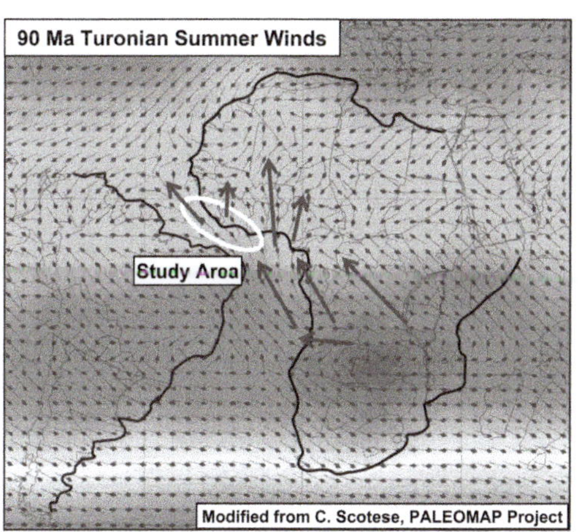

Fig. 21. Mid-Campanian palaeogeography (modified from Scotese 2014, courtesy of C. Scotese). Arid desert area circled by authors.

Fig. 22. Interpreted southeasterly Turonian winds towards study area (modified from Scotese 2014, courtesy of C. Scotese). Lower to mid-Campanian wind maps not available. Summary arrows by authors.

section with a suite of geochemical logs from the top of the Turonian to near the top of the Maastrichtian. These logs plot the core total uranium (ppm) and total organic carbon (TOC) versus depth. The total uranium averages 1–3 ppm in most of the Campanian section. Worldwide crustal averages are about 2.5 ppm. The TOC values of 10% or higher are from the Coniacian–Santonian oceanic anoxic event (OAE 3) interval with corresponding higher total uranium concentrations (8–20 ppm). The upper 50 m of the Campanian section has TOC values of 0.5–2.0%, whereas the lower 20 m is richer with a TOC in the range 1–5%, which is consistent with the slightly higher total uranium content of up to 4.5 ppm.

Studies from the ODP 959 Site D core

Some interesting observations were made from the ODP 959D core site from the OAE-3 time interval by Beckmann *et al.* (2005), who proposed that deposition in the vicinity of the ODP 959 site was cyclic during the Coniacian–Santonian stages,

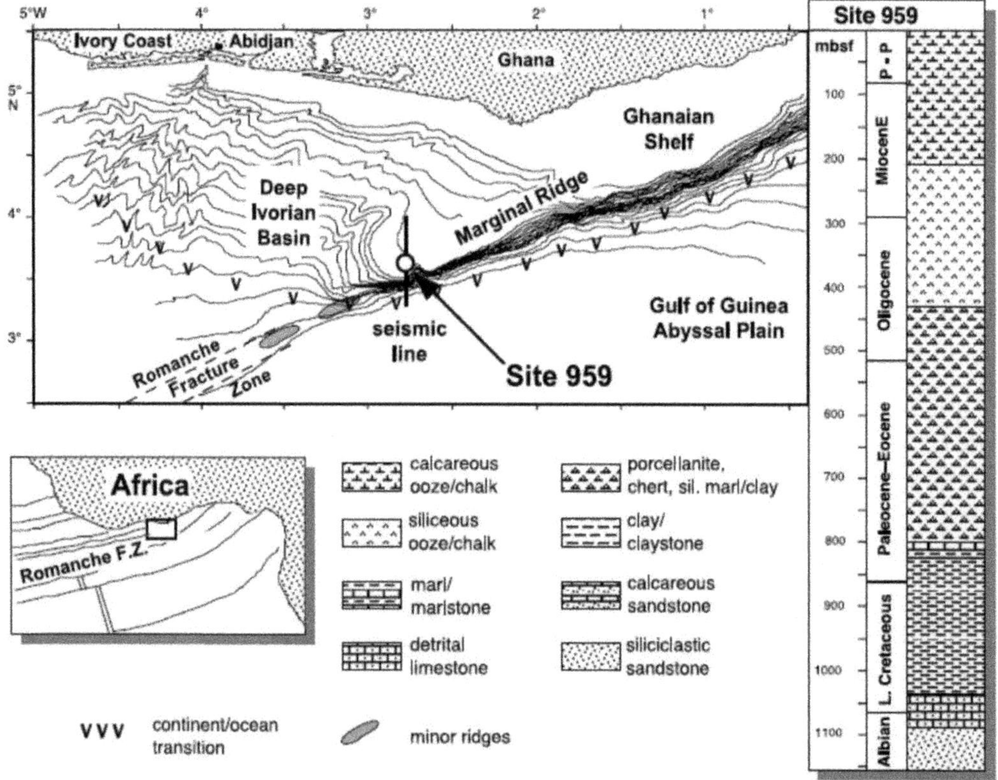

Fig. 23. Location of ODP 959 drill sites about 155 km south of the Ghana coast (from Beckmann *et al.* 2005, after Mascle *et al.* 1996; courtesy of ODP Texas A&M).

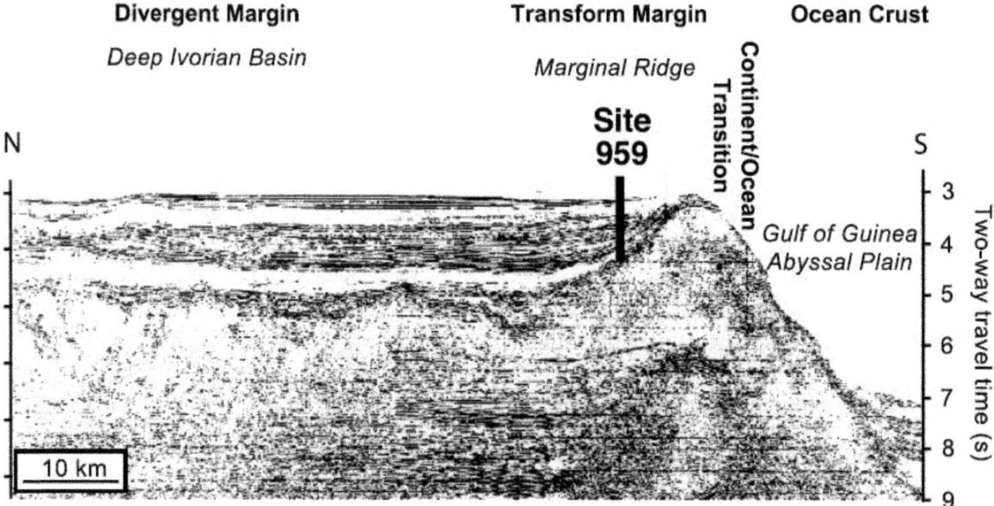

Fig. 24. Seismic time section and the location of Site 959 (from Beckmann *et al.* 2005, after Mascle *et al.* 1996; courtesy of ODP Texas A&M).

with sediments derived from alternating periods of local continental weathering and organic-rich detritus ('run-off') interbedded with aeolian dust transported great distances from southern arid environments. This cyclicity is attributed to shifting of the Inter-Tropical Convergence Zone. Aeolian particulate matter was blown into equatorial West Africa via southeasterly trade winds (Fig. 26) (Beckmann *et al.* 2005). High and low spikes in the TOC data from this interval (Fig. 25, left-hand side) (Luning & Kolonic 2003) suggest a variation between organically-rich sediments and Si/Al-dominated sediments that were organically lean.

The proposal by Beckmann *et al.* (2005) that aeolian dust from south-central arid Africa may have been transported intermittingly into the ODP 959 area during Late Turonian to Santonian time is of great interest. The possibility that aeolian dust could have been deposited into the middle part of the African Transform Margin study area through to the end of the Santonian was encouraging. Could early Campanian deep-water

sedimentation have shifted to a dominance of wind-transported aeolian dust into the study area, with only occasional influxes of organic matter as seen on the TOC plot? This gave a little more credence to using the interpreted Turonian wind map (Fig. 22) as a proxy for early Campanian winds (no interpreted map was available) if, in fact, late Santonian winds were from the SE based on Beckmann *et al.* (2005).

Possible alternative interpretations

An alternative source for the origin of the quartz fraction of the Conundrum facies may be diatom and radiolarian tests that were deposited by pelagic sedimentation. Oceanic upwelling may have driven organic productivity at sea-level and the resulting siliceous sediments deposited on the deep seafloor may have been composed of siliceous ooze. However, as no trace of diatom or radiolarian fragments has been observed under high magnification in this study, this hypothesis remains largely untested and

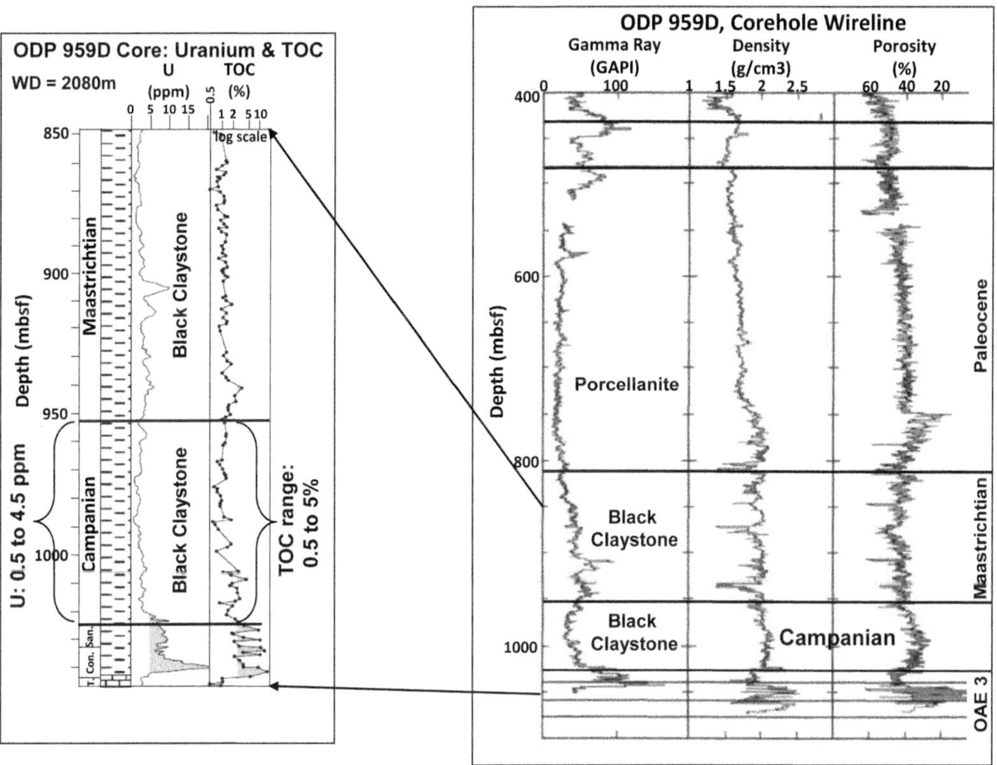

Fig. 25. ODP 959D examples of core uranium, total organic carbon (TOC), wireline gamma ray (GR) and porosity across the Campanian Conundrum section from two different studies. Log on right is from the entire cored interval and log on left is zoomed-in to the bottom 200 m of the core (left modified from Luning & Kolonic 2003 and right from Mascle *et al.* 1996; figure courtesy of ODP Texas A&M and Prof. Thomas Wagner, Newcastle University, Newcastle upon Tyne, UK).

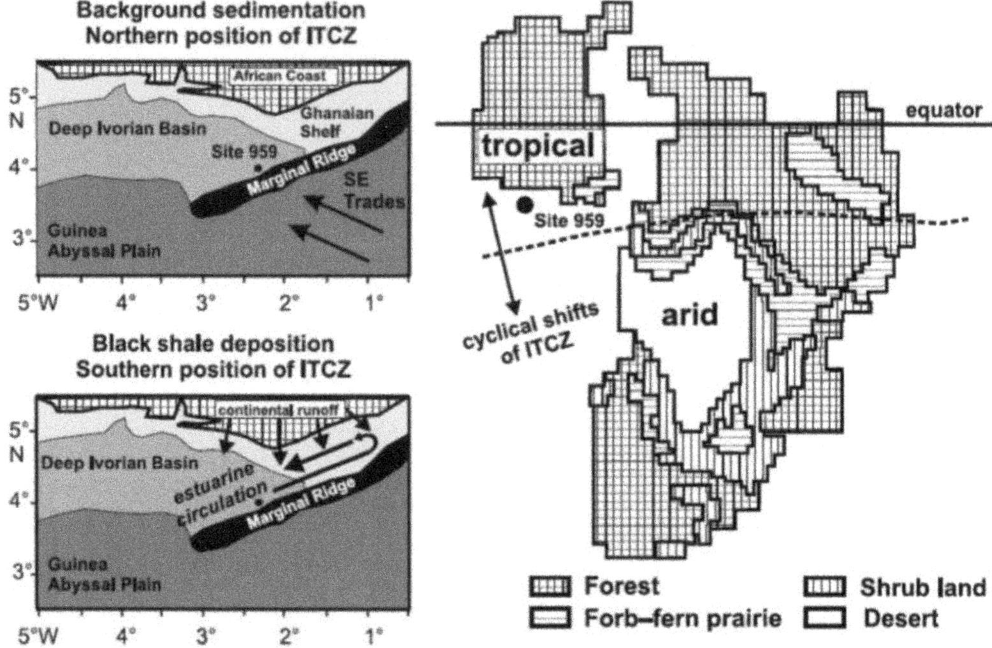

Fig. 26. Model for cyclical shifting of Inter-Tropical Convergence Zone (ITCZ) and the effects on the dominant sediment type and input direction (from Beckmann *et al.* 2005, proposed by Hofmann *et al.* 2003, modelled with GENESIS-EVE by DeConto *et al.* 1999; figure courtesy of the Geological Society of America).

unproved. Isotopic analysis of various trace elements may differentiate an aeolian source of silica from a biogenic source. These two processes may coexist. The aeolian transport of fine particulates may also be the very mechanism that drove oceanic overturn and siliceous productivity. The broad area over which the Conundrum facies is present ($>30\,000$ km^2) and its thickness (90–275 m) favour a repeated pattern of deposition, silica dissolution and silica re-precipitation. The slight elevation in the Campanian of total uranium and TOC in the ODP 959D core could support an upwelling model. Recent internal biostratigraphic studies of a well within the study area indicated that the upper section of the Conundrum lithofacies contained a dinoflagellate, *Manumiella*, indicative of cooler water, which could indicate an upwelling event. More regional biostratigraphic work needs to be carried out before coming to any conclusions about the presence of this dinoflagellate and its implications.

Volcanic ash as a source of silica may be partially discounted because documented volcanic activity in the region is largely confined to the Hauterivian, Apto-Albian and the Miocene through to the present day. However, oceanic upwelling and surface productivity has been postulated at various other times during the Cretaceous for the greater South Atlantic region, notably during the Ceno-Turonian and the Coniacian–Santonian (see Jenkyns 2010). Volcanic degassing is a commonly cited explanation for the initiation of the palaeoclimatic change required to drive sea-level highstands and episodes of oceanic upwelling. The deposition and preservation of wind-borne volcanic ash in the marine setting is possible and may also have punctuated the Campanian stage. Metastable volcanic glass is readily dissolved and re-precipitated as microcrystalline or cryptocrystalline quartz. Coupled with a slightly elevated TOC and uranium content in the ODP 959D core, the Conundrum lithofacies may have a partial volcanic provenance as well as one related to high organic productivity.

Non-preservation of Conundrum lithofacies

There are two geological settings that are not conducive to the preservation of the Conundrum lithofacies. The first geological setting is the continental shelf. The second area is the shelf–slope break at the edge of the continental shelf. The shelfal environment is commonly one of dilution by river- and shore-dominated sedimentation, where deltaic, estuarine and shallow-marine shelf sediments overwhelm any quiet, pelagic or hemipelagic sedimentation. Fair weather and storm waves are common agents of erosion and sediment remobilization that easily obscure the fine-grained fingerprint of suspension settling sedimentation. An in-depth study of shelfal well logs and lithology logs needs to be carried out to confirm this. The deeper-water slope environment is commonly one of sedimentary bypass of riverine and shelf sediments into the seawards bathyal setting. This zone of bypass is aggressively incised by submarine canyons within the equatorial study area. Coarse-grained turbidites and debrites inhabit these active canyons and leave little room for the preservation of the Conundrum lithofacies.

Other aeolian examples

DSDP site 463 is located in the central North Pacific about 3700 km SE of Japan. Studies of the 570 m core taken from the Maastrichtian through Aptian indicated a significant component of aeolian input (Rea & Janecek 1981). The Aptian–Albian and Maastrichtian were dominated by volcanic input, while during the rest of the time the input was primarily aeolian continent-derived dust. Rea & Janecek (1981) believed that this continentally derived aeolian source was about 11 500 km away 90 myr ago. This is over 2.5 times the distance from the Conundrum study area to a potential palaeodesert source.

Conclusions

A cratonic provenance for this unusual, deep-water marine facies is proposed, whereby dust-sized particles were transported by wind to the equatorial margin of West Africa and then settled in the Campanian seafloor through suspension settling. However, additional studies need to be carried out to definitively differentiate the proposed aeolian-sourced silica (primary) from a possible biogenic source (secondary).

The locations of exploration objectives probably explain why so many deep-water wells in the study area have encountered the Conundrum lithofacies. One of the primary exploration targets in the Deep Ivorian and Sierra Leone–Liberian basins has been upper Turonian turbidite, slope–channel complexes. As a result of differential compaction and the resultant compensational stacking, the Mid–Lower Campanian turbidite fan systems tend to flank the major sand-rich upper Turonian systems. Hence many wells targeting the upper Turonian strata penetrate the preserved Conundrum lithofacies as opposed to the sand-rich Campanian fan systems (Figs 8 & 9). However, in some instances, the Campanian has been a primary or secondary objective and therefore rock data from these sand-rich turbidite systems are available.

Irrespective of the origin of this unusual facies, its unique quality demonstrates the need for detailed SWC (or cuttings in the absence of cores) analyses to validate the petrophysical interpretations, especially during the exploration phase of drilling in new areas. Close attention should be paid to sample descriptions by the wellsite geologist. Laboratory analyses suggest that what was mistaken for high-quality reservoir sandstones in early wells was actually impermeable seal rocks. It is very important to recognize the existence of this facies and not to mistake it for an exploration target.

References

AKPATI, B.N. 1978. Geologic structure and evolution of the Keta basin, Ghana, West Africa. *Geological Society of America Bulletin*, **89**, 124–132.

ANTOBREH, A.A., FALEIDE, J.I., TSIKLAS, F. & PLANKE, S. 2009. Rift-shear architecture and tectonic development of the Ghana margin deduced from multi-channel seismic reflection and potential field data. *Marine and Petroleum Geology*, **26**, 345–368.

BECKMANN, B., WAGNER, T. & HOFMANN, P. 2005. Linking Coniacian–Santonian (OAE-3) Black-Shale deposition to African climate variability: a reference section from the eastern tropical Atlantic at orbital time scales (ODP Site 959, off Ivory Coast and Ghana). *In*: HARRIS, N.B. (ed.) *The Deposition of Organic-Carbon-Rich Sediments: Models, Mechanisms, and Consequences*. SEPM, Special Publication, **82**, 125–143.

BROWN, A.S., BIRKHEAD, S.S., MCLEAN, D.J., TOWLE, P.J., WHITE, H.J. & WU, YAFEI, 2014. The Campanian quartz claystone conundrum of the African Transform Margin. *In*: *Proceedings of the HGS-PESGB 13th Conference on African E&P*, September, Houston TX, USA, 39–46 (extended abstract).

BROWNFIELD, M.E. & CHARPENTIER, R.R. 2006. Geology and total petroleum systems of the Gulf of Guinea Province of West Africa. *US Geological Survey Bulletin*, **2207-C**, 15.

CECIL, C.B. 2004. *Eolian Dust and the Origin of Sedimentary Chert*. USGS Open-File Report 2004-1098.

CLIFFORD, A.C. 1986. African oil – past, present, and future. *In*: HALBOUTY, M.T. (ed.) *Future Petroleum Provinces of the World*. AAPG Memoir, **40**, 339–372.

CONN, P. & RODRIGUEZ, G. 2011. New insights into prospectivity of Liberia–Sierra Leone Basin because of improvements in seismic acquisition and processing. *The Leading Edge*, **30**, 656–661.

DE MATOS, R.M.D. 1992. The northwest Brazilian rift system. *Tectonics*, **11**, 766–791.

DECONTO, R.M., HAY, W.W., THOMPSON, S.L. & BERGENGREN, J. 1999. Late Cretaceous climate and vegetation interactions: cold continental interior paradox. *In*: BARRERA, E. & JOHNSON, C. (eds) *The Evolution of Cretaceous Ocean/Climate Systems*. Geological Society of America, Special Publications, **332**, 391–406.

GIRAUD, R. & MAURIN, J.C. 1992. Early Cretaceous rifts of Western and Central Africa: an overview. *Tectonophysics*, **213**, 153–168.

Hofmann, P., Wagner, T. & Beckmann, B. 2003. Millennial- to centennial-scale record of African climate variability and organic carbon accumulation in the Coniacian–Santonian eastern tropical Atlantic (Ocean Drilling Program Site 959, off Ivory Coast and Ghana). *Geology*, **31**, 135–138, http://doi.org/10.1130/0091-7613(2003)031<0135:MTCSRO>2.0.CO;2

Jenkyns, H.C. 2010. Geochemistry of oceanic anoxic events. *Geochemistry, Geophysics, Geosystems*, **11**, q03004, http://doi.org/10.1029/2009GC002788

Kesse, G.O. 1986. Oil and gas possibilities on and offshore Ghana. *In*: Halbouty, M.T. (ed.) *Future Petroleum Provinces of the World*. AAPG Memoir, **40**, 427–444.

Luning, S. & Kolonic, S. 2003. Uranium spectral gamma-ray response as a proxy for organic richness in black shales: applicability and limitations. *Journal of Petroleum Geology*, **26**, 153–174.

Mascle, J. & Blarez, E. 1987. Evidence for transform margin evolution from the Ivory Coast–Ghana transform margin. *Nature*, **326**, 378–381.

Mascle, J., Blarez, E. & Marinho, M. 1988. The shallow structure of the Guinea and Cote D'Ivoire–Ghana transform margins: their bearing on the equatorial Atlantic Mesozoic evolution. *Tectonophysics*, **155**, 193–209.

Mascle, J., Lohmann, B.P. *et al.* 1996. Site 959. *In*: Mascle, J., Lohmann, G.P. & Clift, P.D. (eds) *Proceedings of the Ocean Drilling Program, Initial Reports*, **159**. Ocean Drilling Program, College Station, TX, 66–72, 78–90, 118, 130–132, 616.

Nemčok, M., Henk, A., Allen, R., Sikora, P.J. & Stuart, C. 2012. Continental break-up along strike-slip fault zones; observations from the Equatorial Atlantic. *In*: Mohriak, W.U., Danforth, A., Post, P.J., Brown, D.E., Tari, G.C., Nemčok, M. & Sinha, S.T. (eds) *Conjugate Divergent Margins*. Geological Society, London, Special Publications, **369**, 537–556, http://doi.org/10.1144/SP369.8

Pletsch, T., Erbacher, J. *et al.* 2001. Cretaceous separation of Africa and South America: the view from the West African margin (ODP Leg 159). *Journal of South American Earth Sciences*, **14**, 147–174.

Popoff, M. 1988. Du Gondwana a l'Atlantique sud: les connexions du fosse de la Benoue avec les basins de Nord-Est bresilien jusqu'a l'ouverture di golfe de Guinee au Cretace inferior. *Journal of African Earth Sciences*, **7**, 409–431.

Scotese, C.R. 2014. The PALEOMAP Project PaleoAtlas for ArcGIS, version 2, Volume 2, Cretaceous Plate Tectonic, Paleogeographic and Paleoclimatic Reconstructions, Maps 16–32. PALEOMAP Project, Evanston, IL.

Towle, P., Birkhead, S., Brown, A., Layman, J., McLean, D., Perry, A. & White, H. 2012. The Campanian Quartz Claystone Conundrum of the African Transform Margin: A quartz rich silty claystone with possible implications for climatic conditions in Africa during the Upper Cretaceous. Paper presented at the AAPG Annual Convention and Exhibition, 22–25 April, Long Beach, California, http://www.searchanddiscovery.com/abstracts/html/2012/90142ace/abstracts/tow.htm

Rea, D.K. & Janecek, T.R. 1981. Late Cretaceous history of eolian deposition in the mid-Pacific mountains, central north Pacific Ocean. *Palaeogeography, Palaeoclimatology, Palaeoecology*, **36**, 55–67.

Wozazek, S. & Krawinkel, H. 2002. Development of the Cote D'Ivoire Basin: reading provenance, sediment dispersal, and geodynamic implications from heavy minerals. *International Journal of Earth Sciences*, **91**, 906–921.

Unravelling the along-strike variability of the Angola–Gabon rifted margin: a mapping approach

GWENN PÉRON-PINVIDIC[1]*, GIANRETO MANATSCHAL[2], EMMANUEL MASINI[3], EMILIE SUTRA[4], JEAN MARIE FLAMENT[5], ISABELLE HAUPERT[2] & PATRICK UNTERNEHR[5]

[1]*NGU Geological Survey of Norway, Leiv Eirikssons vei 39, 7040 Trondheim, Norway*

[2]*Strasbourg University, 1 rue Blessig, 67084 Strasbourg, France*

[3]*TOTAL, ISS/STRU, Avenue Larribau, 64018 PAU Cedex, France*

[4]*Paul Scherrer Institut, Laboratory for Energy Systems Analysis, 5232 Villigen PSI, Switzerland*

[5]*Total SA – Exploration Production/Projects Nouveaux, 2 place Jean Millier, La Défense 6, 92078 Paris la Défense Cedex, France*

**Corresponding author (e-mail: gwenn@ngu.no)*

Abstract: We summarize here observations from the South Atlantic Angola–Gabon rifted margin. Our study was based on the interpretation of a selection of deep penetration depth-migrated seismic reflection profiles. We describe here the large-scale dip architecture of the margin under five structural domains (proximal, necking, distal, outer and oceanic) and list their characteristics. We investigated the necking domain further and discuss the architecture of the distal domain as a combination of hyper-extended crust and possible exhumed mantle. The mapping and characterization of these domains, at the margin-scale, allow us to illustrate the along-strike structural and stratigraphic variability of the margin. We interpret this variability as the result of a shift from an upper plate setting to a lower plate setting. This shift is either sharp, typified by a major regional normal fault on the northern flank of a residual hanging-wall block, identified offshore Cabinda–Zaire, or more diffuse to the south. First-order screening of conjugate profiles confirmed the segmentation and structural characteristics of the transfer zones. The dataset studied also allowed the identification of key sections that can be considered as type examples of upper plate and lower plate margins and which allow us to discuss the characteristics of these end-member settings.

Various geological, conceptual, analogue and numerical models have been proposed to explain the crustal structure and the stratigraphic, isostatic, magmatic and thermal evolution of rifted margins (e.g. Whitmarsh *et al.* 2001; Reston 2005; Lavier & Manatschal 2006; Ranero & Pérez-Gussinyé 2010; Pérez-Gussinyé 2013; Brune *et al.* 2014; Manatschal *et al.* 2014). It is regularly proposed that rifted margins, including the so-called magma-poor and magma-rich margins, share many first-order similarities (e.g. Franke 2013). The systematic seawards arrangement of characteristic entities is typical of rifted margins, such as the partitioning into a proximal and a distal margin, the presence of shallow-water platforms and deeper areas, the occurrence of crustal necking corridors, ocean–continent transitions and marginal highs. Such observations have been reported from many margins worldwide, such as the Great Australian Bight (Direen *et al.* 2007; Ball *et al.* 2013), the South China Sea (Franke *et al.* 2011, 2013; Savva *et al.* 2013), the Gulf of Aden (d'Acremont *et al.* 2005; Autin *et al.* 2010), the Argentinean margin (Blaich *et al.* 2009), the Brazilian margin (Blaich *et al.* 2011; Zalán *et al.* 2011; Stica *et al.* 2014), the Armorican margin (Thinon *et al.* 2003; Tugend *et al.* 2014) and onshore analogues exposed in the Alps and the Pyrenees (Manatschal 2004; Masini *et al.* 2014; Tugend *et al.* 2015). This systematic arrangement appears to reflect a constancy of the extensional tectonic processes involved in the formation of rifted margins (Péron-Pinvidic *et al.* 2013).

Another well-accepted observation is that rifted margins can present significant along-strike variations in terms of basement structure, stratigraphy, subsidence, and thermal and magmatic evolution (e.g. the Møre, Vøring and Lofoten segments of the mid-Norwegian margin; Faleide *et al.* 2008, 2010; Tsikalas *et al.* 2008). Major structural entities – such as continental ribbons, microcontinents

From: SABATO CERALDI, T., HODGKINSON, R. A. & BACKE, G. (eds) 2017. *Petroleum Geoscience of the West Africa Margin*. Geological Society, London, Special Publications, **438**, 49–76. First published online December 2, 2015, http://doi.org/10.1144/SP438.1

and inner aborted rifts – can complicate the final architecture and often help to explain this lateral segmentation (Péron-Pinvidic & Manatschal 2010). However, these specific features are not always present. Differences from one margin segment to another are then usually interpreted to be the result of the influence of local pre-rift to synrift parameters, such as inherited structures, the thermal and/ or compositional state of the lithosphere, sedimentation, the evolving rheology or variable extension rates (e.g. Lizarralde *et al.* 2007; Manatschal *et al.* 2014; Naliboff & Buiter 2015).

Three decades ago, Lister *et al.* (1986) had already recognized the existence of end-member type margins. Inspired by knowledge gained from studies of the Basin and Range metamorphic core complexes province of the western USA (e.g. Davis & Coney 1979; Wernicke 1985; Wernicke & Axen 1988; Lister & Davis 1989; Buck 1991; Lavier *et al.* 1999; Rosenbaum *et al.* 2005), Lister *et al.* (1986) introduced the upper plate–lower plate concept (Fig. 1). The main purpose of this concept was to try to better explain the observed structural asymmetry of rifted margins; most of the standard conceptual models at that time (e.g. McKenzie 1978) only proposed symmetrical rift systems.

Metamorphic core complexes correspond to exhumed metamorphic rocks deformed in a ductile manner and capped by high-strain fault zones (detachment faults) that experienced tens of kilometres of normal-sense displacement in response to lithospheric extension (for a review, see Whitney *et al.* 2014). Based on these fundamental concepts and influenced by Wernicke (1981), Lister *et al.* (1986) suggested the applicability of the concept of lithospheric-scale low-angle detachment faults to rifted margins (Fig. 1). In their model, rifting is proposed to be governed by a major detachment system that separates an 'upper plate margin' in the hanging wall from a 'lower plate margin' in the footwall. The lower plate margin is formed from highly structured exhumed rocks overlain by tilted

crustal blocks and extensional allochthons (rafts). In contrast, the upper plate margin is less structured and preserves a continued pre-rift stratigraphic section. The lower plate rocks were already considered as potentially hyper-extended (100–400%; Davis *et al.* 1980) as a result of movements on listric normal faults and/or domino-like rotations of fault blocks bounded initially by high-angle normal faults (Wernicke & Burchfield 1982). Based on this model, Lister *et al.* (1986) proposed that rifted margins display marked, but complementary, asymmetry, with an upper plate margin devoid of structures and often uplifted and a conjugate lower plate margin that is highly structured, faulted and hyperextended (Fig. 1).

At a first-order scale, many observations of deep rifted margins tend to support the key concept of Lister *et al.* (1986). Narrow(er) margins with a sharp crustal taper and no well-developed hyperextended domain can be categorized as upper plate margins (e.g. central East Greenland and Lofoten; Faleide *et al.* 2010), whereas wider margins – presenting a hyper-extended domain, a distal sagtype basin and a more complex architecture – may be lower plate margins (e.g. Vøring, Iberia and Angola; Reston 2009; Faleide *et al.* 2010; Huismans & Beaumont 2011). However, distinguishing between upper and lower plate margins is not straightforward and is often hampered by the resolution of seismic and geophysical datasets.

We propose here a mapping approach to define the margin-scale architecture of the Angola–Gabon rifted system and discuss its along-strike variability using the upper/lower plate concept. We begin by introducing the terminology and methodology used to map the first-order structures and then summarize the geological setting of the study area. Based on a dataset of long-offset seismic reflection profiles (CongoSPAN1, ION), we describe the lateral segmentation of the Angola–Gabon margin based on key sections that may be considered as good examples of upper and lower plate margins.

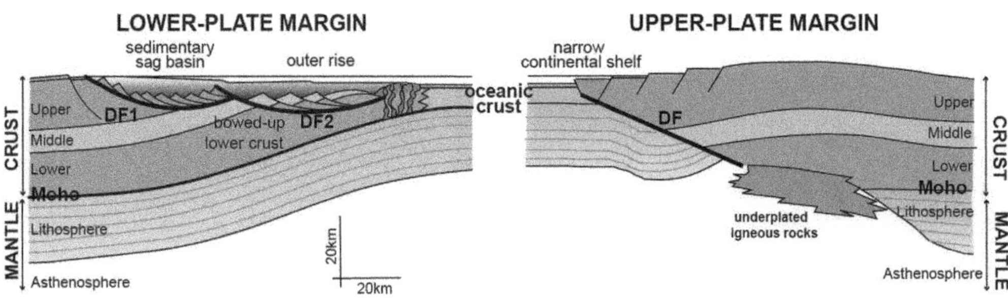

Fig. 1. Schematic representation of the upper plate and lower plate concept after Lister *et al.* (1986). DF, detachment fault.

We propose a template for the complex along-strike variability of these margins and discuss the major structural elements that define their architecture. Finally, based on the model of Lister *et al.* (1986), we review the major characteristics of upper plate and lower plate end-member margin types.

Terminology

Research into the formation and evolution of rifted margins has undergone a major paradigm shift in recent years. Based on Ocean Drilling Project deep drilling results on the Iberia–Newfoundland rift system (e.g. Whitmarsh *et al.* 2001; Tucholke & Sibuet 2007) and newly available high-resolution seismic datasets, it has been shown that rifted margins have a large structural variability that requires new mapping strategies and concepts to define rift structures other than the classical series of tilted blocks and oceanic crust. To map and describe the crustal structures forming the Angola–Gabon margins, we used the terminology developed by Péron-Pinvidic & Manatschal (2010), Péron-Pinvidic *et al.* (2013), Sutra *et al.* (2013) and Tugend *et al.* (2014, 2015). Figure 2 summarizes the terms used in this paper.

Based on these concepts, we mapped the Angola–Gabon margins and defined five rift domains. From the continent to the ocean, these are: (1) the proximal domain; (2) the necking domain; (3) the distal domain; (4) the outer domain; and (5) the oceanic domain. It is important to note that each of these domains can be described by a number of observational criteria (Péron-Pinvidic *et al.* 2013). Each of the domains has been formed by one or a combination of several extensional modes, including stretching, thinning, hyper-extension, exhumation and magmatic accretion. It is of major importance to describe the spatial and temporal superposition of these modes and the distribution of the different domains along the margin. We focus here on a structural description of the margin. A more stratigraphic approach was proposed by Masini *et al.* (2012).

The term 'hanging-wall block' or 'H-block' was first introduced by Lavier & Manatschal (2006) in a dynamic model. Péron-Pinvidic & Manatschal (2010) then defined this as a piece of relatively undeformed continental crust that has preserved its pre-rift stratigraphic cover. Within the framework of an evolving rift system, the H-block can best be compared with a 'keystone'. Depending on various parameters – such as structural heritage, the extension rate, and the initial and ongoing lithospheric thermal configuration – the H-block can be of various shapes, sizes and thicknesses that depend on the preservation of its mid- to lower crustal material

during extension and on the progress of rifting. Very often, this 'keystone' is delaminated during the final evolution of rifting. We therefore refer to the 'residual' and 'delaminated' H-blocks, in contrast with its 'initial' version that can be observed in aborted rift systems (Fig. 2).

Geological background

The South Atlantic Ocean is usually divided into four segments: the Equatorial, the Central, the South and the Falkland segments (Fig. 3). The segmentation is justified in the sense that each of these segments experienced distinct structural and stratigraphic evolution with different magmatic budgets (see the thorough review by Heine *et al.* 2013).

After rift propagation and evolution in the Triassic–Cretaceous, the South Atlantic Ocean separated western Gondwana. Opening started in the southern regions and propagated northwards. In the southern segment, rifting had started by the Late Triassic (220 Ma) and propagated along the Argentina–Namibia rift during the Jurassic before reaching the Central segment by the Early Cretaceous (140 Ma, Berriasian; Heine *et al.* 2013).

We focus here on the margins belonging to the Central segment, which encompass the Brazilian rifted basins from Santos to Sergipe–Alagoas and the African rifted basins from Kwanza to Gabon, between the Ascension and the Rio Grande Fracture Zones (Fig. 3). These basins resulted from lithospheric extension that operated mainly in Early Cretaceous times (Rabinowitz & LaBrecque 1979; Austin & Uchupi 1982; Nürnberg & Müller 1991; Aslanian *et al.* 2009; Torsvik *et al.* 2009; Moulin *et al.* 2010; Heine *et al.* 2013). Their development was rooted on the Pan-African suture, between the Congo, the São Francisco and the Rio de la Plata cratons (Buiter & Torsvik 2014). The pre-rift conditions were therefore complex and heterogeneous at a lithospheric scale. Inherited structures may have played a major role during (early) rifting and may have controlled the lateral variation observed along-strike as well as the segmentation in the margin.

From the Berriasian (140 Ma; Heine *et al.* 2013), a stretching phase affected a wide intracontinental region via series of high-angle normal faults bounding classic half-graben and graben-type basins that registered the deposition of lacustrine sediments (Guiraud & Maurin 1992; Karner *et al.* 1997). Later, in a first stage from the mid-Berriasian (138 Ma; Heine *et al.* 2013) to mid-Barremian (*c.* 127 Ma), while break-up and seafloor spreading started south of the Orange Basin, normal faults related most obviously to the thinning and hyper-extension phases of deformation developed in the Central

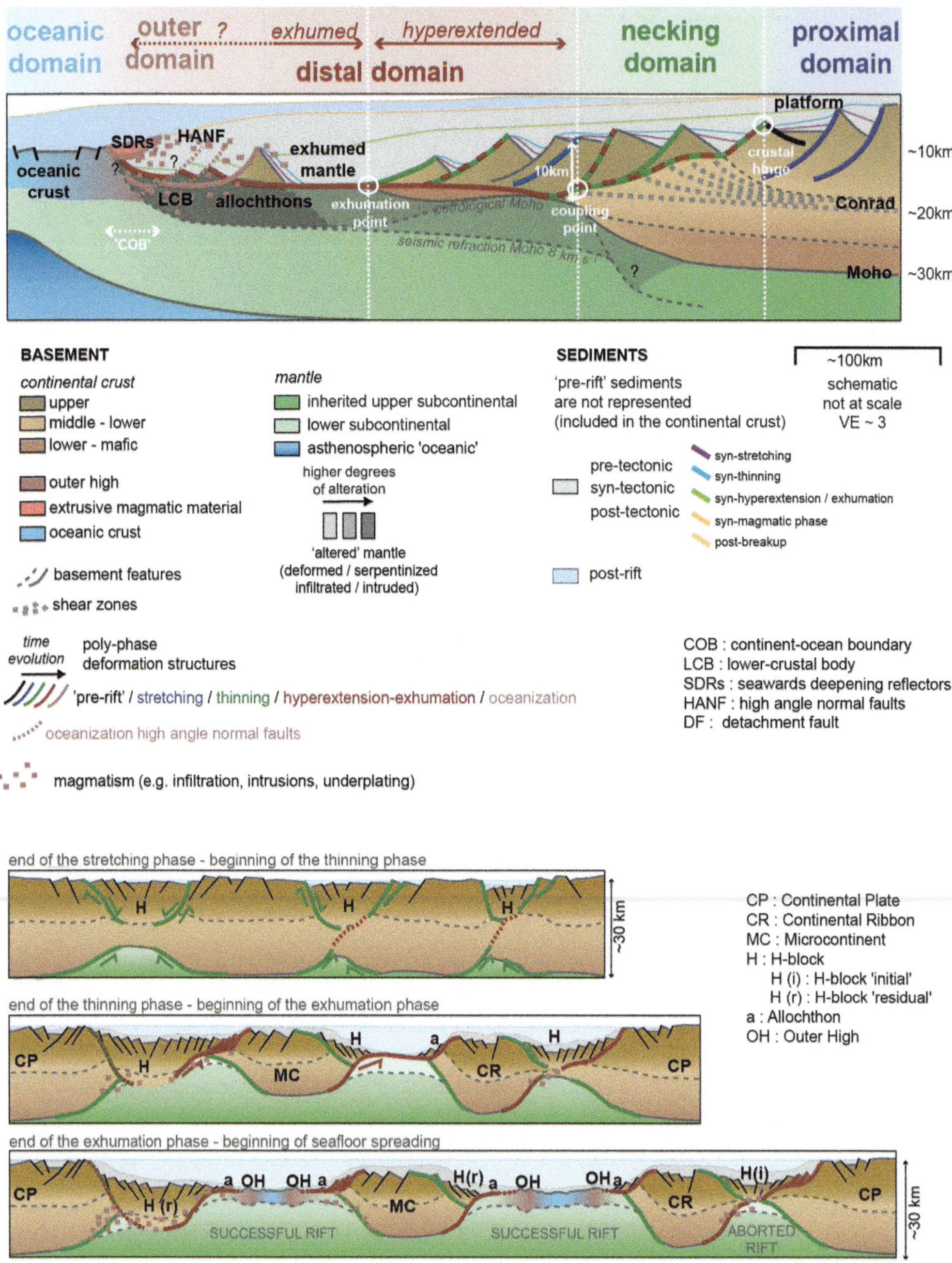

Fig. 2. Schematic sections after Péron-Pinvidic & Manatschal (2010), Péron-Pinvidic *et al.* (2013) and Sutra *et al.* (2013) illustrating the terminology used in this paper. Upper section: schematic representation of a typical rifted margin illustrating the various terms used (after Péron-Pinvidic *et al.* 2013). Lower sections: schematic model of evolution of rifting illustrating the formation of the different categories of crustal blocks. The genesis, evolution, final shape and position in the margin of each block are related to the distinct modes of deformation affecting the margin during rifting (after Péron-Pinvidic & Manatschal 2010).

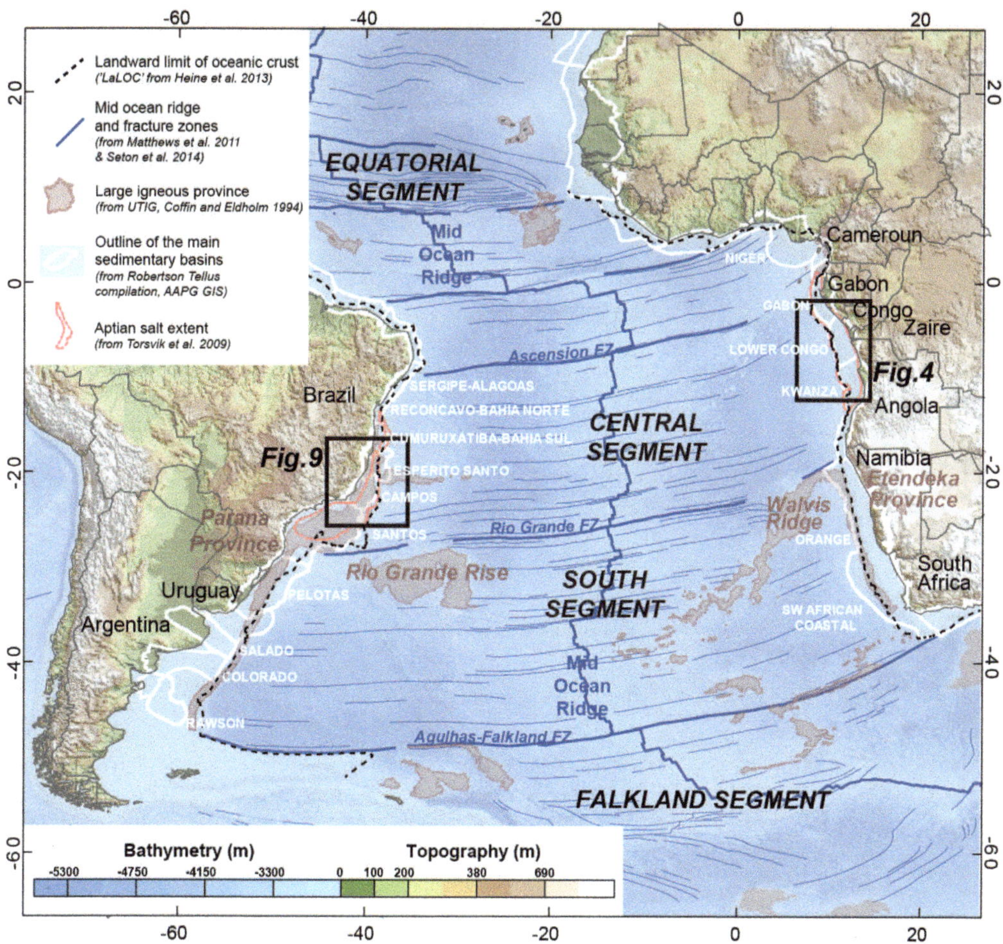

Fig. 3. Regional map of the South Atlantic showing the location of the features discussed in this paper. The two black squares show the location of the more local maps discussed here. The bathymetry is the SRTM30-Plus (Smith & Sandwell 1997; Becker *et al.* 2009). The LaLOC (landward limit of oceanic crust) is from Heine *et al.* (2013). The oceanic structures (mid-ocean ridge and fracture zones) are from Matthews *et al.* (2011) and Seton *et al.* (2014). The major sedimentary basins are outlined in white. These first-order limits are from the Robertson Tellus compilation available on the AAPG GIS webpage. The large igneous province volcanics are from Coffin & Eldholm (1994). The Aptian salt extent outline is from Torsvik *et al.* (2009).

segment, generating tilted blocks and associated half-graben basins. These basins were rapidly filled by an important sedimentary influx and locally present syn-tectonic stratigraphic geometries. A second distinct stage, probably the end of the hyper-extension phase and the exhumation/magmatic accretion phase, from the mid-Barremian to the upper Aptian (*c.* 127–117 Ma), corresponds strati-graphically to unfaulted sedimentary sequences associated with the development of the so-called pre-salt sag basin (Karner *et al.* 2003). A late synrift phase was associated with evaporite/salt deposits, with peak deposition by the mid-Aptian (Karner & Gamboa 2007). The salt layer was later covered by a carbonate platform, deposited by early Albian time, which underwent subsequent gravitational deformation. Heine *et al.* (2013) proposed that the chaotic salt deformation characterizing the Gabon, Kwanza, Espirito Santo, Campos and Santos basins was related to break-up and subsequent rapid sub-sidence, which introduced topographic gradients favouring gravitational sliding and downslope com-pression in the earliest post-rift period. Break-up has been interpreted to have occurred at *c.* 112–110 Ma (Guiraud & Maurin 1992; Davison 1999). However, Heine *et al.* (2013) proposed a revision of that age to 119 Ma between the Campos and Kwanza basins, which is coherent with the 118 Ma age assumed for the adjacent Gabon margin (Karner *et al.* 1997; Dupré *et al.* 2007). The post-rift Albian to Eocene marine sedimentation was then controlled by aggradation of the post-rift systems deposited onto the evaporitic interval and the carbonate platform (Seranne *et al.* 1992). One or several uplift phase(s) affected the sedimentary evolution of the margin. A prominent phase occurred in the Eocene and resulted in a new prograding system related to the development of the Congo–Zaire depocentre (Anka & Seranne 2004).

The magmatic budget, as well as the timing of magma emplacement relative to break-up, changes from south to north, from magma-rich in the south (Pelotas–Namibia) to magma-poor further north. However, magmatic additions are observed everywhere and their age can range from pre-rift (Parana–Etendeka) to synrift to late synrift to post-rift (Abrolhos). All the margins went through a stage of hyper-extension that was overprinted by a more magmatic phase during later stages of rifting.

Margin domains

We studied the Angola–Gabon rifted margin based on the interpretation of long-offset depth-migrated seismic reflection profiles issued from ION's CongoSPAN1 dataset. Our main observations were summarized by Péron-Pinvidic *et al.* (2013). Here, we present the results of a project in which we aimed to map the first-order margin-scale structures and focused more particularly on the along-strike structural and stratigraphic variability of these structures. As a result of the limitations of publication rights, we mainly focus on two profiles that we consider as 'champion lines', exemplifying the laterally contrasting architecture of the margin.

The original study on which this contribution is based dates from 2007. New seismic datasets, both two- and three-dimensional, are now available for this area and these allow more accurate interpretations of the top-basement architecture and the sedimentary and magmatic infill. The aim of this paper is therefore not to propose detailed descriptions, but to highlight the major structural variability at the margin-scale and to discuss the main structural envelopes (seafloor, top-basement and (seismic) Moho), the lateral variability of which is interpreted here in terms of upper to lower plate transfers.

Methodology

Our methodology consisted in investigating the architecture of rifted margins by mapping distinct structural domains. In our view, the key to mapping domains and domain boundaries in seismic reflection profiles are the seafloor, top-basement

and Moho horizons. Their geometries (parallelism, inflection points, convergence) enable the definition of margin domains and the eventual discussion of the processes that led to their formation (Péron-Pinvidic *et al.* 2013; Sutra *et al.* 2013; Tugend *et al.* 2015). For clarity and convenience, we list our observations within the framework of these domains.

Practically, from east to west, two structural limits have been mapped (the necking and continent-wards limit of unequivocal oceanic crust) and five margin domains have been tentatively identified (the proximal, necking, distal, outer and oceanic domains) (Figs 4 & 5). These limits and domains vary along-strike with different crustal, stratigraphic and magmatic characteristics. In addition, along the margin, from south to north, three major structural segments have been defined (see circled numbers on Figs 4 & 5): (1) a southern segment at the level of the Kwanza Basin; (2) a central segment corresponding to the mouth of the Congo River, offshore Cabinda–Zaire; and (3) a northern segment at the level of the Gabon Basin. It is important to note that the southern part of the Angola margin is not constrained by our dataset and therefore not resolved in our maps (see question mark on Fig. 4).

Proximal domain

The proximal domain corresponds to the inboard continental crust that has been stretched at low values of extension (Péron-Pinvidic *et al.* 2013; Tugend *et al.* 2015) (Fig. 2). The brittle upper crust is cut by high-angle normal faults (concave upwards) that bound classic graben and half-graben basins. The related faults sole out at mid- to lower crustal levels (where the ductile regime dominates) and therefore do not affect the Moho (e.g. the Jeanne d'Arc basin and Murre fault; Welsink & Tankard 2012). At depth, other sets of faults are assumed to affect the rigid part of the upper lithospheric mantle and lower mafic crust (Fossen *et al.* 2014). The deformation pattern characterizing the proximal domain is therefore decoupled at the crustal/lithospheric scale: the faults affecting the upper crust and those affecting the upper mantle–lower crust are decoupled along ductile mid- to lower crustal levels. As the bulk crust/lithosphere are only slightly thinned, the subsidence is relatively minor and sediments are mainly restrained within the stretching basins, which can show thick wedge-shaped syn-tectonic sedimentary units. The accommodation space outside these basins is typically very small. The sedimentary record includes continental to shallow water sedimentary systems often associated with sub-aerial exposure markers, with no or only minor aggradation of post-rift sequences (Tugend *et al.* 2015).

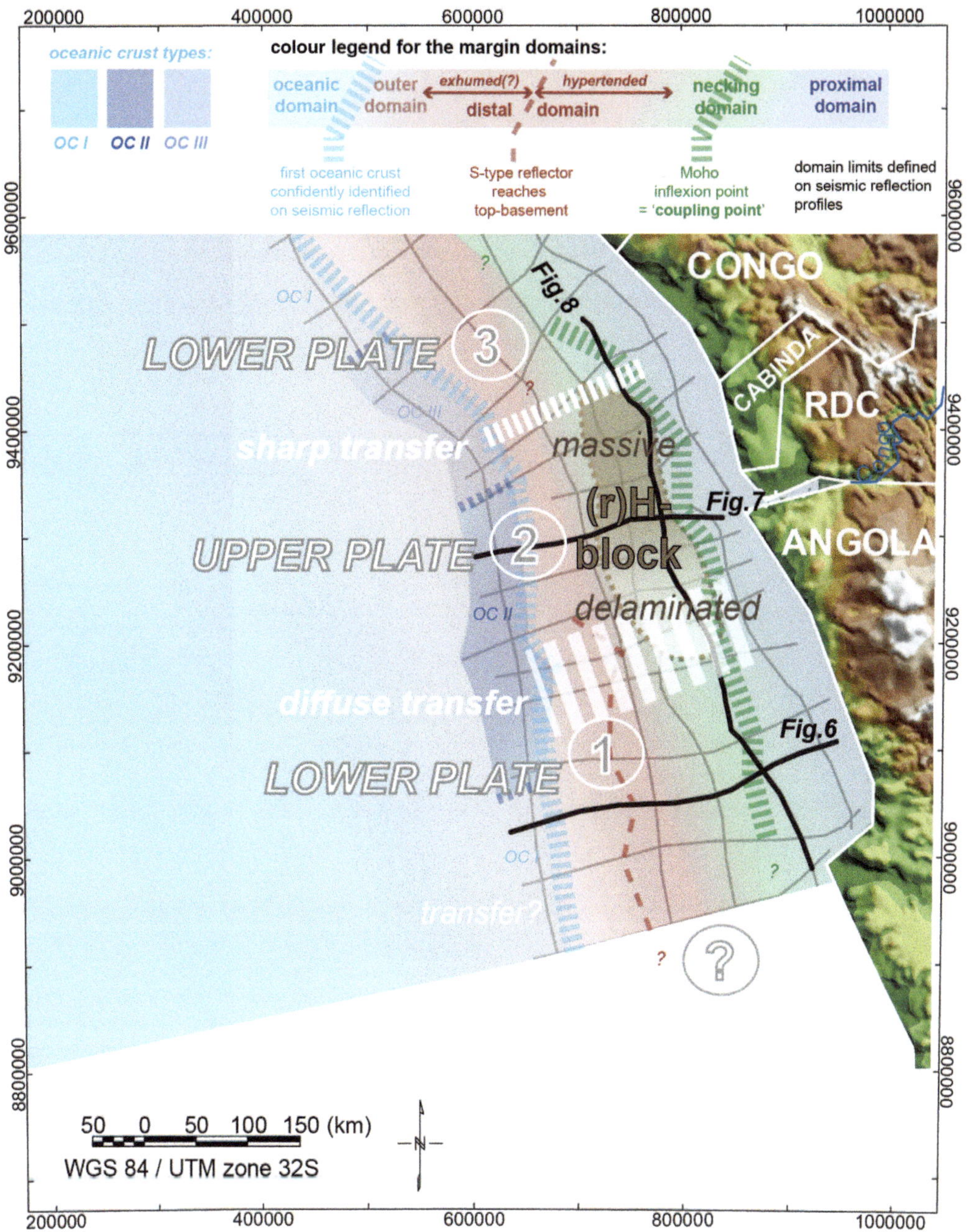

Fig. 4. Local map of the Angola–Gabon margin with delimitation of the structural margin domains. The circled grey numbers define the margin segments identified along-strike, with the alternation of lower plate and upper plate settings, with either a diffuse transfer (thick dashed white segment) or a sharp transfer (thin dashed white segment). The thick black lines locate the major seismic reflection profiles discussed in this paper.

Along the Angola–Gabon dataset, the Moho has been defined as corresponding to a series of deep, high-amplitude reflectors rising westwards from >25 km depth from the eastern extremities of the seismic profiles. The top-basement has been mapped at the base of the pre-salt sag basin sequence. In the proximal domain, these two sets of reflector bands are typically sub-parallel. The transition to

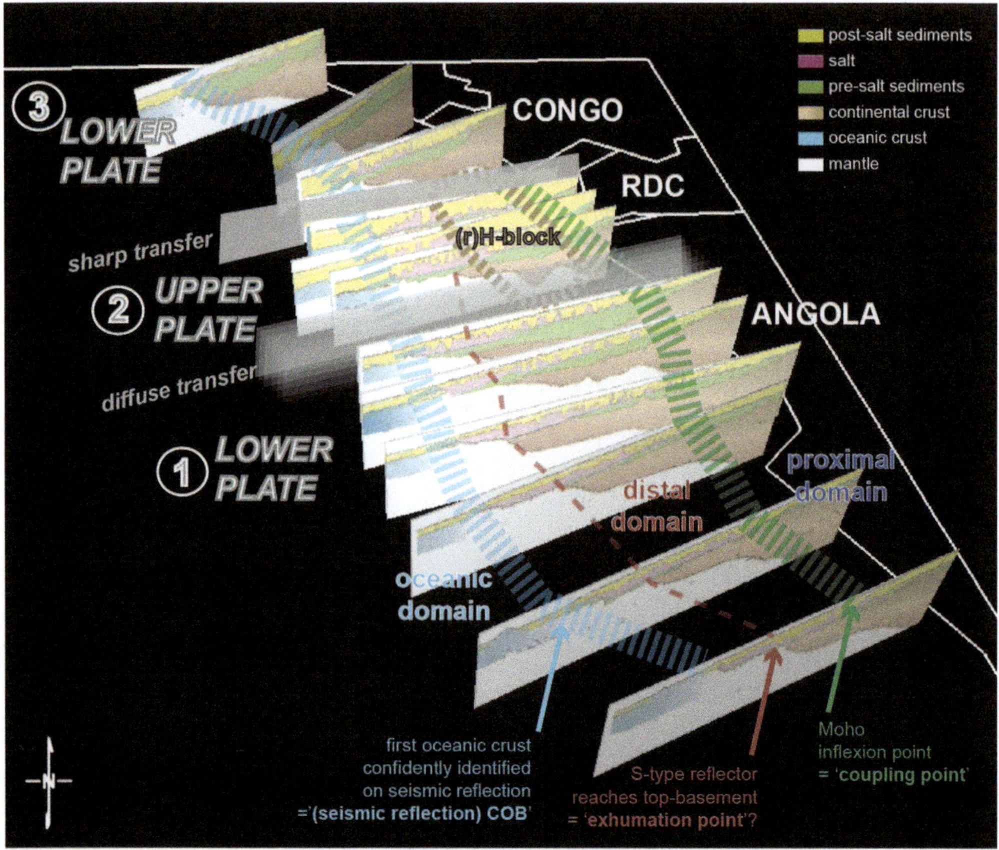

Fig. 5. Perspective view of one part of the seismic reflection profiles interpreted in the study (ION CongoSPAN1). The green, red and blue dashed lines indicate the location of the coupling point, exhumation point and (seismic reflection) continent–ocean boundary (COB), respectively. The circled grey numbers define the margin segments identified along-strike, with the alternation of lower plate and upper plate settings, with either a diffuse or a sharp transfer. The location of the residual (r)H-block is indicated by a dashed brown line.

the necking domain occurs when these begin to converge. The Angola–Gabon proximal domain partly outcrops onshore (e.g. the Kwanza Basin, Guiraud *et al.* 2010) and extends offshore up to the necking point.

In the south, in margin segment 1 (Figs 4 & 5), the continental crust is seismically defined by two distinct layers: a lower reflective layer and an upper layer that is more transparent and homogeneous. These two layers are generally separated by medium- to high-amplitude reflectors presenting variable dips. In the case of the Variscan system (e.g. Iberia), this vertical partition of the basement seismic facies most probably reflects the upper, middle and lower continental crust. It is more uncertain in the case of the South Atlantic. As the reflectivity increases towards the necking, some of the reflectivity may be related to extensional mylonitic shear zones in the ductile parts of the crust (Holliger *et al.* 1993;

Holliger & Levander 1994). As a result of the overlying salt filter, the top-basement is difficult to define precisely. However, concave-upwards faults delimiting half-graben-type basins can be identified, some presenting syn-tectonic stratigraphic geometries (Fig. 6), similar to observations reported at the conjugate Brazilian margin (Blaich *et al.* 2009, 2011; Zalán *et al.* 2011). The base of the crust (Moho) is characterized by a medium- to high-amplitude discontinuous reflector that plunges eastwards >25 km deep under the continental platform, whereas oceanwards it gently points towards the top-basement in the distal domain (Fig. 6).

Northwards, in the central margin segment 2 (Figs 4 & 5), the crust is more homogeneous and the partition into a transparent upper part and an energetic lower part is no longer evident; the Moho is still definable, but presents a lower amplitude signal than that to the south.

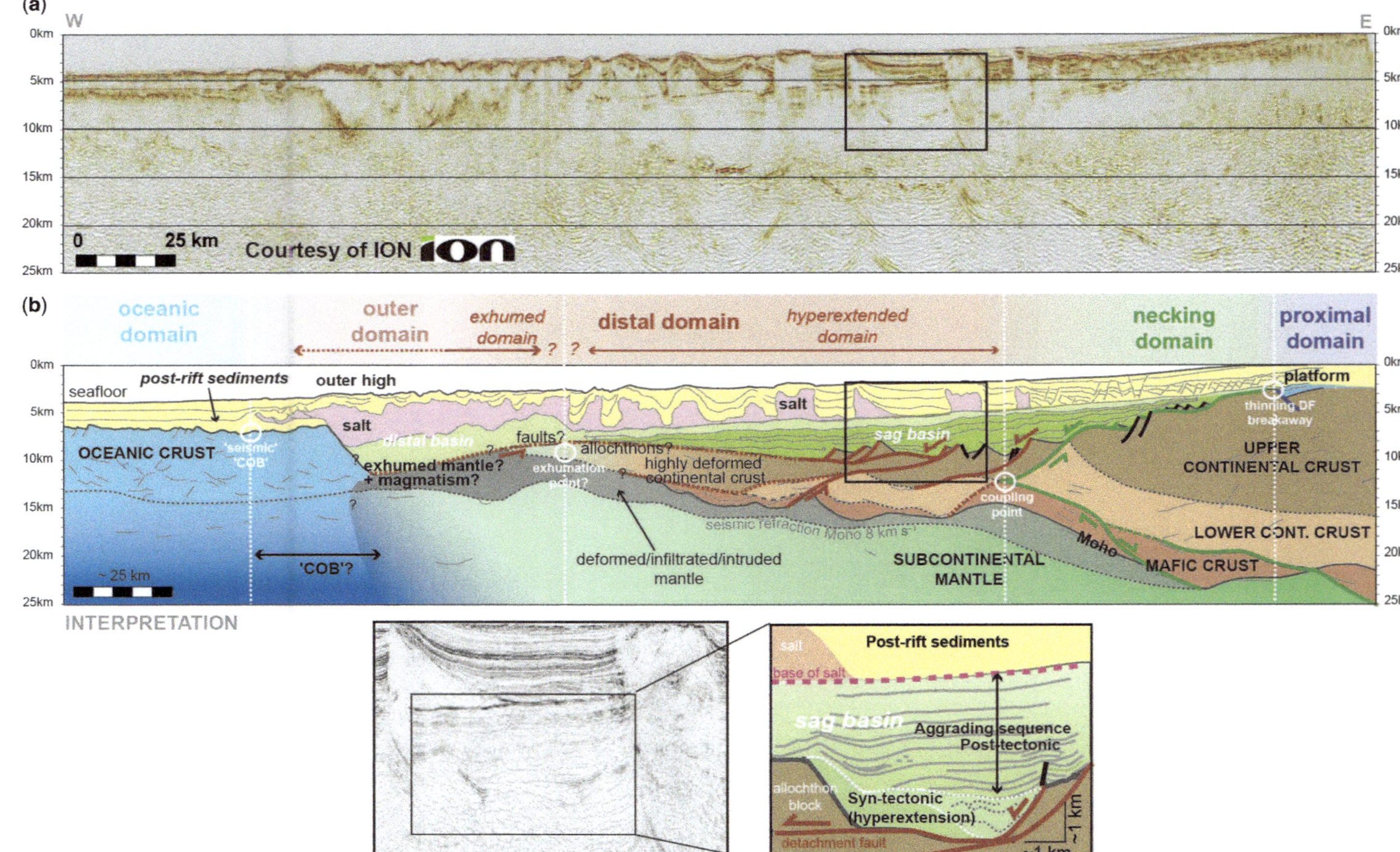

Fig. 6. Seismic profile (ION CongoSPAN1) illustrating the archetype of a lower plate rifted margin: (**a**) depth-migrated seismic reflection profile (ION); and (**b**) interpretation after Unternehr *et al.* (2010), Péron-Pinvidic *et al.* (2013) and Masini *et al.* (2012). See Figure 2 for the detailed legend and Figure 4 for location. ION is acknowledged for publication permission.

In margin segment 3 (Figs 4 & 5), continental basement presents a homogeneous reflective facies. The partitioning into two distinct layers identified in margin segment 1 is not observed and the crustal thickness seems to be more important at the proximal margin. The lower crustal limit is still defined by discontinuous reflectors of variable amplitude.

Necking domain

The necking domain is the domain in which the crust/lithosphere is efficiently thinned. In a similar manner to the proximal domain, the deformation is still decoupled at the crustal scale. However, in contrast, the top-basement and Moho are no longer sub-parallel. Seismically, the necking domain corresponds to a crustal wedge (e.g. the taper geometry of Osmundsen & Redfield 2011) (Fig. 2), with the seismic reflection Moho and the top-basement that have converged defining crustal thinning from *c.* 30 km to less than 10 km (Péron-Pinvidic & Manatschal 2009; Mohn *et al.* 2010; Sutra *et al.* 2013). Both the top-basement and Moho mark an inflexion point defining this drastic crustal thinning so, stratigraphically, the necking domain is also characterized by a marked basin-wards increase in total accommodation space (Sutra & Manatschal 2012). The necking domain is defined between the 'crustal hinge' and the 'coupling point' and has been defined as the point where the first brittle fault transects the entire crust (Sutra *et al.* 2013). It therefore delimits the decoupled domain (inboard) from the coupled domain (outboard) and corresponds to the location where there are no longer any ductile layers in the crust. This very specific location corresponds to the 'crustal embrittlement' constrained by Pérez-Gussinyé *et al.* (2003).

The continental crust is characterized within margin segment 1 by a relatively gentle necking, with crustal thinning from a >20 km- to <10 km-thick crust over *c.* 75 km (Fig. 6). Based on the dataset currently available in the public domain, some rare faulted sedimentary basins are described on the continental slope and are associated with the early synrift phase (Moulin *et al.* 2005). The top-basement is interpreted to correspond to a thinning detachment surface covered by extensional crustal allochthons and cut by high-angle normal faults at some stages (Fig. 6). The intra-crustal reflectivity often increases in intensity towards the coupling point. This may reflect an oceanwards increase in shear deformation accommodated by the ductile layer. Stratigraphically, the progressive oceanwards increase in accommodation space is supposed to be recorded by deltaic sedimentary systems or slope facies, including gravitational systems with a probable oceanwards transition from shallow- to deeper-marine environments (Tugend *et al.* 2015). In the case where the sedimentation rates are similar to the subsidence rates, or when the base level does not correspond to global sea-level, the depositional environments can also be subaerial.

Northwards, in margin segment 2, the Moho presents a very specific V-like geometry (Fig. 7). From the eastern end of the seismic profiles, the Moho rises classically oceanwards from >25 km depth to *c.* 20 km and then, contrary to the other margin segments, it does not gently continue shallowing to join the basement surface. It deepens again over *c.* 50 km, delimiting at depth a well-identified major crustal block. Then, oceanwards, the Moho shows a pronounced westwards rising and rapidly joins the top-basement in the distal margin. The necking domain of margin segment 2 is therefore difficult to define: its continent-wards limit can be identified based on the crustal hinge (increase in accommodation space); however, very often, its oceanwards limit, the coupling point, is not identifiable (Fig. 7).

Northwards, in margin segment 3, the necking domain is very gentle with a (seismic reflection) Moho gently rising from >25 km inboard to reach top-basement in the outer domain, 150 km further west, without any major disruption or inflexion point. Again, the coupling point there is often rather ambiguous. However, the Moho geometry is distinctly different from that of margin segment 2. In addition, in contrast with margin segment 1, major half-graben-type basins are identified in the necking domain with stratigraphic geometries indicating syn-tectonic deposition.

Distal domain

The distal domain corresponds to the margin domain between the crustal necking and the oceanic crust. Conceptually, it encompasses a hyper-extension sub-domain and a possible exhumation sub-domain and corresponds to high extensional β factors associated with an important accommodation space (Fig. 2). The hyper-extended sub-domain stands between the coupling point and the exhumation point (Fig. 2; Sutra *et al.* 2013) and corresponds to continental crust that has been thinned down to <10 km thickness and, importantly, where no remnant ductile layers prevail, allowing faults to cut from the crustal surface into the mantle, indicating that the deformation is coupled on a crustal scale. The exhumation sub-domain is proposed to lie outboard of the last crustal block of the hyper-extension sub-domain (the 'exhumation point' of Sutra *et al.* 2013) (Fig. 2), where the S-type reflector cuts out at the top- (seismic reflection) basement. From there, the basement is proposed to be composed of exhumed and variously serpentinized mantle that can be capped by continent-derived extensional allochthons and/or magmatic additions that may

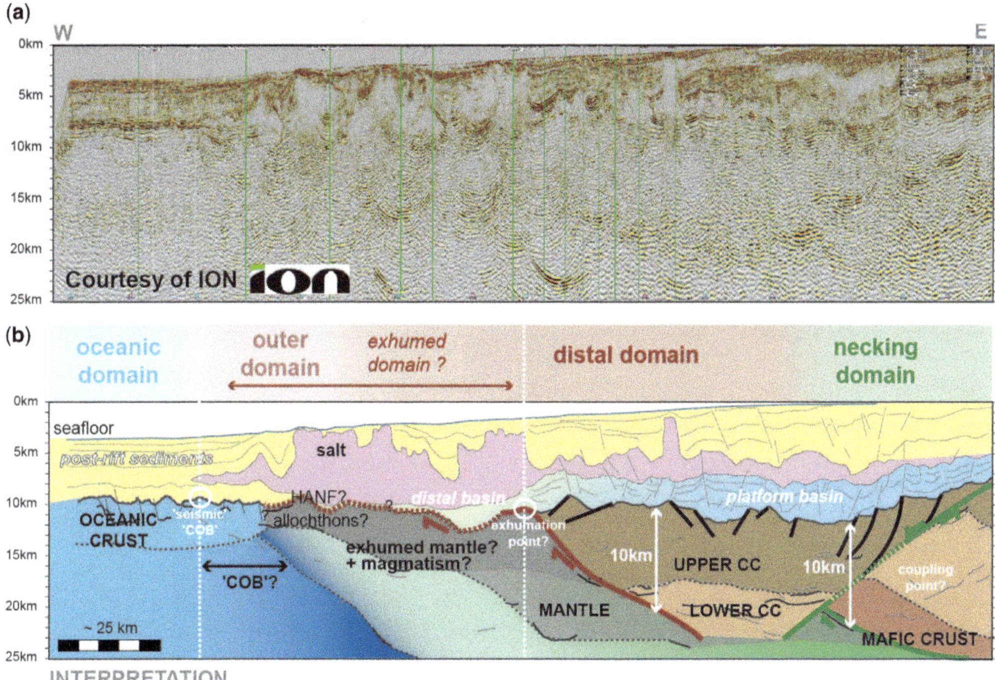

Fig. 7. Seismic profile (ION CongoSPAN1) illustrating the archetype of an upper plate rifted margin: (**a**) depth-migrated seismic reflection profile; and (**b**) interpretation after Unternehr *et al.* (2010) and Péron-Pinvidic *et al.* (2013). See Figure 2 for the detailed legend and Figure 4 for location. ION is acknowledged for publication permission. CC, continental crust.

become more voluminous oceanwards. The top-basement of the exhumation sub-domain may therefore not correspond to simple serpentinized mantle. Complex assemblages of crustal remnants and magmatic products are highly probable.

The Angola–Gabon hyper-extended sub-domain is generally larger than the potential exhumation sub-domain. It is characterized by a well-defined Moho, a top-basement difficult to identify below the salt and sag sequence, and a basement seismic facies oscillating between transparent and more reflective, but globally homogeneous. The (seismic reflection) Moho progressively rises towards the top-basement oceanwards (Fig. 6). The amplitude, geometry and overall structural context of this rising Moho reflector (occurrence in the distal domain, surface emersion) are highly comparable with other reflectors described at other margins worldwide, such as the S reflector for the Iberia margin (Montadert *et al.* 1979; Reston 1996), the M reflector for the Brazilian margin (Blaich *et al.* 2011) or the P reflector for the Porcupine Basin (Reston *et al.* 2001). Therefore it is proposed that the location where this S-type reflection reaches the basement surface coincides with the wedging out of the

continental crust and the exhumation of the mantle at the seafloor (the exhumation point, Fig. 6). Continental crust may exist further oceanwards, but rather as extensional allochthons (of variable size) overlying the 'exhumed' mantle. Magmatic additions are also very likely. The exhumation sub-domain is then characterized by a thick and highly transparent overlying salt sequence (the foot of the gravitational deformation). The top-basement limit is difficult to define, but could be associated with the top of a high-amplitude strip of reflectors characteristic of this segment (Fig. 6). However, if this partition into two distinct sub-domains is well-observed in margin segment 1, they are not distinguishable further northwards. In margin segment 3, the deep margin presents no more high-amplitude reflectors and the Moho is ill-defined and joins the oceanic Moho directly.

The question of exhumation. The existence or non-existence of exhumed mantle in deep water rifted margins is highly debated. Unfortunately, except for the Iberia–Newfoundland, Red Sea or the South Australia systems, no deep drillhole is available to validate or invalidate the hypothesis. In the case of the South Atlantic rift, based on almost the same dataset, various interpretations of the composition of the distal margin basement have been published, including propositions of igneous crust (e.g. Mohriak *et al.* 2008), transitional crust (e.g. Blaich *et al.* 2011), exhumed lower crust (e.g. Aslanian *et al.* 2009) and hyper-extended continental crust associated with zones of exhumed mantle (e.g. Zalán *et al.* 2011). The available dataset may not therefore be conclusive.

It is true that very often the distal margin geometries are difficult to constrain based on seismic reflection alone as a result of various imagery issues. The standard interpretation protocol in that case is to make up the deficiency of the seismic reflection dataset by refraction and/or potential field modelling (e.g. von Nicolai *et al.* 2013). The resulting models bring crucial information on the relative repartition of high and low densities/velocities (and gradients) in the margin (e.g. Contrucci *et al.* 2004). The modelling results are then often interpreted in terms of rock lithology. It has been known from a long time, however, that rock velocity values, densities and magnetic parameters are not conclusive of any lithology (the same value can correspond to very different rock types; Christensen & Mooney 1995). The interpretation of the deep margin architecture in terms of igneous crust, continental crust or exhumed mantle based on these geophysical models alone may therefore not be straightforward. We tend to think that this may be particularly true for the distal and outer margin domains, where the rocks went through so many complex multiphase deformation processes that they probably no longer retain their classic petrophysical parameters.

As for every structural interpretation based on seismic reflection, the interpretation given here is humbly considered as one among many other possible alternatives. We propose considering the possibility of the existence of a zone of exhumed mantle in the deep Angola margin based on the knowledge acquired for other systems, such as the Iberia–Newfoundland or South Australia systems, where similar structural geometries (the band of rising reflectors reaching top-basement) has been proved by drilling or dredging to image mantle exhumation. We thus consider the geometries proposed in our figures as geologically meaningful.

It is also worth adding that 'mantle exhumation' does not necessarily mean that mantle rocks have to form the top of the basement in the deep margin. Crustal allochthons and magmatic material added to the system can represent a significant part of the distal domain, overlying and/or underplating the detachment-capped 'exhumed' mantle. In Alpine ophiolites, for example, mantle exhumation is overprinted by the subsequent emplacement of magmatic products (Manatschal & Muntener 2009). Another independent observation is that at magma-poor, thermally equilibrated margins, the top-basement over the distal margin is regularly deeper than that over the adjacent oceanic crust. The stepping up onto oceanic crust, often referred to as the 'outer high', is commonly explained by the fact that the rock column forming the distal crust is denser and/or thinner than the oceanic domain rock column, an explanation which is compatible with a lower magmatic budget, but also with the exhumation of mantle rocks.

Outer domain

The outer domain is, by definition, the key domain in the distinction of magma-poor and magma-rich margins (Péron-Pinvidic *et al.* 2013). It is located between the (usually) ill-defined basement of the distal domain and the oceanic crust. The composition of the basement is, in most cases, undetermined; similar structures have not been drilled, except from the Iberia–Newfoundland rifted margins.

Its inboard limit is envisaged as a transition rather than as a boundary. It corresponds to the area where the volume of magmatic additions become more important. Magmatic additions can correspond to well-identifiable sills and lavas, but also intrusions in or at the base of the basement (thinned crust or serpentinized mantle), or infiltration and melt channels in the deeper part of the mantle that are more difficult to identify on geophysical datasets. Therefore, contrary to the proximal and necking domains, the inboard limit of the outer domain can rarely be confidently defined with a sharp boundary.

Its outboard limit corresponds to the continent–ocean boundary (COB). Although the COB is a well-accepted term, it is often difficult to map. The definition is often ambiguous as it is strongly subjective and is strongly dependent on the dataset used (see Heine *et al.* 2013). The COB defined in this work corresponds to the first oceanic crust identified on the seismic reflection profiles. It is very often defined westwards of the foot of the salt gravitational body, where it can be confidently identified. It is, however, well-accepted that the first blocks of accreted oceanic crust probably extend further continent-wards (see 'COB?' on figures). In the absence of another high-resolution dataset (refraction, potential field, new seismic datasets or new processing) that could have helped to further define the boundary, we favoured the conservative seismic definition, keeping in mind that the real oceanic boundary may well stand further continent-wards.

The outer domains of the Angola–Gabon and Campos–Esperito Santo margins correspond to the apparent structure-less distal basins (Figs 6 & 7).

Unfortunately, these very specific basins are not straightforward to characterize as they are regularly overlain by allochthonous salt, which masks the underlying basement, preventing the rigorous distinction of the crustal or sedimentary architecture at depth (Torsvik *et al.* 2009; Unternehr *et al.* 2010; Blaich *et al.* 2011). The outboard limit of the Angola–Gabon distal domain is generally well-defined by a pronounced change in the reflectivity of the basement and the presence of sub-horizontal layered sediments covering the oceanic basement (Figs 6 & 7). Seawards deepening reflectors and major volcanoes are identified at the southern tip of the dataset and attest a certain magmatic activity.

Oceanic domain

Oceanic crust presents various geophysical characteristics depending on its magma-poor or magma-rich accretion type (Penrose Conference Participants 1972; Dick *et al.* 2003; Cannat *et al.* 2006). These characteristics can range from low-amplitude, disorganized magnetic anomalies, with no clear V_p contrasts with the adjacent margin distal domain and few intra-basement seismic reflectors (e.g. Iberia) to high-amplitude, well-defined linear magnetic anomalies (e.g. Norway Basin) with clear tripartite V_p characteristics and a proper seismic reflection pattern (e.g. the Angola margin).

With respect to the South Atlantic oceanic crust, the ION CongoSPAN1 dataset images a marked lateral structural variability with very distinct geometries from south to north (Figs 4 & 5). Three main types of oceanic crust have been tentatively defined. Interestingly, these correlate well with our mapped upper plate and lower plate margin segments (see Fig. 4).

(1) Oceanic crust type I corresponds to a 'normal magmatic crust' with a standard crustal thickness of about 6–8 km. It is observed at the southern and northern tips of our dataset. The basement limit is well-defined and rough, covered by a relatively thin *c.* 2–3 km-thick sequence of layered sediments. The oceanic crust itself presents a standard three-layered type architecture: a *c.* 2–3 km-thick upper part with reflective seismic facies; a *c.* 3 km-thick middle layer with a marked transparent seismic facies; and a *c.* 2–3 km-thick more energetic layer characterized by numerous reflectors with medium to high amplitude, disorganized, with variable dips and geometries. This first-order architecture may reflect a standard oceanic layering, according to the Penrose definition (Penrose Conference Participants 1972), including, from top to bottom: a mafic volcanic (basalt) and sheeted dyke complex (oceanic layer 2); a gabbroic complex (oceanic layer 3); and a deformed and intruded ultramafic complex.

(2) Oceanic crust type II probably corresponds to a 'slow-spreading magma-poor crust' with a crustal thickness varying between 3 and 6 km. It is observed southwestwards of the H-block, nearly coincident with the mapped upper plate margin segment (Fig. 4). The basement limit is characterized by numerous fault-bounded tilted blocks. It is covered by a thick sedimentary sequence (from *c.* 3 to *c.* 6 km thick). The crust is structured into three layers, comparable with oceanic crust type I, but with significantly different amplitudes, geometries and thicknesses. The upper layer is typically 1 km thick, but can also taper down to nearly zero. Where it exists, it is reflective and is cut by numerous normal faults showing variable displacements. The middle layer is globally transparent and around 2.5 km thick, but this layer can also be almost absent. The lower layer is significantly different from the other margin segments: it is characterized by reflectors with medium to high amplitude, the lower reflectors defining an easily mappable oceanic Moho. The reflectors are mainly sub-horizontal, although some have higher dips and join the surface cutting through the nearly absent upper layers where the oceanic crust appears to be thinnest. Cannat *et al.* (2006) reported different categories of oceanic spreading depending on the supply of melt to the accreting system. The variability of the geometries and thicknesses of the oceanic layers in our dataset may illustrate such melt variations. Our oceanic crust type II, which shows thinner to absent mafic and volcanic layers, could then represent a section with relatively less magma and volcanism than the other two adjacent types.

(3) Oceanic crust type III corresponds to an intermediate crust, with intermediate seismic characteristics. The top-basement is rough with pronounced topography. The crustal thickness can reach 9 km. It is composed of a *c.* 4 km-thick relatively reflective upper layer overlying a *c.* 4–5 km-thick homogeneous band of high reflectivity, with no clear individual reflector or particular geometry.

Along-strike structural variation

The Angola–Gabon rifted margin presents structural and stratigraphic along-strike variations that we have interpreted as the result of an upper plate to lower plate alternation with sharp or diffuse transfer zones. The dataset available for this study enabled us to map a number of key observable features that helped us to define the first-order structural and stratigraphic characteristics of the upper and lower plate margins. Two profiles from the ION CongoSPAN1 dataset have been identified as key sections exemplifying the upper and lower plate margins.

Figures 6 and 7 show our interpretations of the two key profiles. Figure 8 shows an along-strike profile highlighting the lateral structural variations. Figures 9 and 10 refer to the conjugate first-order observations. Figure 11 proposes a schematic reconstruction of the conjugate Brazilian and African margins at 115 Ma and Figures 12 and 13 are cartoons summarizing the key points of our discussion.

Upper plate setting

We propose that the upper plate setting is exemplified by the ION CongoSPAN1 3600 profile offshore Zaire (Fig. 7; see Fig. 4 for location). Other examples of this setting may be the mid-Norwegian Lofoten segment, the central East Greenland Jameson Land segment, the central East India segment, the Newfoundland margin or the Brianconnais (Tsikalas *et al.* 2005; Péron-Pinvidic & Manatschal 2009; Mohn *et al.* 2010, 2012; Péron-Pinvidic *et al.* 2012; Nemčok *et al.* 2013).

The upper plate necking domain is considered to be relatively narrow and probably typified by structures similar to core complexes with the denudation of mid-crustal material in the footwall of major thinning detachment faults (Fig. 13). In the hanging wall, significant perched basins have been regularly observed (e.g. East India, Nemčok *et al.* 2013). The crustal necking usually flanks an H-block-like structure. Within successful rift systems, the residual H-block is considered as the characteristic feature of an upper plate setting (Fig. 13). It corresponds to a sort of keystone, a block of remnant continental crust flanked by major detachment faults. Depending on the local rifting parameters, the final geometry of the residual H-block can be highly variable, from a safeguarded, easily identifiable triangular-shaped massive block to a dismembered series of standard tilted crustal blocks.

The CongoSPAN1 3600 profile shows a *c.* 20 km-thick continental crust strongly rotated in the footwall of a major detachment structure, interpreted to correspond to one of the thinning detachments bordering a massive H-block (Fig. 7). The H-block flanks are well identified on the seismic reflection profiles by series of high-amplitude reflectors (Fig. 5). The overall geometry is reminiscent of a keystone block with a characteristic inverted triangular shape (Fig. 7). The crustal thickness exceeds 10 km, so the H-block may still present ductile layers at depth. The top of the H-block is

chaotic. The upper crust is cut by numerous normal faults that probably gradually delaminated the H-block, while the exhumation detachment on its outboard flank exhumed deep crustal and mantle material to the surface. Some of the crustal blocks delimited by these normal faults most likely ended as allochthonous rider blocks overlying exhumed basement in the distal margin.

The stratigraphy on top of the H-block is dominated by standard graben and half-graben basins with relatively little sedimentary accommodation space. The basin that caps the residual H-block is highly tectonized (Fig. 7). It is affected by numerous high-angle normal faults, which often correspond to the prolongation of underlying basement faults bordering basement blocks at the top of the residual H-block. These are evidence of continuing tectonic activity that periodically remobilized the crustal block and its sedimentary strata. The thickness of this sequence can reach >5 km. The base of the salt is also disrupted, deformed and sometimes cut by faults, which highlight out-of-sequence faulting or late post-rift isostatic adjustment of the residual H-block.

The upper plate margin, as the main active detachment that exhumed deep material rooted underneath, is supposed to have experienced a specific isostatic evolution with a high position throughout most of the evolution of the rift (Lister *et al.* 1986; Lemoine *et al.* 1987; Mohn *et al.* 2010, 2012; Ranero & Pérez-Gussinyé 2010; Brune *et al.* 2014). It is also subject to various synrift uplift phases depending on the activity of the detachment fault and the underlying movement of material with variable density (e.g. mid- to lower crust, mantle, serpentinized mantle and melts). Later, the post-rift evolution of the upper plate margin is characterized by rapid subsidence once the tectonic activity migrates oceanwards (Masini *et al.* 2012, 2013). Accordingly, the heat flow is supposed to be high during the evolution of the synrift (McKenzie 1978) when deep crustal and mantle material are dragged out from underneath the H-block (Mohn *et al.* 2010). Magma production and emplacement in the form of infiltration, intrusions and/or underplating are also considered to be typical of an upper plate setting if the petrology and temperature of the mantle is adequate (Muntener *et al.* 2010).

Outboard of the residual H-block, a *c.* 60 km-wide zone is interpreted as exhumed mantle with variable amounts of magmatic additions, capped by a thin sedimentary sequence below massive salt deformations (Fig. 7). Unfortunately, no geometry is identifiable in our dataset to help better constrain the characteristics of this distal basin.

At a post-break-up stage, an upper plate margin can present various lengths: it can be very narrow (e.g. the central East Greenland Jameson Land

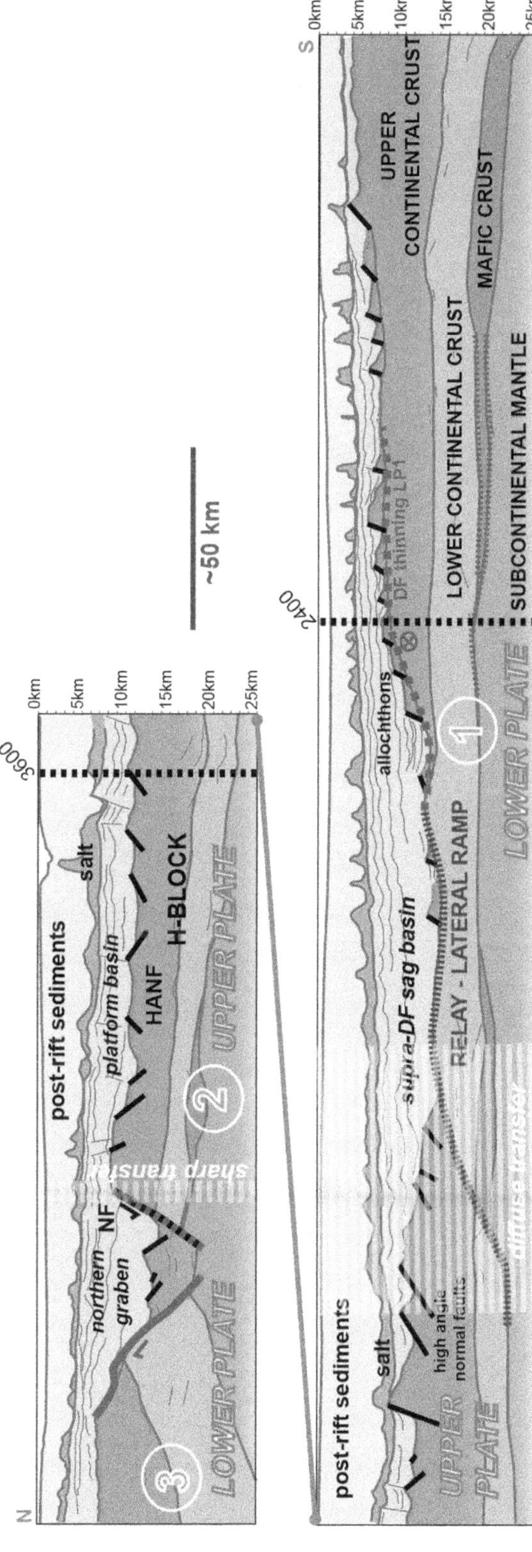

Fig. 8. Line drawing illustrating the along-strike variability of the major structural and stratigraphic entities discussed in this paper. See Figure 4 for location and Figure 2 for legend. The grey numbers circled in white refer to the margin segments described in the text. The dashed light white stripes represent the transfers from an upper plate setting to a lower plate setting. ION is acknowledged for publication permission.

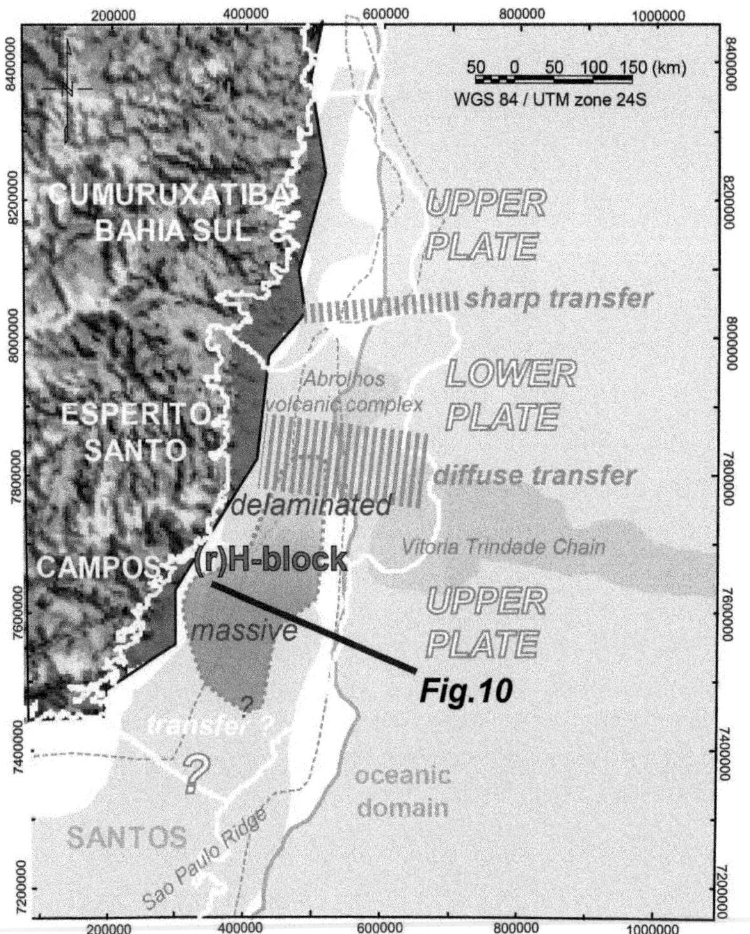

Fig. 9. Local map of the SE Brazilian margin. The Santos, Campos, Esperito Santo and Cumuruxatiba basins are outlined in white. The along-strike alternation from lower plate to upper plate settings is illustrated as in Figure 4 with either a diffuse transfer (thick dashed white segment) or a sharp transfer (thin dashed white segment). The thick black line locates the seismic reflection profile discussed in this paper.

segment, conjugate to the Vøring margin), or very much wider, such as the Newfoundland margin that presents a significant proximal domain (Grand Banks) and a wide distal margin sag-type sequence. The width of a margin actually depends on the location of lithospheric break-up, which is a process considered to be relatively independent of rifting (see Péron-Pinvidic *et al.* 2013). So, although the distal supra-detachment sag basin is considered to be characteristic of the lower plate setting, the break-up line can cross-cut the basin at various locations, from the flank of the upper plate H-block to the lower plate distal domain, which may result in the isolation of a section of the distal sag sequence on the upper plate margin (see blue corridor on Fig. 12). As a result, an upper plate margin can also present a significant distal domain with possibly some hyper-extension and exhumation sub-domains and a related major sag-type sedimentary basin. Moreover, as described by Péron-Pinvidic & Manatschal (2010), additional structural features can complicate the overall geometry, such as continental ribbons with potentially wide and deep aborted inner rift basins, or microcontinents. These are not represented on Figure 13.

Lower plate setting

Lister *et al.* (1986) described the lower plate margin as much wider and structurally complex, with a succession of major detachment surfaces and bowed-up lower crustal material in the deep margin. The

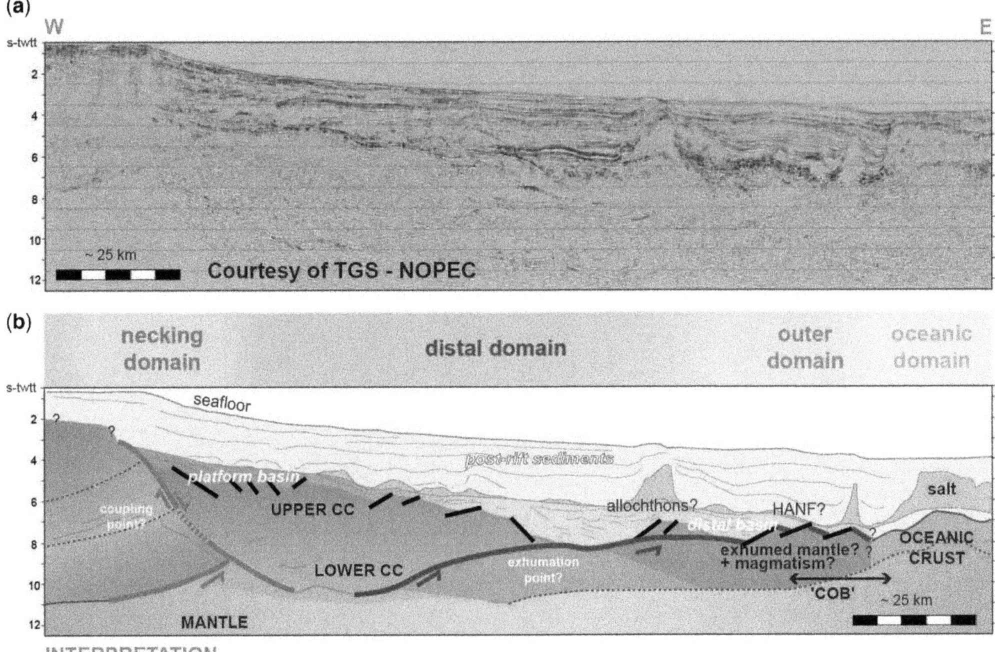

Fig. 10. Line drawing illustrating the upper plate geometry identified in the Brazilian Campos Basin: (**a**) depth-migrated seismic reflection profile; (**b**) interpretation. See Figure 2 for detailed legend and Figures 4 and 5 for location. TGS is acknowledged for publication permission. CC, continental crust.

ION CongoSPAN1 2400 profile is considered to exemplify a lower plate rifted margin. Other examples worldwide are the Iberia margin, the mid-Norwegian Vøring segment and the central South Australia margins (Osmundsen *et al.* 2002; Direen *et al.* 2007; Péron-Pinvidic *et al.* 2007; Faleide *et al.* 2010; Ball *et al.* 2013).

The lower plate necking domain is considered to be probably wider than its upper plate conjugate (e.g. Vøring, Osmundsen *et al.* 2002). The associated thinning detachment fault can present highly complex geometries, with outboard successions of major detachments and intermediate distinct sub-basins (Fig. 13). The distal domain encompasses the hyper-extended and exhumed sub-domains. One of the major characteristics of a lower plate setting is the presence in the distal domain of a supra-detachment sag-type basin, with abundant accommodation space, although sag basins can also occur on upper plate settings. If the sedimentary influx is efficient, there is a potential for thick sag-type aggradational sequences (e.g. Angola), but if the source is almost non-existent, the resulting distal domain can also be significantly starved of sediment and may develop as a deep water margin (e.g. Iberia). Ranero & Pérez-Gussinyé (2010) proposed an example of asymmetry between two

conjugate upper plate and lower plate settings based on the Iberia–Newfoundland system.

The lower plate Angola necking domain is rather wide: the CongoSPAN1 2400 profile shows a >25 km-thick continental crust gently thinned oceanwards over more than 100 km (Fig. 6). The top-basement is interpreted as corresponding to the thinning detachment fault over which some residual crustal allochthons can be recognized. The major detachment is cut by later high-angle normal faults at some stages.

Oceanwards, the lower plate basement is interpreted as a series of major tilted crustal blocks, bordered by detachment faults (Fig. 6). These migrated progressively westwards, cutting into progressively thinner continental crust. This is the hyper-extension domain, where extensional allochthons rest on the detachment surface bounding the graben and are submitted to hyper-extension depositional environments. As a result of the salt blanketing effect, the structure of the deep margin is difficult to define. Based on other examples worldwide, we consider it possible that the mantle has been exhumed at the most distal part of this domain (see Fig. 6).

Stratigraphically, the lower plate proximal domain is similar to the upper plate proximal

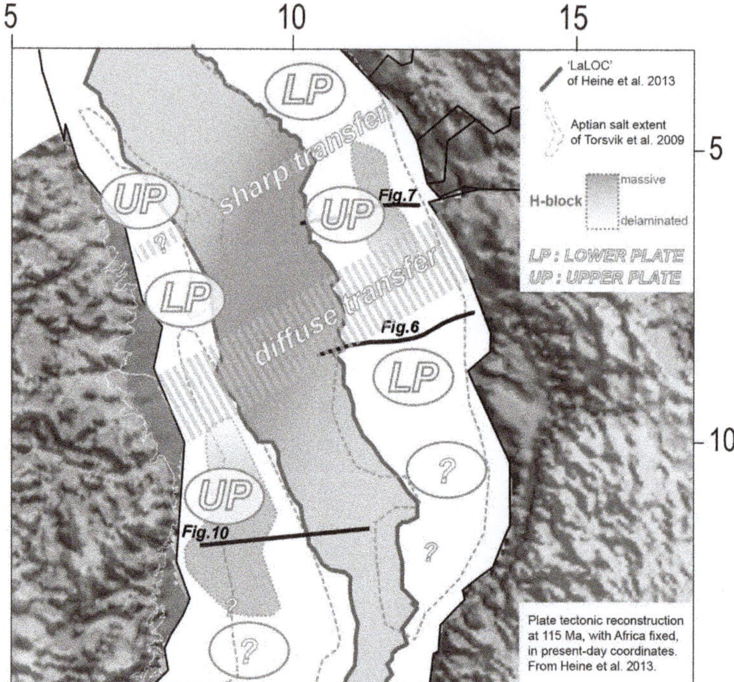

Fig. 11. Local schematic reconstruction of the Angola–Gabon and Campos–Esperito Santo conjugates at 115 Ma, with Africa fixed and in present day coordinates. The reconstruction is based on the review of Heine *et al.* (2013) of the South Atlantic kinematics. The figure shows the coherent conjugate position of the two H-blocks identified on the Brazilian and African conjugates. Note, however, that the southern and northern areas are not constrained.

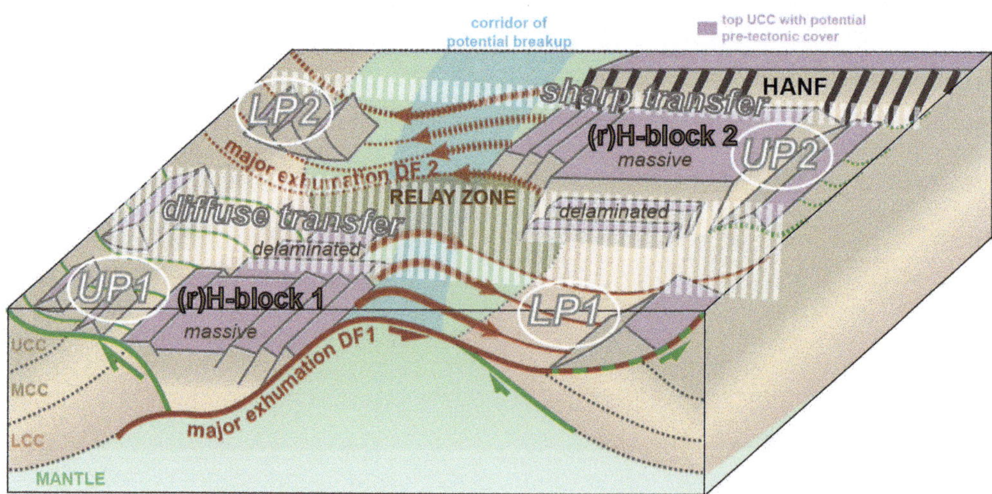

Fig. 12. Perspective three-dimensional schematic diagram illustrating the lateral polarity change from an upper plate to a lower plate margin along a diffuse transfer. Colour coding as Figure 2. Sediments not represented. UCC, upper continental crust; MCC, middle continental crust; LCC, lower continental crust; DF, detachment fault; HANF, high-angle normal fault; UP, upper plate; LP, lower plate.

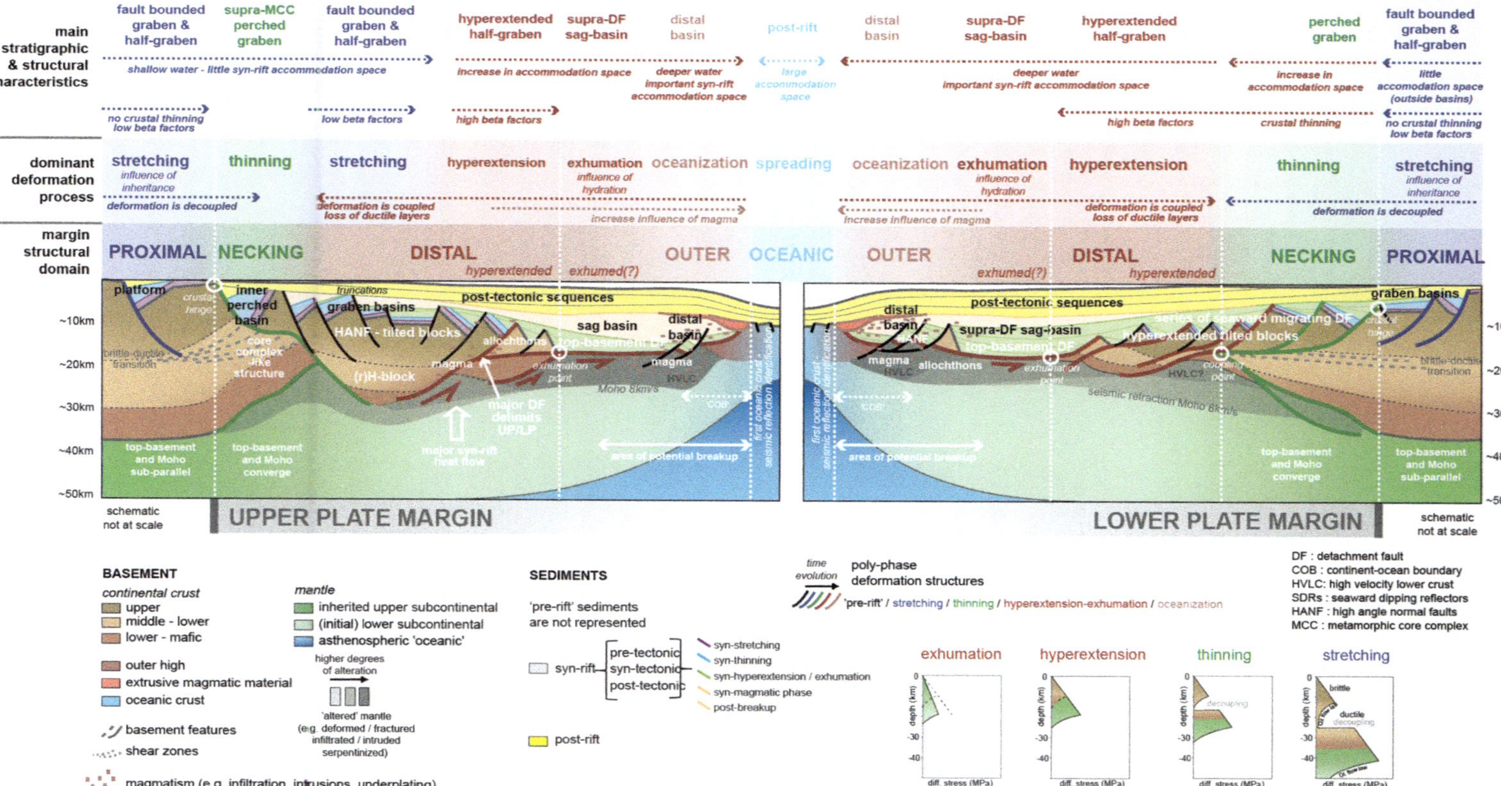

Fig. 13. Schematic diagram summarizing the major structural and stratigraphic characteristics of upper plate and lower plate settings based on the observations made on the Angola–Gabon dataset. Terminology and definitions are (notably) from Ranero & Pérez-Gussinyé (2010), Sutra *et al.* (2013), Péron-Pinvidic *et al.* (2013), Tugend *et al.* (2014), Manatschal *et al.* (2014).

domain, corresponding to standard syn-tectonic wedged-shape packages related to half-graben basins delimited by early rift normal faults, with little accommodation space (Fig. 13). More oceanwards, the distal sag basin presents very distinct geometries, characteristic of hyper-extended and exhumed basement (Masini *et al.* 2012). At the base of the sag basin, supra-detachment half-graben are identified between hyper-extended tilted crustal blocks (Fig. 6). These show growth structures that attest to the syn-tectonic tilting of blocks during deposition. Upwards, the characteristic relatively sub-parallel strata of the sag sequence drape the structured top-basement. These attest to an increasing accommodation space, most probably related to drastic crustal extension and the migration of the tectonic activity outboard (Masini *et al.* 2012). The stratigraphy of the distal sag sequence encompasses three main units (Masini *et al.* 2012): basal, intermediate and top. These distinct facies tracts are intimately linked to the formation and evolution of the distal lower plate with: (1) the initial formation of the regional sag basin with a series of fault-bounded half-graben that can evolve into supra-detachment basins; (2) the widening of the sub-basins; and (3) the migration of the active tectonic deformation further outboard (for details and terminology, see Masini *et al.* 2012, 2013). In contrast with the standard fault-bounded basins characteristic of the proximal margins, the sediments in the distal lower plate margin are not deposited over a tilting H-block, but over a tectonically exhumed footwall with faults formed in sequence and migrating oceanwards (e.g. Ranero & Pérez-Gussinyé 2010). In such supra-detachment basins, the depocentre is supposed to show a lateral migration oceanwards, whereas in classical half-graben basins the depocentre is stable in time and space (Masini *et al.* 2012). A major consequence of the in-sequence migration of the faults is that the use of the term synrift is inadequate to describe the sedimentary sequences in distal margins. Therefore, Masini *et al.* (2013) distinguished between syn-tectonic and sag sequences, the former describing deposition during active faulting and the second describing a sedimentary sequence that is deposited between end of local tectonic activity and break-up, defined as the moment when seafloor spreading starts.

Regarding heat flow, the stretching phase should correspond to heat flow that can be predicted by classic McKenzie-type models, whereas thinning is related to crustal/lithospheric necking that may include thermal erosion of the lithospheric mantle (Muntener *et al.* 2010; Padovano *et al.* 2015). A major change between the evolution of heat flow in the upper plate and lower plate should correspond to the time of deformation coupling of the necking

detachment faults on one flank of the H-block: the time of necking. Afterwards, the lower plate progressively moves away from the active exhumation detachment flanking the H-block, so the heat flow should decrease accordingly, except at the time of break-up. Major subsidence occurs in the lower plate sag basin during the progressive delamination of the upper plate H-block and the exhumation of deep material (Mohn *et al.* 2010). Finally, in a similar manner to the upper plate margin, the length of the lower plate margin should be variable and depend on the line of break-up (Figs 12 and 13).

Lateral evolution and transfers

Structural evolution. We propose a lateral segmentation of the Angola–Gabon rifted margin into (at least) three sectors from south to north: (1) a lower plate segment (Kwanza Basin); (2) an upper plate segment (North Angola to South Congo); and (3) a second lower plate segment (the Gabon Basin) (Figs 4 & 5) (note that the southern and northern tips of this study area are not constrained, so additional segments are possible). Standard lower plate structural characteristics are observed in the southern part of the study area, with gradual crustal thinning, complex top-basement geometries and regional distal sag-type basins (Figs 5 & 6). The geometries become more subtle northwards, with a gradual change into a setting that increasingly resembles an upper plate context. From the ION 2800–3000 profiles, the crustal necking becomes sharper in the footwall of a master detachment fault flanking a crustal block that becomes more massive and clear northwards (e.g. ION 3000). The architecture of the necking domain is, however, more confused and the coupling point is no longer obvious (Fig. 13). At the northern end of the dataset, offshore Gabon, the crustal structure displays other geometries (Fig. 5) with an apparently thicker crust in the proximal domain, gentle crustal thinning in the necking domain and a regional significantly thick sag-type sedimentary sequence (Fig. 5).

The CongoSPAN1 9300 profile is a strike line running over the proximal necking domains (Fig. 4). It illustrates very well these lateral structural and stratigraphic variations (Fig. 8). It cross-cuts the 2400 and 3600 key profiles and highlights the along-strike extent of the residual H-block and the two distinct transfers from upper plate to lower plate settings (Figs 8–12). To the south, the 9300 profile shows a continental crust subdivided into the three characteristic layers, with an upper homogeneous part, a more reflective lower part and an intermediate part that concentrates various reflectors of varying dip and amplitude. Going northwards along the profile, the continental crust is gently thinned over more than 200 km. As the profile laterally cross-cuts the dip structures, we interpret the top-basement as a composite feature including segments of thinning detachment faults, normal faults and a lateral ramp that accommodated the transfer of deformation from the lower plate extensional system to the upper plate system (Figs 8 & 12). Crustal allochthons rest on the extensional composite structure and are draped by supra-detachment sag sequences. North of this zone, the residual H-block can be traced over more than 200 km (Fig. 8). Its northern limit is remarkably neat and well-defined by a major regional normal fault (NF on Fig. 8). This northern limit contrasts with the southern limit, which is more diffuse, probably corresponding to a more gradual dismembering of the crustal block above the master exhumation detachment fault during rift opening.

Stratigraphy. Although the type sediment source, depositional environment and rate of sedimentation are obviously important parameters in defining the sedimentary architecture, the nature and structure of the crust and the tectonic evolution directly define the accommodation space – that is, the space available to record any sedimentary information. The crustal structure of the Angola–Gabon margin shows clear along-strike variations and the observed stratigraphic geometries reflect the lateral segmentation, with evolution from a rather standard lower plate sag-type architecture to more complex geometries over the upper plate H-block (Fig. 8).

Southwards, within margin segment 1, interpreted as a lower plate setting, a major sag-type sedimentary basin is mapped (the 'pre-salt sag', Karner *et al.* 1997). The basin presents a marked thinning in all directions from a depot centre located at the level of *c.* 8° S to 12° E (Figs 5 & 8). Based on refraction models, the sedimentary sequence is likely to be >7 km thick (Contrucci *et al.* 2004). Its northern limit, against the H-block, is chaotic and difficult to characterize. In contrast, its southern limit is better defined and shows a sedimentary wedge defined by converging reflectors that onlap the basement, which shallows southwards (Figs 5 & 8). The sedimentary sequence displays sub-horizontal and sub-parallel reflectors, which are relatively continuous with low amplitudes. The overall geometry has been described as characteristic of a supra-detachment setting (Masini *et al.* 2012) with geometries that record the progressive outboard migration of the deformation and the related important post-tectonic subsidence.

Northwards of the residual H-block, within margin segment 3, the Gabon Basin presents similar kinds of geometries, although the thickness of the sag sequence seems to be more important (Fig. 5). Similarly, two sedimentary intervals are distinguished: a lower interval filling the graben with tilted and truncated syn-tectonic geometries and an upper interval draping the resulting topography with numerous identifiable onlaps. The facies are defined by continuous low-amplitude reflectors.

In between, within margin segment 2, interpreted as an upper plate setting, the stratigraphic geometries are more perturbed. The related sedimentary succession is thinner, pointing to a smaller available accommodation space. In addition, in contrast with the rather unperturbed sag strata, south and north of the studied area, the sedimentary layers on top of the residual H-block are cut, locally eroded and overprinted by numerous faults.

Transfer zones. By definition, transfer faults divide extending crust into segments (Gibbs 1984). Analogous to oceanic transfer faults, Lister *et al.* (1986) proposed that comparable structural features can cause the upper plate to lower plate polarity to change along the strike of the margin. Transfer faults are supposed to correspond to strike-slip faults or shear zones orthogonal to the major detachment faults onto which they terminate. Where major detachment faults change their dip direction across one or more transfer faults, the margin will change along-strike from an upper plate to a lower plate and will exhibit along-strike structural variations.

The Angola–Gabon dataset used in our study allowed the identification of two classes of transfer zone separating an upper plate from a lower plate segment. Structurally, these transfers can be either very sharp, corresponding to a major normal crustal-scale fault that bounds the upper plate residual H-block, which fits the original definition and the oceanic analogy, or far more subtle with vaguer crustal and stratigraphic geometries. The transfer is very diffuse where upper plate delamination occurred, probably over a long time period, above a plunging regional-scale lateral ramp detachment surface.

Key questions related to these transfer structures are the timing of their activity relative to the necking of the margin as well as the influence of the pre-rift inheritance (structural, thermal, compositional). Various key factors may be responsible for the structural variability and segmentation of rift systems, which makes it difficult to determine which is the controlling factor (see Lizarralde *et al.* 2007). Tugend *et al.* (2014) studied the Pyrenean transfer structures that segment the rift system both onshore and offshore and proposed that the transfers may be rooted on pre-rift orogen-related structuration. In that example, the transfers were proposed to originate on pre-rift structures. Based on the example of the Gulf of California, which shows important along-strike variations in rifting style and magmatism, Lizarralde *et al.* (2007) proposed that mantle depletion or fertility could play a significant role in small-scale variations. However, they also highlighted the fact that the sedimentation rate and volume could also play a major part. This means that either pre-rift (the fertility of the mantle) or synrift (sedimentation) parameters could be advocated. In the South Atlantic, Heine *et al.* (2013) showed that the rifting and break-up of Pangaea have been influenced by obliquity during opening and by the rate of extension. At the more local scale of the Angola–Gabon system, these parameters may also have influenced the genesis of our observed transfers, but other (pre- and/or synrift) parameters cannot be ruled out.

Thus the range of possible controlling parameters that could explain the lateral variations in the construction of rifted margins is very large. Detailed local investigations are needed to understand the sensitivity of the rifting processes to single physical parameters. More detailed observations combined with modelling are needed to better understand the control of rift evolution.

The Brazilian conjugate

First-order screening of the conjugate system was performed to verify the coherence and consistency of the margin domains and mapping of the margin segments. Equivalent proximal, distal and oceanic domains can be defined based on previously published work (Franke *et al.* 2007, 2010; Blaich *et al.* 2011; Zalán *et al.* 2011; Mohriak & Leroy 2013; Stica *et al.* 2014). In the context of this study, the major observation is that another residual H-block, comparable with that observed on the Angola margin segment 2, has been identified on the conjugate Brazilian margin over the northern Campos and the southern Esperito Santo basins (Figs 9 & 10).

The TGS 2301 profile has been chosen as one of the key lines in the conjugate dataset to illustrate the upper plate signature theoretically conjugated with the 2400 lower plate signature (Figs 10 & 11). This profile shows crustal geometries and stratigraphic records comparable 0with those reported on the CongoSPAN1 3600 profile: (1) the necking domain illustrates a thick continental crust tilted in the footwall of a major detachment fault, the hanging wall of which corresponds to a massive residual H-block; (2) the residual H-block is flanked by well-defined reflectors interpreted as detachment faults, its top crust is cut by series of normal faults and the capping sedimentary cover is thinned and tectonized and had probably been isostatically remobilized several times; and (3) the distal margin is not well-imaged, but tends to resemble an exhumed domain, potentially covered by remnants of crustal blocks running over the exhumation detachment.

Figure 11 proposes a schematic view of the Angola–Gabon and Campos–Esperito Santo conjugate margins reconstructed at 115 Ma, based on the review of South Atlantic kinematics by Heine *et al.* (2013). The reconstructed map shows the coherent conjugate position of the two residual H-blocks identified in this study. The African residual H-block is well mapped in the north as a massive continental block >10 km thick; its border is identified well as a result of a high-angle normal fault. In contrast, in the south, its border is rather vague, no major structural feature is observed and all the observations tend to show that this border is more transitional than sharp. The same observations can be reported for the Brazilian residual H-block: its northern limit is also difficult to map confidently. We propose that these residual H-blocks used to share a common border during early rift phases. These borders are now rather vague where the H-blocks have been delaminated during subsequent basement exhumation along detachment faults (Fig. 12). By definition, conceptually, one exhumation detachment fault is rooted underneath the Brazilian residual H-block, exhuming material towards the Angola margin segment 1 and, northwards, another detachment is rooted underneath the African H-block, exhuming material towards the Esperito Santo lower plate segment. The northern border of the Brazilian residual H-block and the southern border of the African H-block probably show that the vergence shift in this regional extensional template does not correspond to a sharp, well-defined transfer zone, but rather to a relay zone, a diffuse domain where the surface detachments interacted, overlapped and overprinted (over probably several generations of faults), leading to a complex delamination of the H-blocks and the related stratigraphy (see schematic representation in Fig. 12).

One consequence of these observations is that, even though the global kinematics of the opening between Africa and South America is now well constrained (Heine *et al.* 2013), it can be envisaged that the local movement of material can be more complex, with along-strike movement of material and ramp/relay geometries.

Conclusions

The aim of this study was two-fold: (1) to present some of the results from a regional mapping study of the Angola–Gabon rifted margin; and (2) to discuss the distribution of upper plate/lower plate settings and their lateral evolution along the West African rifted margin.

(1)   The study of a ION dataset of long-offset seismic reflection profiles allowed us to map the

overall margin-scale architecture of the Angola–Gabon rifted system:

- three margin segments have been identified along-strike, delimiting an upper plate margin (central segment, South Congo to North Angola, corresponding to the central part of the Lower Congo Basin) and two lower plate margins (southwards with the Kwanza Basin and northwards with the Gabon Basin);
- the transfer from one setting to the other is either sharp, typified by a major regional normal fault on the northern flank of a residual H-block identified offshore Cabinda–Zaire, or more diffuse at its southern border;
- five margin domains have been defined in the dip direction, from the continental platform to the oceanic crust (proximal, necking, distal, outer and oceanic);
- three types of oceanic crust have been observed (normal magmatic, slow-spreading magma-poor and an intermediate);
- two distinct pre-salt basins have been characterized, either sag-type (lower plate settings) or platform/terrace-type (upper plate setting);
- two key sections have been identified as exemplifying the upper plate/lower plate concept.

(2) Our observations suggest that proximal domains are very similar from one setting to the other and that structural and stratigraphic differences between the upper plate and the lower plate segments arise only from the necking domain oceanwards. The definition of upper plate and lower plate settings is therefore proposed to be applicable only for the necking and distal domains.

The key characteristics of the upper plate margin are:

(1) a relatively narrow necking domain with a sharp crustal taper and a structure similar to a core complex covered by major inner perched basins;

(2) a residual H-block cut by a series of high-angle normal faults into tilted blocks that bound standard graben- and half-graben-type basins;

(3) a continued pre-rift to synrift stratigraphy dominated by relatively shallow-water records, with little sedimentary accommodation space on top of the H-block.

The key characteristics of the lower plate margin are:

(1) a relatively wider necking domain characterized by an outboard succession of detachment faults;

(2) a distal domain encompassing hyper-extended and exhumed sub-domains;

(3) a distal supra-detachment basin, with abundant sedimentary accommodation space that includes three main units – basal (syntectonic), intermediate (widening, migration of the deformation outboard) and top (posttectonic).

The along-strike change in polarity from an upper plate setting to a lower plate setting can be operated over distinct end-member types of transfers, either sharp or diffuse. The structural heritage and the obliqueness and rate of opening are considered to play major roles in the resulting geometries.

It is important to note that, if particular structures can be listed as 'key characteristics' of one setting, this does not mean that we consider them as diagnostic. It is the combination and the relative geometries that are diagnostic. The presence of a structure similar to a core complex is, for instance, not considered to be a conclusive observation for an upper plate setting (e.g. the Hobby High in the Southern Iberia Abyssal Plain is such a structure; it has been drilled and well-imaged on a margin that can easily be interpreted as a lower plate setting). It is the large-scale crustal and stratigraphic geometry – ideally studied at both conjugate margins – that are diagnostic.

Rifted margins are geological features that are too complex for our current models to explain. The huge variety of controlling parameters (e.g. the inheritance – structural, rheological and/or thermal – the extension rate and the magmatic input) are profoundly interdependent and preclude a full understanding of the system. In addition, these features are three-dimensional. Models and concepts are still generally developed for simple orthogonally divergent margins and simple two-dimensional sections may be good approximations of the global structure. However, many settings present complex pre-rift geometries and oblique opening that require more complex three-dimensional modelling.

TOTAL is acknowledged for funding this project. ION and TGS are acknowledged for publication permission regarding some of their seismic profiles. We thank Marta Pérez-Gussinyé and two anonymous reviewers for their useful comments and advice. Christian Heine and Trond Torsvik are acknowledged for their precious help in building Figure 3. They kindly provided the shapefiles used for the figure and gave very useful and pertinent advice. Frants von Platen is also acknowledged. He created the colour scales used for the bathymetry and elevation for the various map figures and kindly allowed us to use them for this paper. Schlumberger is also acknowledged for providing access to the Petrel interpretation software. Their support team is notably sincerely acknowledged for their invaluable help.

References

ANKA, Z. & SERANNE, M. 2004. Reconnaissance study of the ancient Zaire (Congo) deep-sea fan (ZaiAngo Project). *Marine Geology*, **209**, 223–244.

ASLANIAN, D., MOULIN, M. *ET AL.* 2009. Brazilian and African passive margins of the Central Segment of the South Atlantic Ocean: kinematic constraints. *Tectonophysics*, **468**, 98–112, http://doi.org/10.1016/j.tecto.2008.12.016

AUSTIN, J. A. J. & UCHUPI, E. 1982. Continental-oceanic crustal transition off southwest Africa. *AAPG Bulletin*, **66**, 1328–1347.

AUTIN, J., LEROY, S. *ET AL.* 2010. Continental break-up history of a deep magma-poor margin based on seismic reflection data (northeastern Gulf of Aden margin, offshore Oman). *Geophysical Journal International*, **180**, 501–519, http://doi.org/10.1111/j.1365-246X.2009.04424.x

BALL, P., EAGLES, G., EBINGER, C., MCCLAY, K. & TOTTERDELL, J. 2013. The spatial and temporal evolution of strain during the separation of Australia and Antarctica. *Geochemistry, Geophysics, Geosystems*, **14**, 2771–2799, http://doi.org/10.1002/ggge.20160

BECKER, J. J., SANDWELL, D. T. *ET AL.* 2009. Global bathymetry and elevation data at 30 arc seconds resolution: SRTM30_PLUS. *Marine Geodesy*, **32**, 355–371, http://doi.org/10.1080/01490410903297766

BLAICH, O. A., FALEIDE, J. I., TSIKALAS, F., FRANKE, D. & LEON, E. 2009. Crustal-scale architecture and segmentation of the Argentine margin and its conjugate off South Africa. *Geophysical Journal International*, **178**, 85–105.

BLAICH, O. A., FALEIDE, J. I. & TSIKALAS, F. 2011. Crustal breakup and continent–ocean transition at South Atlantic conjugate margins. *Journal of Geophysical Research*, **116**, B01402, http://doi.org/10.1029/2010JB007686

BRUNE, S., HEINE, C., PÉREZ-GUSSINYÉ, M. & SOBOLEV, S. V. 2014. Rift migration explains continental margin asymmetry and crustal hyper-extension. *Nature Communications*, **5**, article 4014, http://doi.org/10.1038/ncomms5014

BUCK, W. R. 1991. Modes of continental lithospheric extension. *Journal of Geophysical Research*, **96**, 20161–120178.

BUITER, S. J. H. & TORSVIK, T. H. 2014. A review of Wilson cycle plate margins: a role for mantle plumes in continental break-up along sutures? *Gondwana Research*, **26**, 627–653, http://doi.org/10.1016/j.gr.2014.02.007

CANNAT, M., SAUTER, D. *ET AL.* 2006. Modes of seafloor generation at a melt-poor ultraslow-spreading ridge. *Geology*, **34**, 605–608.

CHRISTENSEN, N. I. & MOONEY, W. D. 1995. Seismic velocity structure and composition of the continental crust: a global view. *Journal of Geophysical Research – Solid Earth*, **100**, 9761–9788.

COFFIN, M. F. & ELDHOLM, O. 1994. Large igneous provinces – crustal structure, dimensions, and external consequences. *Reviews of Geophysics*, **32**, 1–36.

CONTRUCCI, I., MATIAS, L. *ET AL.* 2004. Deep structure of the West African continental margin (Congo, Zaire, Angola), between 5 degrees S and 8 degrees S, from reflection/refraction seismics and gravity data. *Geophysical Journal International*, **158**, 529–553.

D'ACREMONT, E., LEROY, S. *ET AL.* 2005. Structure and evolution of the eastern Gulf of Aden conjugate margins from seismic reflection data. *Geophysical Journal International*, **160**, 869–890.

DAVIS, G. A., ANDERSON, J. L., FROST, E. G. & SHACKELFORD, T. J. 1980. Mylonitization and detachment faulting in the Whipple–Bucksin–Rawhide Mountains terrane, southeastern California and western Arizona. *In*: CRITTENDEN, M. D., Jr, CONEY, P. J. & DAVIS, G. H. (eds) *Cordilleran Metamorphic Core Complexes*. Geological Society of America, Memoirs, **153**, 79–130, http://doi.org/10.1130/MEM153-p79

DAVIS, G. H. & CONEY, P. J. 1979. Geologic development of the Cordilleran metamorphic core complexes. *Geology*, **7**, 120–124.

DAVISON, I. 1999. Tectonics and hydrocarbon distribution along the Brazilian South Atlantic margin. *In*: CAMERON, N. R., BATE, R. H. & CLURE, V. S. (eds) *The Oil and Gas Habitats of the South Atlantic*. Geological Society, London, Special Publications, **153**, 133–151, http://doi.org/10.1144/GSL.SP.1999.153.01.09

DICK, H. J. B., LIN, J. & SCHOUTEN, H. 2003. An ultraslow-spreading class of ocean ridge. *Nature*, **426**, 405–412.

DIREEN, N. G., BORISSOVA, I., STAGG, H. M. J., COLWELL, B. & SYMONDS, P. A. 2007. Nature of the continent–ocean transition zone along the southern Australian continental margin: a comparison of the Naturaliste Plateau, SW Australia, and the central Great Australian Bight sectors. *In*: KARNER, G. D., MANATSCHAL, G. & PINHEIRO, L. M. (eds) *Imaging, Mapping and Modelling Continental Lithosphere Extension and Breakup*. Geological Society, London, Special Publications, **282**, 239–263, http://doi.org/10.1144/SP282.12

DUPRÉ, S., BERTOTTI, G. & CLOETINGH, S. 2007. Tectonic history along the South Gabon Basin: anomalous early post-rift subsidence. *Marine and Petroleum Geology*, **24**, 151–172, http://doi.org/10.1016/j.marpetgeo.2006.11.003

FALEIDE, J. I., TSIKALAS, F. *ET AL.* 2008. Structure and evolution of the continental margin off Norway and the Barents Sea. *Episodes*, **31**, 82–91.

FALEIDE, J. I., BJØRLYKKE, K. & GABRIELSEN, R. H. 2010. Geology of the Norwegian continental shelf. *In*: BJØRLYKKE, K. (ed.) *Petroleum Geoscience: From Sedimentary Environments to Rock Physics*. Springer, Heidelberg, 467–499.

FOSSEN, H., GABRIELSEN, R. H., FALEIDE, J. I. & HURICH, C. A. 2014. Crustal stretching in the Scandinavian Caledonides as revealed by deep seismic data. *Geology*, **42**, 791–794, http://doi.org/10.1130/g35842.1

FRANKE, D. 2013. Rifting, lithosphere breakup and volcanism: comparison of magma-poor and volcanic rifted margins. *Marine and Petroleum Geology*, **43**, 63–87, http://doi.org/10.1016/j.marpetgeo.2012.11.003

FRANKE, D., NEBEN, S., LADAGE, S., SCHRECKENBERGER, B. & HINZ, K. 2007. Margin segmentation and volcano-tectonic architecture along the volcanic margin off Argentina/Uruguay, South Atlantic. *Marine Geology*, **244**, 46–67.

FRANKE, D., LADAGE, S. *ET AL.* 2010. Birth of a volcanic margin off Argentina, South Atlantic. *Geochemistry Geophysics Geosystems*, **11**, Q0AB04, http://doi.org/10.1029/2009gc002715

FRANKE, D., BARCKHAUSEN, U. *ET AL.* 2011. The continent–ocean transition at the southeastern margin of the South China Sea. *Marine and Petroleum Geology*, **28**, 1187–1204, http://doi.org/10.1016/j.marpetgeo.2011.01.004

FRANKE, D., SAVVA, D. *ET AL.* 2013. The final rifting evolution in the South China Sea. *Marine and Petroleum Geology*, **58B**, 704–720, http://doi.org/10.1016/j.marpetgeo.2013.11.020

GIBBS, A. 1984. Structural evolution of extensional basin margins. *Journal of the Geological Society, London*, **141**, 609–620, http://doi.org/10.1144/gsjgs.141.4.0609

GUIRAUD, M., BUTA-NETO, A. & QUESNE, D. 2010. Segmentation and differential post-rift uplift at the Angola margin as recorded by the transform-rifted Benguela and oblique-to-orthogonal-rifted Kwanza basins. *Marine and Petroleum Geology*, **27**, 1040–1068.

GUIRAUD, R. & MAURIN, J.-C. 1992. Early Cretaceous rifts of Western and Central Africa: an overview. *Tectonophysics*, **213**, 153–168.

HEINE, C., ZOETHOUT, J. & MULLER, D. 2013. Kinematics of the South Atlantic rift. *Solid Earth*, **4**, 215–253.

HOLLIGER, K. & LEVANDER, A. 1994. Structure and seismic response of extended continental crust: stochastic analysis of the Strona-Ceneri and Ivrea zones, Italy. *Geology*, **22**, 79–82, http://doi.org/10.1130/0091-7613(1994)022<79:sasroe>2.3.co;2

HOLLIGER, K., LEVANDER, A. R. & GOFF, J. A. 1993. Stochastic modeling of the reflective lower crust: petrophysical and geological evidence from the Ivera Zone (northern Italy). *Journal of Geophysical Research: Solid Earth*, **98**, 11 967–11 980, http://doi.org/10.1029/93jb00351

HUISMANS, R. & BEAUMONT, C. 2011. Depth-dependent extension, two-stage breakup and cratonic underplating at rifted margins. *Nature*, **473**, 74–78, http://doi.org/10.1038/nature09988

KARNER, G. D. & GAMBOA, L. A. P. 2007. Timing and origin of the South Atlantic pre-salt sag basins and their capping evaporites. *In*: SCHREIBER, B., LUGLI, S. & BABEL, M. (eds) *Evaporites Through Space and Time*. Geological Society, London, Special Publications, **285**, 15–35, http://doi.org/10.1144/SP285.2

KARNER, G. D., DRISCOLL, N. W., MCGINNIS, J. P., BRUMBAUGH, W. D. & CAMERON, N. R. 1997. Tectonic significance of syn-rift sediment packages across the Gabon-Cabinda continental margin. *Marine and Petroleum Geology*, **14**, 973–1000.

KARNER, G. D., DRISCOLL, N. W. & BARKER, D. H. N. 2003. Syn-rift regional subsidence across the West African continental margin: the role of lower plate ductile extension. *In*: ARTHUR, T. J., MACGREGOR, D. S. & CAMERON, N. R. (eds) *Petroleum Geology of Africa: New Themes and Developing Technologies*. Geological Society, London, Special Publications, **207**, 105–129, http://doi.org/10.1144/GSL.SP.2003.207.6

LAVIER, L. L. & MANATSCHAL, G. 2006. A mechanism to thin the continental lithosphere at magma-poor margins. *Nature*, **440**, 324–328.

LAVIER, L. L., BUCK, W. R. & POLIAKOV, A. N. B. 1999. Selfconsistent rolling-hinge model for the evolution of large-offset, low-angle normal faults. *Geology*, **27**, 1127–1130.

LEMOINE, M., TRICART, P. & BOILLOT, G. 1987. Ultramafic and gabbroic ocean-floor of the Ligurian Tethys (Alps, Corsica, Apennines) – in search of a genetic model. *Geology*, **15**, 622–625.

LISTER, G. S. & DAVIS, G. A. 1989. The origin of metamorphic core complexes and detachment faults formed during Tertiary continental extension in the northern Colorado River region, U.S.A. *Geology*, **11**, 65–94.

LISTER, G. S., ETHERIDGE, M. A. & SYMONDS, P. A. 1986. Detachment faulting and the evolution of passive continental margins. *Geology*, **14**, 246–250.

LIZARRALDE, D., AXEN, G. J. *ET AL.* 2007. Variation in styles of rifting in the Gulf of California. *Nature*, **448**, 466–469.

MANATSCHAL, G. 2004. New models for evolution of magma-poor rifted margins based on a review of data and concepts from West Iberia and the Alps. *International Journal of Earth Sciences*, **93**, 432–466.

MANATSCHAL, G. & MUNTENER, O. 2009. A type sequence across an ancient magma-poor ocean-continent transition: the example of the western Alpine Tethys ophiolites. *Tectonophysics*, **473**, 4–19, http://doi.org/10.1016/j.tecto.2008.07.021

MANATSCHAL, G., LAVIER, L. & CHENIN, P. 2014. The role of inheritance in structuring hyperextended rift systems: some considerations based on observations and numerical modeling. *Gondwana Research*, **27**, 104–164, http://doi.org/10.1016/j.gr.2014.08.006

MASINI, E., MANATSCHAL, G., MOHN, G. & UNTERNEHR, P. 2012. Anatomy and tectono-sedimentary evolution of a rift-related detachment system: the example of the Err detachment (central Alps, SE Switzerland). *Geological Society of America Bulletin*, **124**, 1535–1551.

MASINI, E., MANATSCHAL, G. & MOHN, G. 2013. The Alpine Tethys rifted margins: reconciling old and new ideas to understand the stratigraphic architecture of magma-poor rifted margins. *Sedimentology*, **60**, 174–196, http://doi.org/10.1111/sed.12017

MASINI, E., MANATSCHAL, G., TUGEND, J., MOHN, G. & FLAMENT, J. M. 2014. The tectono-sedimentary evolution of a hyper-extended rift basin: the example of the Arzacq–Mauléon rift system (Western Pyrenees, SW France). *International Journal of Earth Sciences*, **103**, 1569–1596, http://doi.org/10.1007/s00531-014-1023-8

MATTHEWS, K. J., MÜLLER, R. D., WESSEL, P. & WHITTAKER, J. M. 2011. The tectonic fabric of the ocean basins. *Journal of Geophysical Research – Solid Earth*, **116**, B12109, http://doi.org/10.1029/2011jb008413

MCKENZIE, D. P. 1978. Some remarks on the development of sedimentary basins. *Earth and Planetary Science Letters*, **40**, 20–32.

MOHN, G., MANATSCHAL, G., MÜNTENER, O., BELTRANDO, M. & MASINI, E. 2010. Unravelling the interaction between tectonic and sedimentary processes during lithospheric thinning in the Alpine Tethys margins. *International Journal of Earth Sciences*, **99**, 75–101.

MOHN, G., MANATSCHAL, G., BELTRANDO, M., MASINI, E. & KUSZNIR, N. J. 2012. Necking of continental crust in magma-poor rifted margins: evidence from the fossil Alpine Tethys margins. *Tectonics*, **31**, 1–5, http://doi.org/10.1029/2011TC002961

MOHRIAK, W. U. & LEROY, S. 2013. Architecture of rifted continental margins and break-up evolution: insights from the South Atlantic, North Atlantic and Red Sea–Gulf of Aden conjugate margins. *In*: MOHRIAK, W. U., DANFORTH, A., POST, P. J., BROWN, D. E., TARI, G. C., NEMČOK, M. & SINHA, S. T. (eds) *Conjugate Divergent Margins*. Geological Society London, Special Publications, **369**, 497–535, http://doi.org/10.1144/SP369.17

MOHRIAK, W., NEMCOK, M. & ENCISO, G. 2008. South Atlantic divergent margin evolution: rift-border uplift and salt tectonics in the basins of SE Brazil. *In*: PANKHURST, R. J., TROUW, R. A. J., DE BRITO NEVES, B. B. & DE WIT, M. J. (eds) *West Gondwana: Pre-Cenozoic Correlations Across the South Atlantic Region*. Geological Society, London, Special Publications, **294**, 365–398, http://doi.org/10.1144/sp294.19

MONTADERT, A., ROBERTS, D. G., DE CHARPAL, O. & GUENNOC, P. 1979. Rifting and subsidence of the northern continental margin of the Bay of Biscay. *Reports of the Deep Sea Drilling Project*, **48**. US Government Printing Office, Washington, DC, 1025–1060.

MOULIN, M., ASLANIAN, D. *ET AL.* 2005. Geological constraints on the evolution of the Angolan margin based on reflection and refraction seismic data (ZaiAngo project). *Geophysical Journal International*, **162**, 793–810, http://doi.org/10.1111/j.1365-246X.2005.02668.x

MOULIN, M., ASLANIAN, D. & UNTERNEHR, P. 2010. A new starting point for the South and Equatorial Atlantic Ocean. *Earth-Science Reviews*, **98**, 1–37, http://doi.org/10.1016/j.earscirev.2009.08.001

MUNTENER, O., MANATSCHAL, G., DESMURS, L. & PETTKE, T. 2010. Plagioclase peridotites in ocean-continent transitions: refertilized mantle domains generated by melt stagnation in the shallow mantle lithosphere. *Journal of Petrology*, **51**, 255–294, http://doi.org/10.1093/petrology/egp087

NALIBOFF, J. & BUITER, S. J. H. 2015. Rift reactivation and migration during multiphase extension. *Earth and Planetary Science Letters*, **421**, 58–67, http://doi.org/10.1016/j.epsl.2015.03.050

NEMČOK, M., SINHA, S. T. *ET AL.* 2013. East Indian margin evolution and crustal architecture: integration of deep reflection seismic interpretation and gravity modelling. *In*: MOHRIAK, W. U., DANFORTH, A., POST, P. J., BROWN, D. E., TARI, G. C., NEMČOK, M. & SINHA, S. T. (eds) *Conjugate Divergent Margins*. Geological Society, London, Special Publications, **369**, 477–496, http://doi.org/10.1144/sp369.6

NÜRNBERG, D. & MÜLLER, R. D. 1991. The tectonic evolution of the South Atlantic from Late Jurassic to present. *Tectonophysics*, **191**, 27–53.

OSMUNDSEN, P. T. & REDFIELD, T. 2011. Crustal taper and topography at passive continental margins. *Terra Nova*, **23**, 349–361, http://doi.org/10.1111/j.1365-3121.2011.01014.x

OSMUNDSEN, P. T., SOMMARUGA, A., SKILBREI, J. R. & OLESEN, O. 2002. Deep structure of the Mid Norway rifted margin. *Norwegian Journal of Geology*, **82**, 205–224.

PADOVANO, M., PICCARDO, G. B. & VISSERS, R. L. M. 2015. Tectonic and magmatic evolution of the mantle lithosphere during rifting of a fossil slow–ultraslow spreading basin: insights from the Erro-Tobbio peridotite (Votri Massif, NW Italy). *In*: GIBSON, G. M., ROURE, F. & MANATSCHAL, G. (eds) *Sedimentary Basins and Crustal Processes at Continental Margins: From Modern Hyper-extended Margins to Deformed Ancient Analogues*. Geological Society, London, Special Publications, **413**, 205–238, http://doi.org/10.1144/SP413.7

PENROSE CONFERENCE PARTICIPANTS 1972. Penrose Field Conference: Ophiolites. *Geotimes*, **17**, 24–25.

PÉREZ-GUSSINYÉ, M. 2013. A tectonic model for hyperextension at magma-poor rifted margins: an example from the West Iberia–Newfoundland conjugate margins. *In*: MOHRIAK, W. U., DANFORTH, A., POST, P. J., BROWN, D. E., TARI, G. C., NEMČOK, M. & SINHA, S. T. (eds) *Conjugate Divergent Margins*. Geological Society, London, Special Publications, **369**, 403–427, http://doi.org/10.1144/sp369.19

PÉREZ-GUSSINYÉ, M., RANERO, C. R., RESTON, T. J. & SAWYER, D. 2003. Mechanisms of extension at non-volcanic margins: evidence from the Galicia interior basin, west of Iberia. *Journal of Geophysical Research – Solid Earth*, **108**, http://doi.org/10.1029/2001jb000901

PÉRON-PINVIDIC, G. & MANATSCHAL, G. 2009. The final rifting evolution at deep magma-poor passive margins from Iberia-Newfoundland: a new point of view. *International Journal of Earth Sciences*, **98**, 1581–1597, http://doi.org/10.1007/s00531-008-0337-9

PÉRON-PINVIDIC, G. & MANATSCHAL, G. 2010. From microcontinents to extensional allochthons: witnesses of how continents break apart? *Petroleum Geoscience*, **16**, 189–197, http://doi.org/10.1144/1354-079309-903

PÉRON-PINVIDIC, G., MANATSCHAL, G., MINSHULL, T. A. & SAWYER, D. S. 2007. Tectonosedimentary evolution of the deep Iberia-Newfoundland margins: evidence for a complex breakup history. *Tectonics*, **26**, Tc2011.

PÉRON-PINVIDIC, G., GERNIGON, L., GAINA, C. & BALL, P. 2012. Insights from the Jan Mayen system in the Norwegian–Greenland Sea – II. *Architecture of a microcontinent. Geophysical Journal International*, **191**, 413–435, http://doi.org/10.1111/j.1365-246X.2012.05623.x

PÉRON-PINVIDIC, G., MANATSCHAL, G. & OSMUNDSEN, P. T. 2013. Structural comparison of archetypal Atlantic rifted margins: a review of observations and concepts. *Marine and Petroleum Geology*, **43**, 21–47, http://doi.org/10.1016/j.marpetgeo.2013.02.002

RABINOWITZ, P. D. & LABRECQUE, J. L. 1979. The Mesozoic South Atlantic Ocean and evolution of its continental margins. *Journal of Geophysical Research*, **84**, 5973–6002.

RANERO, C. R. & PÉREZ-GUSSINYÉ, M. 2010. Sequential faulting explains the asymmetry and extension discrepancy of conjugate margins. *Nature*, **468**, 294–297, http://doi.org/10.1038/nature09520

RESTON, T. J. 1996. The S reflector west of Galicia: the seismic signature of a detachment fault. *Geophysical Journal International*, **127**, 230–244.

RESTON, T. J. 2005. Polyphase faulting during the development of the west Galicia rifted margin. *Earth and Planetary Science Letters*, **237**, 561–576.

RESTON, T. J. 2009. The structure, evolution and symmetry of the magma-poor rifted margins of the North and Central Atlantic: a synthesis. *Tectonophysics*, **468**, 6–27.

RESTON, T. J., PENNELL, J., STUBENRAUCH, A., WALKER, I. & PÉREZ-GUSSINYÉ, M. 2001. Detachment faulting, mantle serpentinization, and serpentinite- mud volcanism beneath the Porcupine Basin, southwest of Ireland. *Geology*, **29**, 587–590.

ROSENBAUM, G., REGENAUER-LIEB, K. & WEINBERG, R. F. 2005. Continental extension: from core complexes to rigid block faulting. *Geology*, **33**, 609–612.

SAVVA, D., MERESSE, F. ET AL. 2013. Seismic evidence of hyper-stretched crust and mantle exhumation offshore Vietnam. *Tectonophysics*, **608**, 72–83, http://doi.org/10.1016/j.tecto.2013.07.010

SERANNE, M., SEGURET, M. & FAUCHIER, M. 1992. Seismic super-units and post-rift evolution of the continental passive margin of southern Gabon. *Bulletin de la Société géologique de France*, **163**, 135–146.

SETON, M., WHITTAKER, J. M. ET AL. 2014. Community infrastructure and repository for marine magnetic identifications. *Geochemistry, Geophysics, Geosystems*, **15**, 1629–1641, http://doi.org/10.1002/2013gc005176

SMITH, W. H. & SANDWELL, D. T. 1997. Global seafloor topography from satellite altimetry and ship depth soundings. *Science*, **277**, 1957–1962.

STICA, J. M., ZALÁN, P. V. & FERRARI, A. L. 2014. The evolution of rifting on the volcanic margin of the Pelotas Basin and the contextualization of the Paraná–Etendeka LIP in the separation of Gondwana in the South Atlantic. *Marine and Petroleum Geology*, **50**, 1–21, http://doi.org/10.1016/j.marpetgeo.2013.10.015

SUTRA, E. & MANATSCHAL, G. 2012. How does the continental crust thin in a hyperextended rifted margin? Insights from the Iberia margin. *Geology*, **40**, 139–142, http://doi.org/10.1130/G32786.1

SUTRA, E., MANATSCHAL, G., MOHN, G. & UNTERNEHR, P. 2013. Quantification and restoration of extensional deformation along the Western Iberia and Newfoundland rifted margins. *Geochemistry, Geophysics, Geosystems*, **14**, 2575–2597, http://doi.org/10.1002/ggge.20135

THINON, I., MATIAS, L., REHAULT, J. P., HIRN, A., FIDALGO-GONZALEZ, L. & AVEDIK, F. 2003. Deep structure of the Armorican Basin (Bay of Biscay): a review of Norgasis seismic reflection and refraction data. *Journal of the Geological Society, London*, **160**, 99–116, http://doi.org/10.1144/0016-764901-103

TORSVIK, T. H., ROUSSE, S., LABAILS, C. & SMETHURST, M. A. 2009. A new scheme for the opening of the South Atlantic Ocean and the dissection of an Aptian salt basin. *Geophysical Journal International*, **177**, 1315–1333, http://doi.org/10.1111/j.1365-246X.2009.04137.x

TSIKALAS, F., ELDHOLM, O. & FALEIDE, J. I. 2005. Crustal structure of the Lofoten-Vesterålen continental margin, off Norway. *Tectonophysics*, **404**, 151–174, http://doi.org/10.1016/j.tecto.2005.04.002

TSIKALAS, F., FALEIDE, J. I. & KUSZNIR, N. J. 2008. Along-strike variations in rifted margin crustal architecture and lithosphere thinning between northern Voring and Lofoten margin segments off mid-Norway. *Tectonophysics*, **458**, 68–81.

TUCHOLKE, B. E. & SIBUET, J. C. 2007. Leg 210 synthesis: tectonic, magmatic, and sedimentary evolution of the Newfoundland-Iberia rift. *In*: TUCHOLKE, B. E., SIBUET, J. C. & KLAUS, A. (eds) *Proceedings of the ODP, Scientific Results, 210*. Ocean Drilling Program, College Station, TX, 1–56.

TUGEND, J., MANATSCHAL, G., KUSZNIR, N. J., MASINI, E., MOHN, G. & THINON, I. 2014. Formation and deformation of hyperextended rift systems: insights from rift domain mapping in the Bay of Biscay-Pyrenees. *Tectonics*, **33**, 2014TC003529, http://doi.org/10.1002/2014tc003529

TUGEND, J., MANATSCHAL, G., KUSZNIR, N. J. & MASINI, E. 2015. Characterizing and identifying structural domains at rifted continental margins: application to the Bay of Biscay margins and its Western Pyrenean fossil remnants. *In*: GIBSON, G. M., ROURE, F. & MANATSCHAL, G. (eds) *Sedimentary Basins and Crustal Processes at Continental Margins: From Modern Hyper-extended Margins to Deformed Ancient Analogues*. Geological Society, London, Special Publications, **413**, 171–203, http://doi.org/10.1144/sp413.3

UNTERNEHR, P., PÉRON-PINVIDIC, G., MANATSCHAL, G. & SUTRA, E. 2010. Hyper-extended crust in the south Atlantic: in search for a model. *Petroleum Geoscience*, **16**, 207–215, http://doi.org/10.1144/1354-079309-904

VON NICOLAI, C., SCHECK-WENDEROTH, M., WARSITZKA, M., SCHØDT, N. & ANDERSEN, J. 2013. The deep structure of the South Atlantic Kwanza Basin – insights from 3D structural and gravimetric modelling. *Tectonophysics*, **604**, 139–152, http://doi.org/10.1016/j.tecto.2013.06.016

WELSINK, H. & TANKARD, A. 2012. Extensional tectonics and stratigraphy of the Mesozoic Jeanne d'Arc basin, Grand Banks of Newfoundland. *In*: ROBERTS, D. G. & BALLY, A. W. (eds) *Regional Geology and Tectonics: Phanerozoic Rift Systems and Sedimentary Basins*. Elsevier, Amsterdam, 336–381.

WERNICKE, B. 1981. Low-angle normal faults in the Basin and Range province: nappe tectonics in an extending orogen. *Nature*, **291**, 645–648.

WERNICKE, B. 1985. Uniform-sense normal simple shear of the continental lithosphere. *Canadian Journal of Earth Sciences*, **22**, 108–125.

WERNICKE, B. & AXEN, G. J. 1988. On the role of isostasy in the evolution of normal fault systems. *Geology*, **16**, 848–851.

WERNICKE, B. & BURCHFIELD, B. D. 1982. Modes of extensional tectonics. *Journal of Structural Geology*, **4**, 105–115.

WHITMARSH, R. B., MANATSCHAL, G. & MINSHULL, T. A. 2001. Evolution of magma-poor continental margins from rifting to seafloor spreading. *Nature*, **413**, 150–154.

                G. PÉRON-PINVIDIC *ET AL.*

WHITNEY, D. L., TEYSSIER, C., REY, P. & BUCK, R. 2014. Continental and oceanic core complexes. *GSA Bulletin*, **125**, 273–298, http://doi.org/10.1130/B30754.1

ZALÁN, P. V., SEVERINO, M. G., RIGOTI, C. A., MAGNAVITA, L. P., OLIVEIRA, J. A. B. & VIANA, A. R. 2011. An entirely new 3-D view of the crustal and mantle structure of a South Atlantic passive margin – Santos, Campos and Espírito Santo Basins, Brazil. *AAPG Annual Conference and Exhibition*, 10–13 April 2011, Houston, TX, USA.

Evolution of the South Atlantic lacustrine deposits in response to Early Cretaceous rifting, subsidence and lake hydrology

TERESA SABATO CERALDI* & DARRYL GREEN

BP Exploration, Sunbury on Thames TW16 7LN, UK

Corresponding author (e-mail: teresa.ceraldi@uk.bp.com)

Abstract: The discovery of the giant Lula Field (Santos Basin, Brazil) has focused exploration on the pre-salt lacustrine carbonate play in both Brazil and Angola. Understanding the main controls on deposition and the link to the unique tectonic setting of the South Atlantic during the Early Cretaceous is paramount in successful exploration. Observations from conjugate regional seismic lines between the Kwanza and South Campos basins define the key megasequences as pre-rift, rift, sag, salt basin, post-salt carbonates and post-salt clastic sediments. The pre-salt stratigraphy can be described by three significant phases. In the first phase, the synrift lakes (Valanginian–Hauterivian), representing the onset of continental break-up and the formation of synrift graben, created several deep basins leading to overfilled freshwater lakes. In the second phase, Sag 1 (Barremian–Aptian), there was widespread subsidence created by syn-kinematic stretching of the continental crust and/or continuous rifting. Lacustrine sediments composed of coarse Pelecypod coquina with moderate faunal diversity are found, suggesting a fresh to brackish, balanced-filled, interconnected lake. In the third phase, Sag 2 (Aptian), there was continuous subsidence and isolation of the lake. Base level falls below sea-level resulted in an under-filled, alkaline and hypersaline lake with widespread microbialite growth. Based on the integration of seismic data, well data, biostratigraphy and tectonic models, a regional predictive model has been developed for reservoirs in phases 2 and 3.

There has recently been renewed interest in petroleum exploration in the pre-salt regions of the basins on both sides of the South Atlantic and a new type of 'lacustrine carbonate' play has been identified. Previous discoveries were insufficiently researched and described and therefore little is known about how the lacustrine systems are related to the evolution of the opening of the South Atlantic. An understanding of the systems that lie beneath the thick layer of salt may provide insights into the details of the formation and history of the early opening of the South Atlantic.

The rifting and the Early Cretaceous continental break-up of the South Atlantic – i.e. the opening of the South Atlantic Ocean – has been reported by a number of research groups with a particular focus on the Angola–Brazil conjugate (Karner & Driscoll 1999*a*, *b*; Watcharanantakul & Morley 2000; Meredith & Egan 2002; Kusznir & Karner 2007; Aslanian *et al.* 2009; Moulin *et al.* 2010; Unternehr *et al.* 2010). Many researchers have proposed a polyphase rifting model with two to three rifting events starting from the Barremian up to the Aptian (Karner & Driscoll 1999*a*, *b*; Kusznir & Karner 2007). However, the lack of significant extensional faulting visible in the seismic data during the latest stages, commonly referred to as the Sag basin, is not consistent with the observed amount of crustal thinning and subsidence (Sibuet 1992; Driscoll & Karner 1998; Contrucci *et al.* 2004; Davis &

Kusznir 2004; Kusznir & Karner 2007); this phenomenon is described as the extension discrepancy. This discrepancy has been explained by some researchers as depth-dependent stretching (or thinning) (Karner & Driscoll 1999*a*, *b*; Watcharanantakul & Morley 2000; Meredith & Egan 2002; Kusznir & Karner 2007), whereas others postulate that not all extension is recognized in the seismic data (Marrett & Allmendinger 1992; Reston 2009). This problem has been approached and described by Reston (2007, 2009, 2010), who also proposed that synrift subsidence might be underestimated by referencing to the local sea (or lake) water depth rather than the global ocean level.

Many attempts have been made to provide a chronostratigraphic framework for the different basins in Angola and Brazil (Braccini *et al.* 1997; Harris 2000; Moreira *et al.* 2007). However, the lacustrine biostratigraphic record is based on several different criteria, such as ostracods, pollens and spores. There are only tenuous links to absolute ages and an accurate correlation to consistent, reliable, age-dated lithostratigraphic formations or biozones is missing. There are significant time gaps and many intervals are barren of any useful biostratigraphic fauna; it is therefore difficult to correlate these independent methods.

Poropat & Colin (2012) have reported an important margin-wide chronostratigraphic correlation scheme based on biostratigraphic records and have

From: SABATO CERALDI, T., HODGKINSON, R. A. & BACKE, G. (eds) 2017. *Petroleum Geoscience of the West Africa Margin*. Geological Society, London, Special Publications, **438**, 77–98.
First published online June 21, 2016, http://doi.org/10.1144/SP438.10

also provided a chronostratigraphic correlation between the two margins. Chaboureau *et al.* (2013) attempted to assign absolute age dates to the pre-salt stratigraphic units using chemostratigraphy and volcanic ages.

The rift (early Hauterivian) phase led to a series of rift valleys between South America and Africa that were filled with terrigenous and carbonate deposits of probably fluvial–lacustrine origin (Horschutz & Scuta 1992). The stratigraphic unit above the fluvial–lacustrine deposits is formed by fluvial–lacustrine and Pelecypod–Ostracode-rich carbonates, which are evident in the Campos and Santos basins, in the Lagoa Feia and Itapema formations, respectively (Carvalho *et al.* 2000). This facies has been productive since the late 1990s on both the Brazilian and Angolan margins as a coquina facies (Lower Congo, Toca Formation) (Braccini *et al.* 1997; Harris 2000; Moreira *et al.* 2007). The timing of the occurrence of this coquina facies seems to be similar on both margins and is dated as Late Barremian to Early Aptian, although there is uncertainty about the exact age dates and the potential for these two occurrences to be linked.

Several researchers (McHargue 1990; Braccini *et al.* 1997; Guardado *et al.* 2000; Harris 2000) have referred to an important source rock interval associated with, or capping, the coquina-rich units (latest Barremian) on both margins. The age, distribution and formation mechanism of this source rock is poorly understood on the regional scale. In Cabinda, the uppermost Bucomazi (late Bucomazi–synrift II) is suggested to have abundant source rocks, which seem to drape the synrift clastic sediments (Burwood 1999).

Aptian deposition along the Angolan margin (McHargue 1990; Braccini *et al.* 1997; Burwood 1999; Guardado *et al.* 2000; Harris 2000) has been documented as the Chela Formation, which consists of mostly lacustrine sandstones and shales. The deposition of predominantly sandy to silty lacustrine clastic sediments has also been recorded in the proximal settings of the Campos Basin (Lagoa Feia) (Carvalho *et al.* 2000; Guardado *et al.* 2000).

The Aptian succession has received renewed interest as a result of the emergence of a microbialite facies discovered in recent well penetrations; the earliest is in the Tope (Lula) Field discovered in 2006. These rocks were assigned an Aptian age (Barra Velha Formation) in the Santos Basin by Moreira *et al.* (2007) and, more recently, by Petersohn & Abelha (2013). This unit is often referred to as the Sag (Mello *et al.* 2011). In Angola, a microbialite facies has been reported from the Kwanza Basin, although the age was not specified (Cazier *et al.* 2014).

Thick salt deposits attributed to the latest Aptian have been recorded on both margins (Jackson & Hudec 2009). As rifting propagated northwards, the Walvis Ridge subsided and began to lose its role as a regional barrier (Stark 1991; Dingle 1999; Eichenseer *et al.* 1999). The timing of formation related to rifting or thermal sag is still being debated (Dupre *et al.* 2007; Jackson & Hudec 2009; Montaron & Tapponnier 2010; Warren 2010; Quirk *et al.* 2012).

Wright (2012) summarized a number of key observations and potential microbial carbonate depositional models that may apply to the lacustrine deposits (Sag) associated with volcanic-dominated rift basins. However, these only described a large number of possibilities and areal distributions and did not discuss the formation of the separate coquina-rich interval.

Although lacustrine depositional models have existed for some time, very few papers have focused on the potential for the occurrence of significant carbonate reservoirs. However, the discovery of giant pre-salt lacustrine fields in Brazil has attracted renewed interest. In addition, the pre-salt lacustrine fields need to be integrated into the tectonic history and chronostratigraphic framework. The reservoir architecture, composition, distribution and diagenesis of these lacustrine reservoirs can only be understood in the light of the unique tectonic and geological setting. Regional seismic lines and biostratigraphic, well log, core and isotopic data from four basins and two continents are reviewed here to formulate an integrated model for the tectonic and depositional history of the pre-salt section.

Geological setting

The pre-salt of the South Atlantic is a result of slow tectonic and depositional processes involving continental rifting, seafloor spreading and sedimentation. These processes were associated with the split between South America and Africa during the Cretaceous break-up of Gondwanaland (Karner & Driscoll 1999*a*, *b*; Watcharanantakul & Morley 2000; Meredith & Egan 2002; Kusznir & Karner 2007; Aslanian *et al.* 2009; Moulin *et al.* 2010; Unternehr *et al.* 2010). The southern extent of the pre-salt basins in Africa and Brazil is marked by the volcanic lineament of the Walvis Ridge and the Rio Grande Rise, respectively (Fig. 1a) (Thompson *et al.* 2015). This feature served as a barrier in the Aptian, allowing the isolation from seawater of several rift basins to the north. The development of fluvial–deltaic and lacustrine carbonates with little to no influence from marine water followed (Braccini *et al.* 1997; Harris 2000; Moreira *et al.* 2007). The topographic highs (Walvis Ridge and Rio Grande Rise) began to subside in the latest Aptian, allowing the periodic or episodic ingress of marine water and

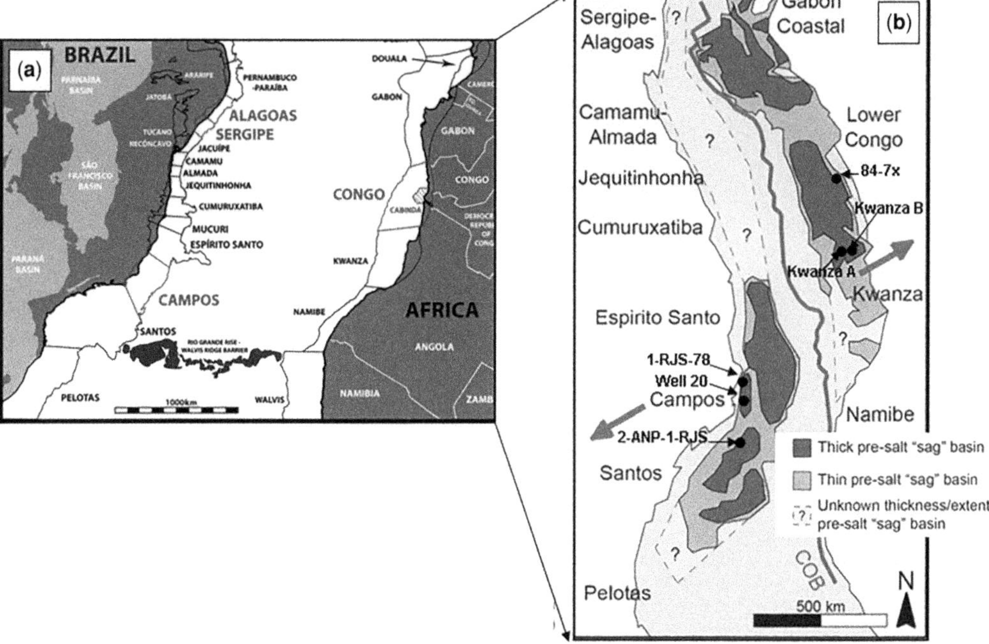

Fig. 1. Geological setting of the pre-salt basins of the South Atlantic shown as a schematic diagram of the proto-South Atlantic. (**a**) The numerous basins on both margins are delineated by black outlines. The ocean–continent boundary is delimited by the seaward boundary of the basins, which corresponds to the 3000 m bathymetric line. The Walvis Ridge and Rio Grande Rise form the prominent southern extent of the rift basin on the African and Brazilian sides, respectively (Thompson *et al.* 2015). (**b**) Restoration showing the main Sag basins with salt thicknesses during the Early–Late Cretaceous (Karner & Gamboa 2007). The approximate well positions used in Figures 4 and 8 are shown.

resulting in the deposition of thick and widespread evaporite deposits. Fully marine conditions returned in the Albian, resulting in the deposition of extensive marine carbonates. The post-Albian succession recorded rising sea-levels and the deposition of siliciclastic sediments.

The pre-salt areas with significant post-salt seals along the African margin stretch from Cabinda in northern Angola to the Benguela Basin of southern Angola (Fig. 1b). The main Brazilian pre-salt margin stretches from the Campos Basin to the Santos Basin of Brazil. The pre-salt extent ranges from current water depths of less than 100 m to 3000 m in depth. The width of the pre-salt margin in Brazil is several hundreds of kilometres, whereas the width of the Angolan pre-salt margin is generally <200 km; both the African and Brazilian margins are >1000 km in length.

The study area reported here consists of the main pre-salt basins: the Campos and Santos basins of the Brazilian margin and the Kwanza and Cabinda basins of the Angolan margin. The pre-salt basins have undergone major exploration and are the source of most of the recent well data. Exploration

has focused on the areas of the basins sealed by salt, which provides a robust top seal that can trap hydrocarbons.

Regional seismic lines and well data: defining sequences and timing

The observations made on the regional conjugate seismic lines shown in Figure 2 have been used to define the key megasequence boundaries within the sedimentary record on both the Angolan and Brazilian passive margins. The megasequences can be described as follows: pre-rift, rift (Valanginian–Hauterivian to Early Barremian), sag (Early Barremian to Late Aptian), salt basin (Late Aptian), post-rift, post-salt carbonates (latest Aptian to Late Albian) and post-rift, post-salt clastic sediments (Cenomanian to present day).

The lines chosen were GXT 2400 and a TGS regional line because the seismic data quality of these lines is fairly good and they offer a good representation of the tectonostratigraphic sequences for each margin.

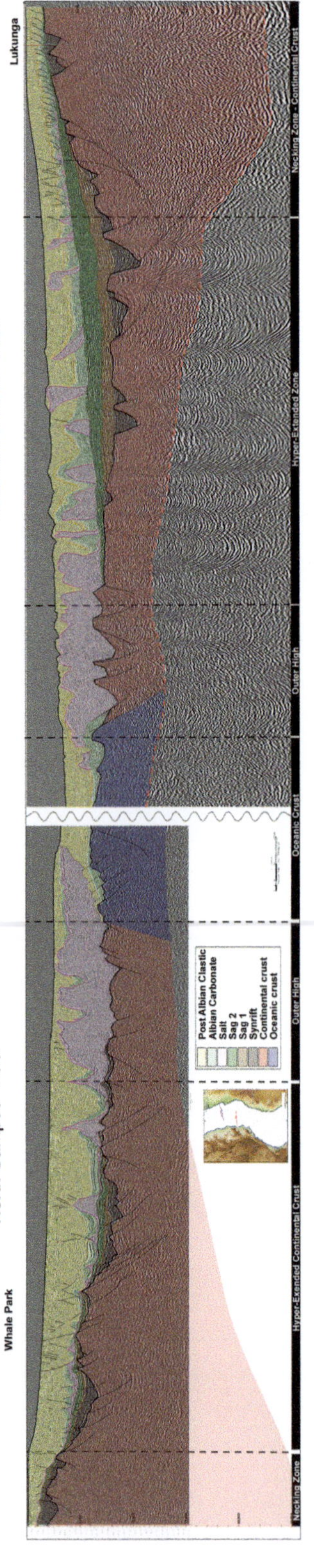

Fig. 2. Campos –Kwanza conjugate regional time lines. Key pre-salt megasequences: Synrift, Sag 1, Sag 2. The synrift sequence is focused in deep graben, shows classic growth geometries and is generally filled by terrigenous sediments. The Sag 1 is the unit formed by sub-parallel reflectors with little evidence of active faulting and is the coquina-bearing unit. The Sag 2 unit is also formed by sub-parallel reflectors with bilateral onlaps on the flanks of the sub-basins and it is the microbialite bearing unit. Interpretation of the Campos seismic line from L. Neal. Data courtesy of ION and TGS.

The pre-rift section is very difficult to identify and is only visible locally on the Angola margin (line GXT 2400). For example, in the Lukunga well location it is possible to see a series of sub-parallel reflectors underneath the synrift graben in the seismic data. The pre-rift stratigraphy in Angola and Brazil is filled by continental and fluvial deposits. The land mass of Gondwanaland was still intact at this time, before the onset of rifting and crustal stretching (pre-Valanginian).

Both the Brazilian and the Angolan margins were extensional and the incipient synrift phase is represented by deep graben formed mainly by normal faulting. These graben contain reflectors showing classic growth fault geometries. They can be very deep and their sedimentary fill can reach thicknesses of 2–3 km. The wells that penetrate these units (e.g. Lukunga and various onshore wells) show that these graben are filled by siltstones and shales in a lacustrine-type depositional setting with occasional continental and alluvial deposits (Fig. 3).

The dating of the synrift sediments from the biostratigraphic record gives an approximate Hauterivian to Barremian rifting age, Palynozone CVII–CIV, Ostracode zone AS2–AS6, RT2–6 (Grosdidier *et al.* 1996; Braccini *et al.* 1997; Bueno 2004). The synrift graben are separated by horst blocks, as seen in the Brazilian conjugate line in Figure 2, which cross the Whale Park high. The thinning of the pre-salt sediments on the crest of these horsts suggests that these features remained as persistent topographic and bathymetric highs throughout the Aptian. Most of the major fields with resources in the pre-salt carbonates are found on these outboard highs: the Lula Field, Whale Park, Libra and Franco in Brazil. Early exploration in Angola focused on the margin, whereas current exploration success in the Kwanza Basin tested outboard of the margin.

The top of the synrift is represented by a seismically expressed unconformable surface and is overlain by a unit formed by sub-parallel, sub-horizontal reflectors, which are commonly referred to as the 'Sag' unit (Karner & Driscoll 1999*a*, *b*; Marton *et al.* 2000; Karner *et al.* 2003; Lentini *et al.* 2010).

This Sag unit appears on seismic data as a series of sub-parallel reflectors with bilateral onlaps on the flanks of the Angolan and Brazilian marginal basins. Thicknesses of up to 5 km in the Lower Congo Basin and in the Kwanza Basin are characterized by a lack of synsedimentary faulting.

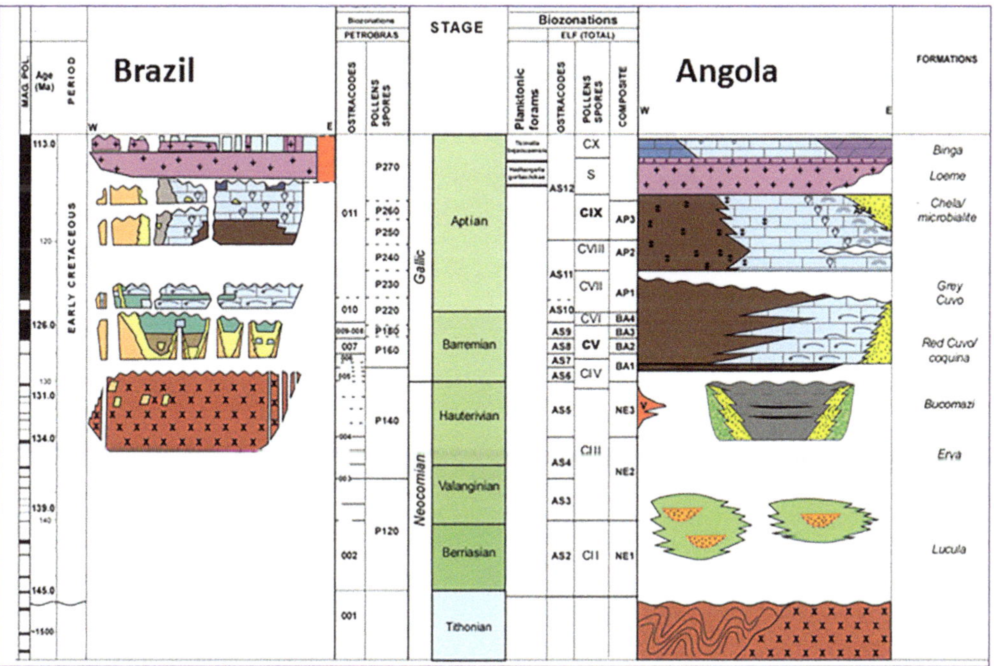

Fig. 3. Stratigraphic columns for Angola and Brazil. The Brazilian margin column was modified from Moreira *et al.* (2007). The Angolan margin column was derived from biostratigraphy, sequence stratigraphy and seismic stratigraphy from numerous wells along the Angolan margin. The proposed equivalent stages, formations, biozones and ages are correlated across both margins.

The outboard sag onlap is better imaged on the Angola line (Fig. 2). The sedimentary sequence thins and onlaps onto what is sometimes referred to as the 'outer high' (Quirk *et al.* 2013); the same type of stratigraphic relationship is visible on the Campos margin. The Sag sequence can be divided into two sub-units (Sag 1 and Sag 2) separated by an unconformity (Fig. 2). This surface appears to be erosional on the flanks of the basin (the areas where we see the onlap relationship). In the centre of the basin this surface becomes paraconformable and is overlain by a bright negative reflection that can also be seismically tied to a shale section with a high total organic content (TOC) content (Kwanza wells Falcao and Baleia; Fig. 1). The geometry of the reflections in the seismic line suggests that the localized rifting in the Angola and Campos basins had ceased (no fault propagates through this interval) and the discrete lakes that had formed during the rifting phase had started to merge.

The shale interval that separates the Sag 1 and Sag 2 units can be mapped fairly confidently on regional seismic lines all along the Kwanza and Lower Congo basins and can be extended north into Cabinda and Gabon. Seismic reflectivity in the Sag 1 unit is generally expressed as a series of repeating peaks and troughs related to the interbeds of shell-rich carbonates and organic shales. A similar reflection characteristic can be identified in the Campos Basin and is associated with the Jiquia-aged shale units, which, on the Brazilian margin, separate the coquina-bearing unit (Coqueiros Formation) and the unit containing microbialites (Macabu Formation.). The Sag 2 unit appears to have a very similar character to the Sag 1 unit. It is also formed by sub-parallel reflections with bilateral onlap onto the flanks of the sub-basins.

The Sag tectonostratigraphic unit is formed of lacustrine deposits and contains at least two different types of carbonate reservoirs: the coquina-rich bioclastic unit and the microbialite–spherulite-dominated unit. As seen in the well cross-section (Fig. 4), derived from cuttings and core information, the Sag 1 unit is formed by carbonates with abundant and moderately diverse non-marine mollusc and Ostracode assemblages, commonly displayed as concentrated shell hash (coquina) beds. This limited diversity of fauna (commonly only one species, indicating a stressed environment) and the presence of stevensite ooids associated with some coquinas, indicates that the lakes were at times slightly brackish and even extremely saline (Carvalho *et al.* 2000; Harris 2000; Altenhofen 2013). Biostratigraphic control suggests that this zone is Barremian to earliest Aptian in age, corresponding to Palynozone CVI–CVII and Ostracode Zone AS9–AS10, RT 09–RT 010 (Braccini *et al.* 1997; Poropat & Colin 2012; Chaboureau *et al.* 2013). Existing well data

from both the Campos Basin (Brazil) and the Kwanza Basin (Angola) indicate that the facies are variable.

This uppermost shale interval in Brazil occurs within RT 010 and within Biozones AS10 and CVII in Angola, which are roughly time-equivalent to the conjugate margin (Poropat & Colin 2012). This intra-Sag shale interval (Upper Bucomazi and Jiquia equivalent) has a high TOC, averaging 4%, and the interval is believed to be one of the main source rocks for the pre-salt and post-salt plays. Isotope data and biomarkers suggest a restricted (saline) depositional environment (McHargue 1990; Braccini *et al.* 1997; Guardado *et al.* 2000; Harris 2000).

The nearshore margins of both conjugates portray what is observed to be a large regional unconformity with the omission of one or more biostratigraphic zones (Braccini *et al.* 1997; Guardado *et al.* 2000; Moreira *et al.* 2007; Poropat & Colin 2012). The nature and the geomorphological setting of the Early Aptian hiatus – either subaquatic or subaerial – are still unknown (Chaboureau *et al.* 2013). However, similar evidence for an unconformity and missing time interval (the pre-Alagoas unconformity) is suggested from the existing Santos Basin chronostratigraphy (Moreira *et al.* 2007). Work published on the basins in the Lower Congo highlight a regional unconformity termed the pre-Chela unconformity (Braccini *et al.* 1997; Harris 2000).

Biostratigraphic control suggests that the Sag 2 unit is Middle to Late Aptian in age, corresponding to Palynozone CVIII–CIX, Ostracode Zone AS11–AS12 and RT011 between wells; the underlying faunal types (depositional environments and hydrology) were similar at this time.

The Sag 2 unit is overlain by a variable thickness evaporitic sequence. The evaporite package is a reflective sequence of deformed halite, interbedded autochthonous anhydrite and clastic sediments. The margins of both conjugate margins display thin to absent salt with minimal halokinesis. The evaporites increase in thickness (to the same degree or apparent original thickness) away from the margin. Significant halokinesis and downslope translation of the evaporites is evident on both conjugate margins. These evaporites can be seen in the shallow sections and act as a detachment for localized compressional thrusts and extensional listric fault features. The evaporites end or toe-out as they approach the assumed proto-continental–ocean boundary.

The post-salt, post-rift carbonate megasequence starts in the very Late Aptian and extends into the Albian (Biozone CX) and is characterized by a change to fully marine conditions with the deposition of grainstones and wackestones in a shallow marine environment and deeper water mudstones (Eichenseer *et al.* 1999).

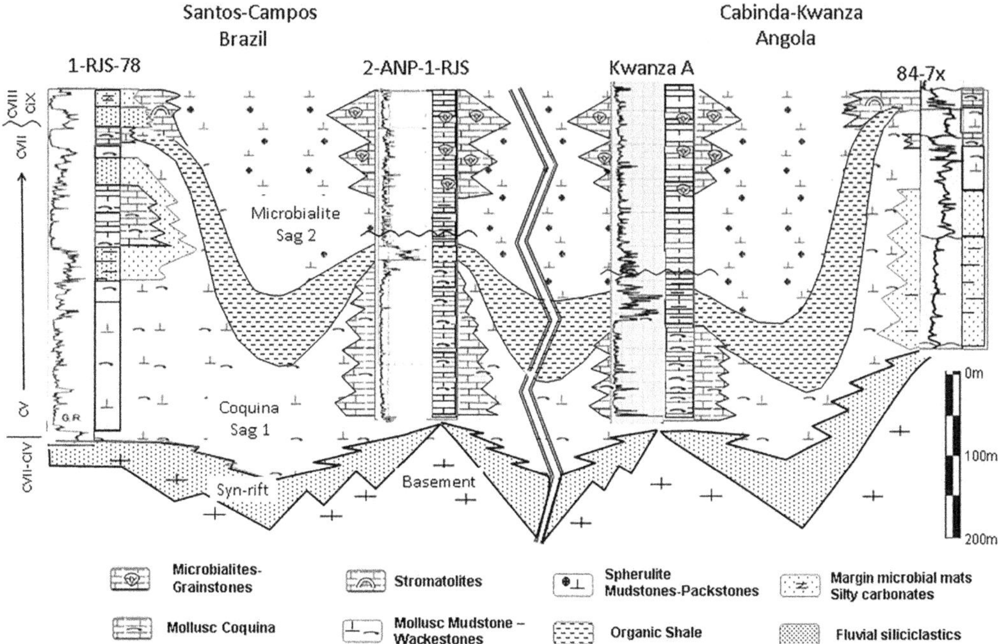

Fig. 4. Schematic cross-section across the Campos and Santos basins of Brazil and the Kwanza and Cabinda basins of Angola. Well 1-RJS-78 adapted from Guardado *et al.* (2000), well 2-ANP-1-RJS from Petersohn & Abelha (2013) and well 84-7x from Harris (2000) and Braccini *et al.* (1997). Two main phases characterize the sections: CVIII–CIX (microbialite section) and CV–CVII (coquina section). The margins show either condensed sections or fluvial–lacustrine-influenced carbonates. The inferred off-structure and deepest parts of the palaeo-lake are silty and shaly with sparse grains (coquina) or spherulites and stevensite dominating. The large basement-rooted structures develop coquina grainstones (coquina section) and microbialites (microbialite section).

At the end of the Albian the sedimentary record changed drastically on both the West Africa margin and the South American margin. With rising sea-levels the system switched from a carbonate-dominated to a clastic-dominated environment with intermittent pulses of clastic material into the basin. A detailed depositional model for the Sag 2 micro-bialite reservoir is described in the following sections.

Facies associations and isotopic data

Facies

Well log, core and seismic data from several wells from both the Campos Basin and the Kwanza and Benguela basins in Angola serve to formulate a depositional/palaeogeographical model for the pre-salt of both margins. We cannot encompass or include all the features or observations that occur in the pre-salt (or demonstrate with well and core data), but we attempt to detail what is believed to be the underlying depositional environments interpreted for the majority of the pre-salt.

The distribution and quality of lacustrine pre-salt carbonates of both the coquina-rich Sag 1 and microbialite-rich Sag 2 units are a reflection of the palaeogeography and lake chemistry. The changes in lake chemistry and hydrology are unique and had a major influence on deposition relative to the carbonate depositional models derived for the marine system.

The Upper Barremian (Sag 1) section reflects the potential fresh–brackish-water lake chemistry, with reservoir zones displaying variable amounts of Pelecypod–Ostracode shell debris. Large centi-metre-scale pelecypods preserved in grainstones–packstones with well-developed shelter and inter-particle porosity display the best reservoir porosi-ties. The shell material consists of disarticulated and abraded shell material in both a convex and concave-up orientation. Additional poorer quality reservoir facies consist of Pelecypod mudstones–wackestones and silty (peloid?) and Ostracode mudstones.

The main characteristics of the latest Barremian to Aptian Sag 2 unit are thin (centimetre- to metre-scale) cyclic beds of microbial mats, variable 'shrub' morphologies (microbialites, dendrolites and thrombolites), microbial crusts and spherulites–stevensite wackestones–packstones. We have categorized and grouped the large number of facies into a more manageable number and will briefly describe the following facies types (Fig. 5).

(1) The siliciclastic-dominated facies are poorly sampled and interpretations are based on sparse rock data with abundant log data. Log patterns commonly show fining upwards packages of calcite-cemented sandstones interbedded with siltstones and shales. Many sandstones are feldspar-rich with angular to subrounded quartz clasts. These are occasionally interbedded with dolomitic and silty wackestones.

(2) Mixed fine-grained siliciclastic sediments occur as dark silty to shaly, fine-laminated planar mudstones. Soft sediment and fluid escape structures are common, as are desiccation cracks, some ripple laminations and terrigenous plant material. Light brown wavy laminated, crenulated microbial mats with fine-grained laminated–agglutinated stromatolites occur in many wells. Deep mud cracks occur in some cores, consisting of very thin, light-coloured (frequently dolomitized), wavy laminated deposits with fenestral porosity and mud cracks.

(3) Dark to light brown grainstones to packstones containing unsupported or floating small 'tuft' or 'bud' shrubs or spherulites, usually associated with more wavy laminated, fine-grained, crust-like habits. The buds consist of rounded or subrounded disorganized clots or clusters of millimetre-scale carbonate nodules. Elongate 'fingers' of irregular columns or pustules constrained in centimetre-scale beds make up the tufts. Accessory grain types (ooids, peloids, spherules, Ostracodes, siliciclastic silt/fine sand) are similar to those found in the tuft/shrub sub-facies. The tops of beds commonly display more buds and tufts, whereas the bottoms of the beds are more massive and compact.

(4) Coarse-grained grainstones and packstones of intraclasts, detrital spherulites, shrub debris, rare peloids and ooids are associated with Facies 3 above and Facies 5 below. Intraclast–detrital spherule–ooid packstones–grainstones consist of coarse- to very coarse-grained intraclasts, reworked spherules (detrital spherules), ooids, composite grains and minor (<2%) peloids and skeletal fragments This sub-facies is generally very poorly sorted, with grains ranging from fine sand size to very large, well-rounded, angular or platy clasts up to 10 cm across. Layers are typically 5–30 cm thick and commonly associated or interstratified with framestone facies.

(5) A variety of framework shrub forms are identified across both margins. These can occur as dendrolitic shrub-like forms occurring in an irregular spaced pattern and with a vertical relief or synoptic profiles ranging up to tens of centimetres. Well-developed framework porosity dominates these rocks. These are commonly associated with thin coarse-grained layers of spherulites–intraclast grainstones in repeating patterns. Occurring within the same core or field are coarser, agglutinated thrombolytic forms with fine lamina evident within the growth pillars. Again, a preserved framework porosity is evident in most instances. Coarse-grained stromatolitic forms also are common with irregular crenulated tops and laminations.

(6) Dark, planar laminated to weakly cross-laminated or massive spherulites. Wackestones–packstones are a very abundant facies in wells drilled in deeper water. Clasts exist as round to subrounded and coalesced spherulites with compromised growth zones, floating in a fine dark mud and/or clay matrix. The matrix commonly displays some silicification or dolomitization; the lamina follow the grain boundaries and wrap around the grains. These generally occur with Facies 7 below or with Facies 5 above. Beds are often 5–50 cm thick.

(7) Dark, millimetre-scale laminated spherulitic mudstones and organic-rich shales. There are sometimes green clay horizons and intraclasts near the bottom of the section. Nodular chert is common and is frequently associated with Ostracode-rich shale lamina. Thin layers of *in situ* spherule mudstone–wackestone are also commonly interlayered with dark micritic crusts in some intervals.

Isotopic data

Carbon and oxygen isotopes are powerful and insightful tools to help discern the lake chemistry and potential models for microbialite formation. These tools can help unravel the temperature and composition of the parent fluid of carbonates and also serve to look for potential cross-basin and even cross-conjugate margin correlations.

Primary calcite (micro-sampled shrubs of a limited number of samples) from Angolan–Brazilian pre-salt wells (Sag 2) showed that calcite has a $\delta^{18}O$ range of 4.6 to $-6.2‰$ (Pee Dee Belemnite, PDB) and a $\delta^{13}C$ range of 0.3 to $-4.3‰$ (PDB) (Fig. 6). Isotope values from the lower coquina (Sag 1) unit have wider ranges of $\delta^{18}O$ 1.4

Fig. 5. Core photos from two key wells: 2-ANP-1-RJS of the Santos Basin and a well from offshore Kwanza Basin. Many of the same facies are similar across the two basins. (**a**, **b**) Silty–spherulite–Ostracode mudstone with chert nodules (Kwanza Basin (a) and 2-ANP-1-RJS (b)), facies 7. (**c**, **d**) Argillaceous spherulite wackestones to packstones (Kwanza Basin (c) and 2-ANP-1-RJS (d)), facies 6. (**e**, **f**) Shrub framestones, dendritic and arborescent (Kwanza Basin (e) and 2-ANP-1-RJS (f)), facies 5. (**g**) Intraclast–detrital spherulite grainstones, facies 4. (**h**, **i**) Microbial crusts or buds and tuft small shrub boundstones–framestones (Kwanza Basin (h) and 2-ANP-1-RJS (i)), facies 3. (**j**, **k**) Fine crenulated dolomudstones with fenestral pores and potential mud cracks (j) and laminated silty wackestones and packstones with fine moderately sorted volcanic clasts (k), Kwanza Basin, facies 2. Illustrations from Terra *et al.* (2010).

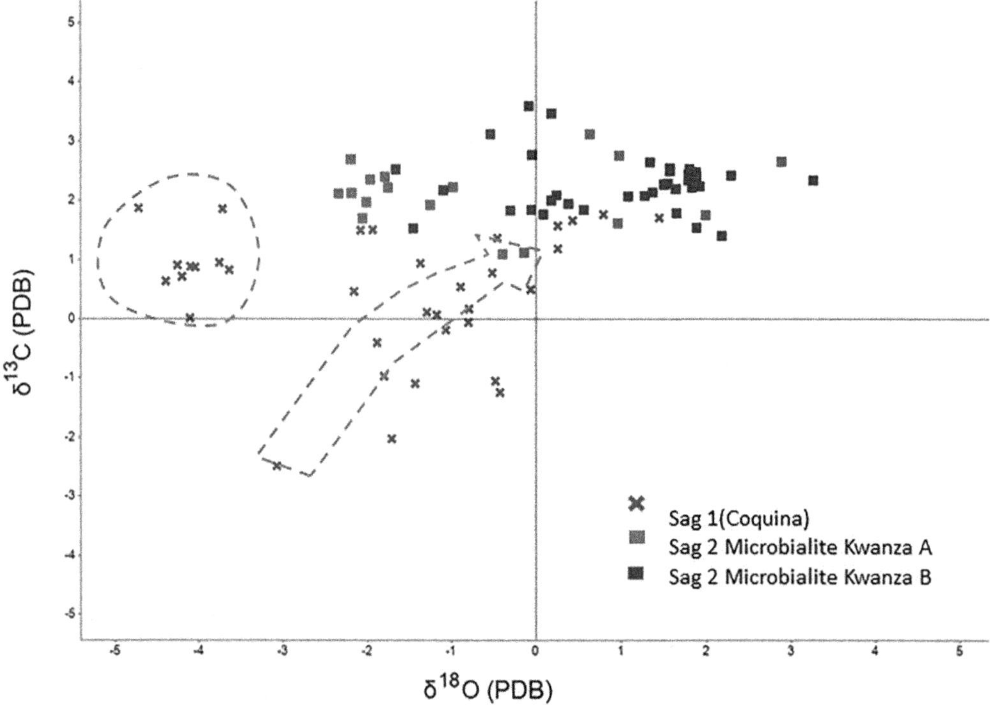

Fig. 6. Stable isotope analysis of the coquina section (Sag 1) Kwanza Basin. The range in values is $\delta^{18}O$ 1.44 to $-6.64‰$ (PDB) and $\delta^{13}C$ 1.77 to $-3.01‰$ (PDB). No covariant trend exists and the data suggest either a mixture of variable salinity lakes (evaporation variable) or some diagenetic overprint of some samples. Dashed arrow shows an evaporation trend with enrichment in $\delta^{18}O$ and $\delta^{13}C$. Dashed polygon could represent a diagenetic trend as a result of recrystallization or cementation during burial.

to $-6.6‰$ (PDB) and $\delta^{13}C$ 1.7 to $-3.0‰$ (PDB) (Fig. 7). A trend can be defined for the Sag 2 dataset, but with a rather non-variant $\delta^{13}C$ and a large variation in $\delta^{18}O$. A widespread of values is seen for the Sag 1 unit, with limited evidence of a reasonable covariant trend.

Depositional models and implications from isotopes

Depositional models

Few lacustrine carbonate models are available and there are few analogies from the marine system (Tucker & Wright 1990; Platt & Wright 1991; Wright 2012). Without detailed water depth indicators (biostratigraphy) and with limited palaeo-water depth indications, the origin of these cycles has to be inferred. We interpret these to be cycles that represent changes in accommodation space in conjunction with changes in lake chemistry. Wright (2012) suggests that these cycles can simply reflect the

changing lake chemistry during times of changes in the water level of the lake.

The Sag 1 unit appears to more predictable because mollusc packstones–grainstones (coquinas) dominate the inferred shoreline or wave-resistant features and the shoreline deposits described for the marine system and lacustrine settings in the Lower Congo and Campos Basin (Carvalho *et al.* 2000; Harris 2000; Jahnert & Collins 2012). Intrabasin highs away from the lake margin can develop thick grainstone coquinas with well-developed intra-granular or shelter porosity. The disarticulated shells, random orientation and low mud content suggest wave base deposition (high energy) or beach shoreline ridges. The presence of mud in Pelecypod wackestones–packstones suggests deposition below the wave base, but perhaps within the storm wave base. Lake level cycles and potential exposure are expressed with deeper water and muddier, low-energy deposits during lake level rises and potential dissolution and dolomitization during lower and more evaporative lake levels. Poorer reservoir zones probably reflect lower energy, deeper water

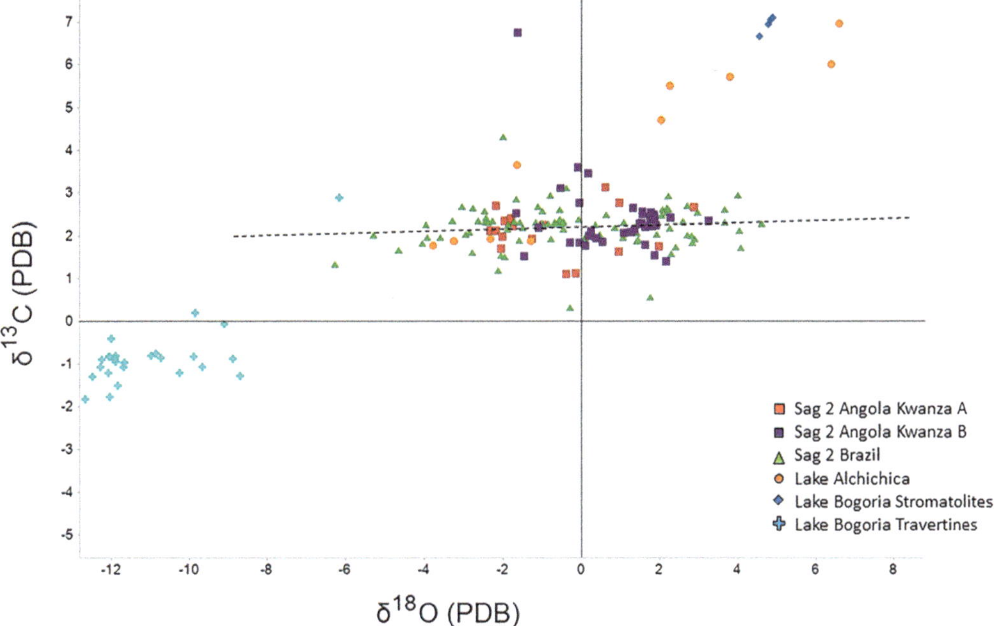

Fig. 7. Stable isotope ranges for the microbialite section of three wells, one in the Campos Basin (Lepley & Piccoli 2013) and two in the Kwanza Basin (Kwanza A and Kwanza B): $\delta^{18}O$ 4.61 to $-6.27‰$ (PDB) and $\delta^{13}C$ 0.32 to $-4.32‰$ (PDB). $\delta^{13}C$ varies moderately, whereas $\delta^{18}O$ has a wide range, leading to a poor covariant trend, but an obvious trend line. Samples from the Campos and Kwanza basins fit on a similar trend line and with similar ranges in both $\delta^{18}O$ and $\delta^{13}C$. Lake Bogoria travertines (Renaut *et al.* 2013) plot in a region well off the South Atlantic samples with more negative $\delta^{18}O$ and $\delta^{13}C$. Stromatolites from Lake Bogoria and Lake Alchichica (Kazmierczak *et al.* 2011) plot with generally more positive $\delta^{18}O$ and $\delta^{13}C$ values with a good covariant trend with a steep slope, indicating small, closed lakes.

deposition or deposition during low-energy storms within the wave base.

Wells near the lake margin shorelines have variable energy conditions and the coquinas are commonly interbedded with clastic debris input at the margin of the lake (Fig. 4). A poorer overall reservoir potential is observed as a result of fine clastic sediments filling the intergranular pores. Here the coquinas were probably competing for accommodation space with the clastic input derived from the continent, restricting the thickness of the coquinas. However, these areas seem to be more prone to dolomitization and exposure, both reservoir-enhancing diagenetic processes.

Wide variations in facies in the Sag 2 stacking patterns within basins are evident, with the lake margin patterns being differentiated from those in the inferred deeper profundal and intrabasinal palaeohigh trends (Fig. 8).

Wells penetrating the shallow basin margins where clastic input from fluvial sources dominates display fluvial–deltaic clastic or mixed clastic sediments (Facies 1–2) (Fig. 2, 1-RJS 78 and 84-7X).

These facies make up what has often been referred to as the Chela Formation on the African margin (McHargue 1990; Braccini *et al.* 1997; Burwood 1999; Guardado *et al.* 2000; Harris 2000). Basinwards, this fluvial–siliciclastic-dominated facies transitions to a more carbonate-prone facies with a shallow carbonate depositional fabric. These outer margins commonly show fine-scale cycles of silty stevensite mudstones or silty spherulite wackestones (Facies 2) and, on occasion, thin beds of microbial mats (Facies 2). The lack of fauna still points to the lake being highly alkaline and saline. Average water depths were probably only a few metres because mud cracks and microbial mats are common. Depending on the proximity to the clastic input and the potential for terrigenous dilution, there are varying amounts of silty Ostracode mudstones–wackestones, spherulite–stevensite wackestones–packstones and microbial mats or thin stromatolitic units. This facies is generally characterized along the margin by millimetre-scale fine laminations with little evidence of wave energy. The energy levels of these nearshore deposits may have been low in many areas.

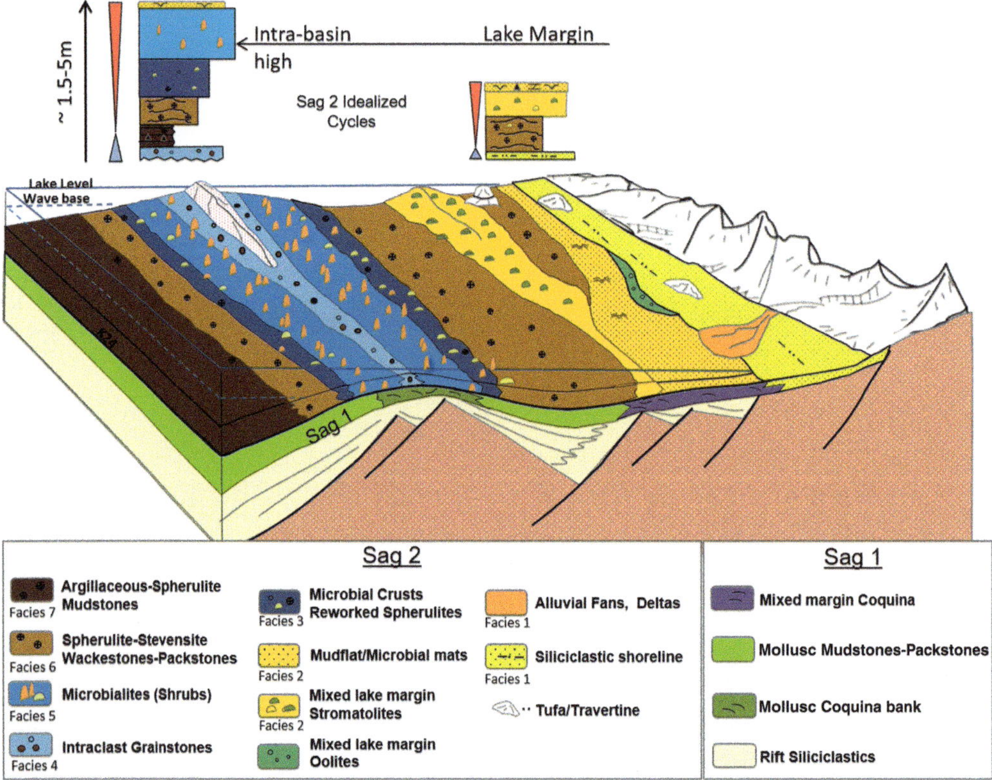

Fig. 8. Depositional model (Sag 2) showing the transition from the lake margin to deeper lake areas (synrift graben) and eventually to intra-basin highs (synrift horsts). The low accommodation margins are a mixture of siliciclastic sediments and carbonates. Carbonates near the margin are interpreted to be predominantly fine stromatolites and microbial mats and are commonly dolomitized. The deeper portions (inferred by the position off the palaeo-structural and present day highs) of the lake are often characterized by variably silty and spherulite–stevensite wackestones–packstones. Intrabasinal highs show a range of microbialite fabrics, ooid–intraclast grainstones, microbial crusts and detrital spherulite boundstones to wackestones. The distal lake centres are populated with argillaceous spherulite mudstones to wackestones. Idealized stacking patterns for both the margin and the intra-basin highs of the Sag 2 show small-scale (1.5–3 m) stacking patterns with sharp facies boundaries. The Sag 1 facies are shown as a thin layer (post-depositional) below the sediment surface.

The outboard wells away from the margin commonly display a thin-bedded (centimetre- to decimetre-scale) pattern of chert-rich organic mudstones (Facies 7), spherulite-rich wackestone–packstone layers (Facies 7), microbial shrub layers (Facies 3 and 5) topped with thin-bedded microbial mats (Facies 3) and grainstones (Facies 4) (Fig. 2). Wells drilled away from the margin are essentially free of clastic influences.

The thickest and most prevalent occurrence of the shrub facies (Facies 5) is coincident with the crests of the inferred palaeo-structure (evident in the current structure mapping), suggesting that this facies may represent shallow deposition influenced by the wave base and potentially also dependent on the photic zone. Facies 4 consists of poor to moderate sorting of probably shrub debris, intraclasts and rare ooids and ostracods, reflecting probable shallow reworking during low lake levels. Facies 2 may represent the shallowest lake level periods based on the thin-bedded and frequently mud-cracked and dolomitized tops. A hypothesized ideal cycle is one of lower energy Facies 6 and 7 with common chert nodules progressing to the small bud and tuft facies and large shrub facies (Facies 3 and 5, respectively) and, ultimately, to Facies 4 (intraclast grainstones) and Facies 2 (microbial mats) – representing a classical shallowing-upwards succession.

The spherulite and stevensite (Facies 6) components may mark the transition to the storm wave base and ultimately below storm wave base (Facies 7). The predominance of either lime mud or stevensite clays in the matrices suggest limited wave action and deposition in quiet waters with the clays settling out. Wells drilled away from obvious palaeo-structure and the thickest Sag 2 sections are predominantly spherulite–stevensite (Facies 6) mixed with silty marls, more evidence that these may represent the low-energy lake centre. Facies 3 often appears at a transition upwards from Facies 6 and 7, as dense crenulated or wavy-bedded crusts or a more open pore network of small shrubs and spherulites (some reworked). Occurrence above or capping Facies 6 indicates a potential moderate depth lake bottom, perhaps at storm wave base or deeper. Similar sublittoral to shallow profundal crusts and stromatolites were documented in the Green River Basin (Sarg *et al.* 2013), although the determination of the palaeogeography and palaeo-structure in an active rift margin setting is difficult. However, well profiles and seismic evidence indicate that outboard rift block highs (outboard highs) were the most prone to develop thick microbia-lite–shrub growth. Figure 5 shows the interpreted distribution and reservoir quality of well-developed shrubs in a position away from any discernible clastic input. An elevated margin or rift shoulder probably led to the development of carbonate micro-bialite growth, perhaps related to the position within the shallow photic zone.

The best reservoir sections pertain to Facies 3 through Facies 5, where the intra- and interparticle porosity of primary framestone is preserved (Fig. 5e and f). Evidence for leached or removed stevensite displayed as remnant silica or dolomite with a lath-like habit is also apparent in the frame-work fabric of Facies 5. Reservoir intervals related to the proposed stevensite removal are also apparent in some spherulite-rich layers (Facies 6), where dolomite appears to have incompletely replaced ste-vensite. These facies show unique lath-like dolomite crystals that appear to have retained the habit of the precursor clay (stevensite) (Terra *et al.* 2010; Wright 2012).

The largely volcanic catchment lithology in the South Atlantic suggests that the lakes received silica-rich inflows from streams and springs. A potential model of silica precipitation could be linked to a rise in lake levels (pluvial events) and a lowering of the pH (Wright 2012) could lead to silica precipitation and the development of chert nodules in association with anoxic lake bottom mudstones. Increases in pH through evaporation could initiate authigenic stevensite precipitation on the lake bottom and spherulite growth coincident with early burial is associated with the stevensite. Increased evaporation with decreased silica activity and decreasing accommodation space could have led to the development of shrubs in shallow water over the palaeohighs. Continued late highstands and reduced accommodation space led to the development of shallow microbial mats and expo-sure along the high.

The origin of the microbial shrub facies (Facies 5) has been questioned previously (Wright 2012). Della Porta & Barilaro (2011) concluded that the various shrub morphologies are not exclusive and can form in a number of different environments, either abiotic or biotic. The abiotic versus biotic origin of microbial carbonates needs to be addressed.

Abundant core data indicate that stromatolite–microbial mats occur in wells along the lake margin. These can be associated with microbial shrub layers (dendritic fan, dendritic branching and laminates) in some wells (Fig. 5). The upright to inclined growth patterns in arborescent features suggest a response to sunlight or to the current energy, perhaps indicat-ing a microbial origin. The intimate association of the framestone lithofacies with the coarse, grain-rich and commonly intraclastic lithofacies clearly indicates that the framestone facies formed in high-energy, wave-agitated environments. Many wells away from the structural highs are almost devoid of large shrubs (but with abundant spherulites) and we propose that they resemble growth frameworks. If the lake cycle was only affected by the chemistry or changes in lake level, then the presence and pat-tern of the cycles would be more ubiquitous or persistent.

New well data from the Angola margin (Cazier *et al.* 2014) interprets the reservoir zone to be a microbial boundstone that accumulated around palaeohighs with a similar lake chemistry to the modern day East African lakes. Baskin *et al.* (2012) proposed that horst block highs could be the focus point of microbial mound build-ups, even in the Great Salt Lake in Utah. Muniz (2013) provided data suggesting a microbial origin with preserved filaments in thin section.

One of the most compelling mineral phases prev-alent in many units in the pre-salt from both the Angolan and Brazilian margins is the trioctahedral magnesium silicate stevensite. Magnesium-rich lake waters at a pH >9 (Jones 1986) best explains this occurrence, whether extracellular polymeric substances (EPS) and microbial activity played a part (Souza-Egipsy *et al.* 2005; Burne *et al.* 2014). Stevensite is commonly associated with spherule calcite and less commonly with shrub microbialites (Facies 6 and 5, respectively). In most instances this is a rather poor reservoir facies, as the compaction and growth of spherulites inhibits the preservation of porosity. However, many wells display par-tially or wholly dolomitized matrices in some spherulite-rich units. Wright (2012) also suggested that spherulites could indicate the formation of a viscous gel as a result of their association and occurrence (displacing stevensite lathes). Stevensite indicates an alkaline, high salinity and high pH solution and, as a result of the easily suspended nature of the clay, is most likely to be deposited in deeper, quiet water areas of the lake (Tosca & Wright 2014).

Isotopic data implications

Interpreting the isotope data involves making some assumptions about the parent water temperature and composition, not uncommon in any isotope study. First, the parent lake water composition has to be deduced for lakes developed in the late Cretaceous. We assume that the Angola–Brazil systems were at a similar tropical latitude to each other (as today) during the late Cretaceous, as both basins were tempered by open oceans at these lower latitudes (Scotese 1997). From a summary of current rain water isotopes (Bowen & Wilkinson 2002), we estimated that rain water would probably have ranged from $\delta^{18}O$ -4 to $-6‰$ (the current range offshore Angola and Brazil). Scotese (1997) reported that the Cretaceous period for Brazil and Angola was probably more arid than today, so the current day values are a maximum. Assuming an average low-latitude temperature of $25°C$ and using the temperature fractionation curves of Friedman & O'Neil (1977), calcite in a freshwater lake should have values of $\delta^{18}O$ -6 to $-8‰$ (PDB).

The lower coquina-rich zone plots over a wide range of potential parent water compositions, ranging from those consistent with our estimate of fresh lake water to those that would indicate a considerable enrichment in $\delta^{18}O$. Some of these trends probably indicate burial and late diagenetic changes as several generations of calcite neomorphism and replacement are observed. Equally important may be the systematic changes in isotopic input into the lake and the degree of overfilling and balanced-fill (essentially climatic changes). Periods of rising lake levels and increased water input reflect lighter $\delta^{18}O$ isotope ratios, whereas periods of reduced lake levels and water input allow evaporation and the enrichment of $\delta^{18}O$. The negative values of $\delta^{13}C$ argue for a contribution from continental near-surface soils and the decay of plant material.

The Sag 1 isotope data, which have no covariance, suggest a balanced to overfilled lake. This is also supported by the inference of limited evaporation (enriched $\delta^{18}O$) from the palaeo-thermometry. We suggest that the values showing depleted $\delta^{13}C$ and $\delta^{18}O$ values could indicate heating or increased temperatures, which could only apply to either the mollusc shell or interparticle cements. However, the weak trend towards values from the Sag 2 unit may just indicate the mixing of fresh and more evaporative lake conditions.

Talbot (1990) suggested that the isotope trends can help to differentiate lake hydrology, residence time and the potential connections between lakes. The trend line of the Sag 2 unit shows a strong correlation, but low covariance (Fig. 7). Talbot (1990) reasoned that a strong covariance is recorded in many closed lakes, but also suggested that the shallower the slope of the $\delta^{13}C$ and $\delta^{18}O$ cross-plot trend, the larger the area to depth ratio of the lake. Large lakes with large surface areas allow the $\delta^{13}C$ values to equilibrate with atmospheric CO_2, limiting the $\delta^{13}C$ variation. However, the correlation line has little slope, which suggests a rather large area to depth ratio of the lake and, although the covariance is poor, the grouping of the data on a well-defined regression line suggests a closed lake. The large range in $\delta^{18}O$ values also supports a closed lake, as the lake levels and variable freshwater input affects the $\delta^{18}O$ of the lake water. Variable evaporation is shown by the data; however, the more positive $\delta^{18}O$ values suggest extensive evaporation. We propose that the shallow slope of the $\delta^{13}C$ and $\delta^{18}O$ cross-plot (Fig. 7) leads to a poor covariance, but still provides good evidence for the lake being very both large (high area to depth ratio) and closed. The similar values and trend lines for two key wells from different basins and two conjugate margins provides strong evidence that the two basins were probably connected.

It is difficult to predict the abundance or role that tufa and travertine deposition contributed to the pre-salt play. However, little evidence of hot or less saline fluids (travertine) is a key finding of the limited isotopic studies to date. Figure 7 plots the isotope values from travertine deposits of Lake Bogoria (Renaut *et al.* 2013). Depleted $\delta^{13}C$ and $\delta^{18}O$ values with low covariance can be expected in travertines (as found at Lake Bogoria) and have also been found in other studies of travertine/tufa deposits around the world (Andrews 2006).

It is difficult to determine for certain that the Sag 2 isotopes are a primary signature. Resetting during burial or even hydrothermal activity could explain the poor covariance (higher temperatures affecting $\delta^{18}O$ with minor resetting of $\delta^{13}C$). However, we would expect the $\delta^{18}O$ values to be significantly lighter and not in the range evidenced by alkaline lakes today. In addition, the similar stratigraphic hierarchy shown between wells from opposite margins (Brazil and Angola) lends support to the view that, at Sag 2 time, the lakes were connected (Fig. 9).

Future isotope work from different basins in the Angola–Brazil pre-salt could be used to determine whether the basins within or across the conjugate margins were linked. If the data share a similar covariant trend, then this suggests that the lakes were linked; individual covariant trends would indicate several closed lakes and the qualitative area to

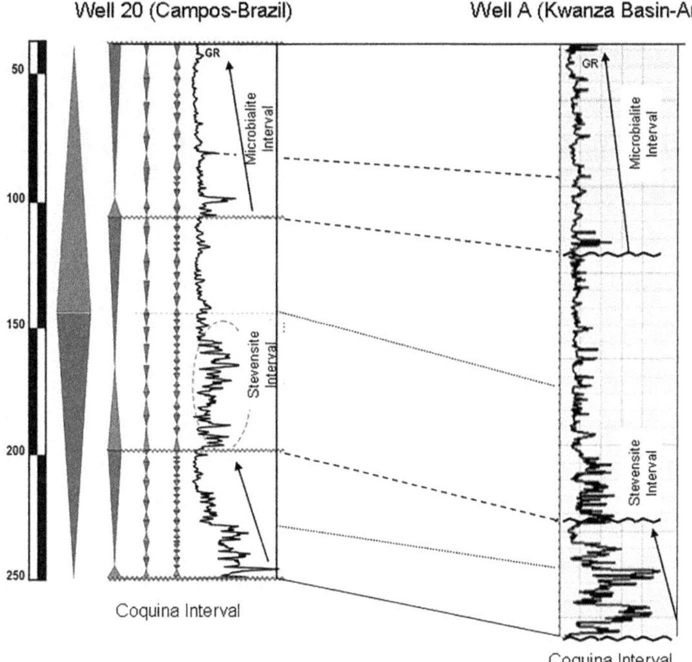

Fig. 9. Well log profile from a Campos Basin well on the Brazil margin (well 20) and a Kwanza Basin well (Kwanza A) on the Angolan margin (Muniz 2013). Both wells display a very similar GR log profile (depositional trend) above the coquina-rich intervals (Sag 1). Both wells report a stevensite-rich (stevensite–spherulite) zone near the base, with a microbialite zone near the top. The GR trends, small deflections and interval thicknesses can be correlated with good confidence.

depth ratio. Poor covariant trends with a range in isotopes could suggest local travertine conditions.

Subsidence and lake hydrology

Lake hydrology is controlled mainly by the relative balance between the accommodation space created by tectonic subsidence and the supply of sediments and water (Fig. 10). Overfilled lakes occur when the rate of supply of water and sediment exceeds the rate of tectonic subsidence in an open basin with a river output to the ocean. These are generally freshwater basins with local fluvial–deltaic inputs (Bohacs *et al.* 2000; Harris *et al.* 2013). Balanced-filled basins occur when the water and sediment supply are in balance with the subsidence rate and the lake water inflow is periodically matched by the outflow (Bohacs *et al.* 2000). The lake hydrology and water chemistry fluctuates; climatically driven lake fluctuations are common (Harris *et al.* 2013). Under-filled lake basins occur when the subsidence rate is consistently higher than the water and sediment supply, creating a closed hydrology that results in a high solute content and evaporative conditions (Bohacs

et al. 2000; Harris *et al.* 2013). The lake hydrology and subsidence history can be inferred from these data and the models of Bohacs *et al.* (2000). A similar picture is emerging across the basins and conjugates, but perhaps will not suffice for all areas of the pre-salt.

From the interpretations of seismic line 2400 and other seismic lines along the Angolan and Brazilian margins, the rifting phase was characterized by the formation of deep graben. These were reportedly filled with mainly clastic sediments, such as gravity flows and conglomerates, in a lacustrine setting (Braccini *et al.* 1997), possibly in a similar setting to East Africa at the present day (Harris *et al.* 2013). At the onset of rifting, the surface elevation in all these lakes was above sea-level, just as in the modern day East Africa rift lakes. It is difficult to infer an accurate lake hydrology at this time, but it is possible that the chemistry of each individual lake was variable depending on whether each basin was in an overfilled or under-filled condition. The modern East Africa lakes have a variable chemistry ranging from hypersaline and hyperalkaline (Lake Bogoria, Lake Manyara) to freshwater (Lake Malawi, Lake Victoria) (Casanova 1986; Cohen &

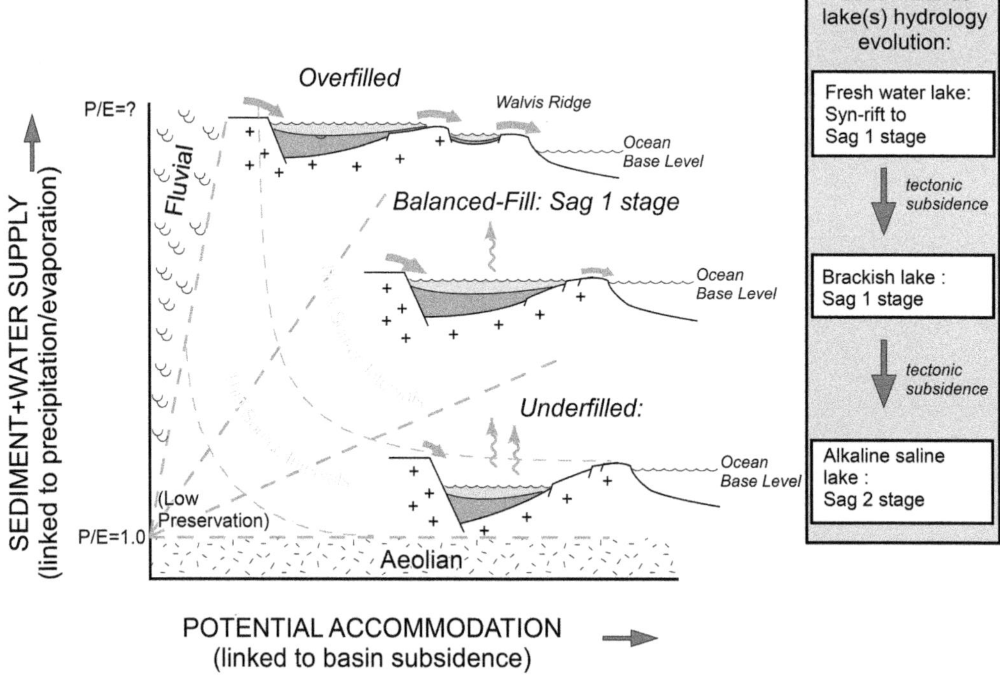

Fig. 10. Synthesis of the evolution of the South Atlantic lake(s) using the lake classification scheme of Carroll & Bohacs (1999) and Bohacs *et al.* (2000). The lake hydrology evolves from freshwater during the Sag 1 stage (coquina deposition) to alkaline in Sag 2 (microbialites–spherulite deposition). This transition may have been caused by the South Atlantic depression falling below sea-level because of continued syntectonic subsidence. P/E is the sediment + water supply relative to potential accommodation space.

Thouin 1987; Hillaire-Marcel & Casanova 1987; Hicks *et al.* 2012; Harris *et al.* 2013; Renaut *et al.* 2013).

There are very few well penetrations of the true synrift section or, at least, penetrations with solid biostratigraphic dates. However, the few well data that exist and the seismic data (progradational wedges off the horst blocks) suggest a typical early fluvial–lacustrine series of basins dominated by fine silty marls to sandstones. Well penetrations in some of these graben in the Kwanza and Santos basins encountered pelecypod shells (coquinas) within the synrift section with variable degrees of biodiversity. The presence of shells implies at least some transitional event of freshwater to brackish conditions; no microbialite or stromatolite has been encountered in this section. From these observations it is assumed that the basin was above sea-level and was dominantly freshwater. This is classified as an overfilled lake, where the rate of water and sediment input exceeds the rate of tectonic subsidence in an open basin with a river output (Bohacs *et al.* 2000; Harris *et al.* 2013).

The Sag 1 unit has a moderately diverse faunal assemblage of non-marine molluscs and ostracods and the isotope data suggest varying degrees of fresh to saline water signatures. The isotopes may also reflect fairly large fluctuations in lake level and several episodes of exposure and diagenetic alteration of the coquinas. The seismic reflections in this unit can be traced in a continuous fashion across many parts of each basin and across different basins (for example, the Kwanza and Lower Congo basins). The seismic reflectors are sub-parallel and have bilateral onlaps on the edge of the basins, implying that the lakes started to coalesce and became deeper. We infer that the lake(s) at this stage fitted the balanced-filled model with the rate of sediment supply in balance with the rate of subsidence and the water inflow generally matched by the outflow (Bohacs *et al.* 2000). The lake hydrology and water chemistry fluctuated and climatically driven lake fluctuations were common (Harris *et al.* 2013). However, the balanced lake indications also suggest that, to allow a balanced inflow and outflow, the surface elevation was above sea-level. The isotope data suggest a complex history of basin hydrology with times of more saline conditions (enriched $\delta^{18}O$) and evaporation followed by periods of high water and sediment influx and fresher conditions (depleted $\delta^{18}O$).

A distinctive seismic and lithological boundary exists between the Sag 1 unit containing the predominantly bivalve coquina rock types and the Sag 2 microbial-rich zone. The boundary is variably a shale to marl zone with a high TOC and appears on the seismic data and in some wells as a sharp angular unconformity. The overlying section is partly or completely missing on the flanks of the basin and local highs; however, it becomes a paraconformable surface in the centre of the basin.

We believe that this transition represented a critical moment in the evolution of the South Atlantic lake and it is this marker that records the transition from a fresh to brackish water open lake to a closed alkaline lake. The Sag 1 stage has also been interpreted to have transitioned from a more open freshwater lake (pelecypod-bearing) to a more restricted and brackish lake (gastropod-bearing) near the top (Muniz 2013; Thompson *et al.* 2015). This transition may have also been the cause of an widespread anoxic event throughout the lake, as supported by the high TOC content typical of this shale and marl layer occurring in the CVII Biozone (AS 10, Top BA4) in Angola and RT-010 in Brazil. However, Harris *et al.* (1994) suggested that the preservation of organic material, particularly in the late rift section, was a result of high rates of organic productivity and chemical sedimentation. Guardado *et al.* (2000) suggest that intermittent seawater incursions led to cyanobacteria blooms, eventual anoxia and the deposition of organic-rich shales. However, the assumed equivalent late Bucomazi source rocks in Angola were characterized as saline ephemeral lacustrine in nature, without an explanation of their mode of occurrence.

We can hypothesize that, during the process of increasing tectonic subsidence to below sea-level, the lake briefly became deeper, followed by a period of high productivity and stagnation (Fig. 10, early Sag stage 2). This change in lake chemistry was responsible for the disappearance of the pelecypods and other biota, causing an increase in algal productivity. In addition, the isolation of the lake resulted in the concentration of organic matter. All of these factors contributed to the anoxic conditions and the deposition of the organic shale.

The overlying microbialite section records a general decrease in the biodiversity of the fauna and the disappearance of all mollusc fauna. The presence of stevensite, indicating an increase in pH, supports lake isolation or stressed conditions. In addition, the $\delta^{18}O$ of the calcitic components indicates increased salinity (higher evaporation

rates as a result of isolation) relative to the presumed freshwater lake.

There is a striking similarity in the stage 2 Sag of the sedimentary record between Angola and Brazil (Figs 7 & 9). The existence of a low diversity, stressed fauna and highly alkaline conditions as evidenced by the occurrence of stevensite and similar isotopic trends suggests an extensive and continuous linked South Atlantic lake. This is best explained by under-filled conditions with the lakes below sea-level, cut off from the ocean by the emerging Walvis Ridge. This could have been caused by the surface of the lakes subsiding below sea-level as a result of continued tectonic subsidence. This subsidence forced the whole system to become closed and the lake to become under-filled, alkaline and saline, similar to present day lakes below sea-level (e.g. the Caspian Sea and Death Valley). Under-filled lakes can also occur above sea-level, but the persistence of the rock record (on both margins) at this time, showing highly alkaline conditions, infers a basin-wide event best explained by isolation occurring as the basin subsided below sea-level (an air-filled hole). It is difficult to justify a large number of isolated lakes, all in alkaline under-filled conditions. In a multiple-lakes alkaline scenario some or many of the lakes would have had an output to the ocean and therefore a freshwater hydrology.

Further evidence that the latest Sag may have been below sea-level is the presence of up to 2000 m of salt (halite and bitter salt). Most researchers agree that salt up to 2 km in thickness was deposited over a short period of geological time (*c.* 2–4 Ma). The models that have been proposed to account for this thick salt are significantly different. Reston (2009, 2010) and Jackson & Hudec (2009) and Crosby *et al.* (2011) suggest that the thick salt can be attributed to incursions of the ocean over the Walvis Ridge into a large and potentially deep (hundreds of metres) air-filled hole below sea-level. Other researchers envisage ocean incursions into a shallow depression near sea-level, with subsidence creating the space for the accumulation of the thick salt (Davison 2007; Dupre *et al.* 2007; Quirk *et al.* 2012)

The formation of bitter salts and thick salt are best explained with a deep basin salt model. The breaching of the Walvis Ridge, associated with global sea-level rise, probably allowed ocean water to flow into the enormous pre-existing depression. Potash salts form during the very last stages of the evaporation of seawater and 1000 m of ocean water can only precipitate 2.8 m of bitter salts and therefore the presence of Aptian potash salt (Hardie & Eugster 1980; James & Kendall 1992; Warren 2010) in the Congo and Gabon basins (Congo, Zaire and Angola) suggests that at least 1000 m of water-filled accommodation space was needed to precipitate the 2–3 m of these salts found in Angola.

It is conceivable that the basin may have been several hundreds of metres below sea-level during the initial, and potentially periodic, seawater encroachment from the Walvis Ridge. Slow subsidence and progressively more vigorous seawater incursions would have occurred until the Walvis Ridge subsided enough to fully connect the ocean basin south of the ridge to the isolated basin to the north during the Early Albian.

An alternative model for the system is that the South Atlantic lake might have been above sea-level and that the under-filled condition of the system occurred as a result of a lack of water outflow to the ocean, similar to the present day examples of under-filled lakes (e.g. the Great Salt Lake). However, a mechanism to isolate two margins and several basins while above sea-level is more difficult to imagine. In this scenario, salt would need to have been deposited in the pre-existing shallow lacustrine basin and rapid subsidence to have occurred over a few million years to accommodate the up to 2 km of salt. The limitations of this model are the rapid subsidence rates needed at a time when the basin should be in slow thermal subsidence; the thick (metre-scale) bitter salts are difficult to explain using this model (Karner & Gamboa 2007; Jackson & Hudec 2009; Montaron & Tapponnier 2010; Warren 2010).

Conclusions

The observations from a conjugate pair of regional seismic lines between the Kwanza and South Campos basins help to define the key megasequences as pre-rift, rift, sag, salt basin, post-rift, post-salt carbonates and post-rift, post-salt clastic sediments. The subsidence and evolution of the lake within the pre-salt interval can be described in three phases.

(1) Synrift lakes. The onset of continental breakup and synrift graben, with the formation of several deep, freshwater (overfilled) lakes above sea-level, representing Palynozones AS2–AS5 and C2–C4 (Valanginian–Barremian) (Braccini *et al.* 1997; Poropat & Colin 2012; Chaboureau *et al.* 2013).

(2) Sag 1; late rift. Syn-kinematic stretching of the continental crust and/or continuous rifting. Abundant fauna and faunal diversity suggest that the lakes probably evolved into a wide interconnected system (fresh–brackish and balanced-filled) during this phase (Palynozone CVI–CVII and Ostracode Zones AS9–AS10, RT 09–RT 010) (Braccini *et al.* 1997; Poropat & Colin 2012; Chaboureau *et al.* 2013).

(3) Sag 2. The basin becomes hydrologically isolated (an under-filled lake) as the base level falls below sea-level. The limited diversity of fauna and widespread occurrence of microbialite, with a prevalence of authigenic stevensite, suggests a hypersaline alkaline lake. The lakes from both margins were probably connected at this point (Palynozone CVIII–CIX, Ostracode Zone AS11–AS12, RT011) (Braccini *et al.* 1997; Poropat & Colin 2012; Chaboureau *et al.* 2013).

The Sag 1 stage was one of fresh to brackish lakes creating a unique depositional system dominated by carbonate deposition and associated with a moderately diverse fauna of molluscs. Coarse coquina grainstones, deposited as shoreface sediments along the lake margin and on palaeohighs, constitute the best reservoir during the Sag 1 stage.

Microbial deposits predominated in the Sag 2 stage as the lake became too saline to support most biota. Microbial deposits dominated along the basin margin, but were most prolific near basement-rooted palaeohighs (outboard highs) away from the basin margin. Authigenic stevensite and calcite spherulites are commonly associated with the microbialite textures and seem to represent the deeper and lower energy areas of the lake. This unique assemblage suggests widespread and often highly evaporative (high pH) lake conditions throughout the Sag 2 stage. The evidence to date – from the fabrics, the extensive distribution and the isotope data – does not support extensive travertine deposits.

The transition from the Sag 1 to Sag 2 stages is marked by a dramatic change in the lake hydrology from a fresh to brackish water, balance-filled lake to an alkaline and hypersaline, under-filled and connected lake. We believe that this change was driven by the tectonic and subsidence evolution of the system and was caused by the continuous subsidence of the basin, eventually resulting in the surface of the lake sinking below sea-level and forcing the whole system to become hydrologically isolated.

Both the South American and African margins display the same age stratigraphy and depositional fabrics, a similar isotope composition for the primary carbonates, and a similar log character and thickness for the Sag 2 section. This evidence supports our theory of a connected lacustrine system at this time. The observation of large Late Aptian salt thicknesses (up to 2 km) above the lacustrine sections also suggests that this phase of basin evolution was significantly below sea-level and represents an under-filled lake.

The last stage of this process is represented by the breaching of the physical barrier (Walvis Ridge) that separated this basin from the ocean This was linked to a global rise in sea-level and allowed the cyclical ingression of marine water into a depression that was located well below sea-level.

The authors thank BP, Jasper Peijs, Jay Thorseth, Liz Jolley, Nicki Adams, Laura Mackay and Rob Satter in addition to our partners, Sonangol P&P and Cobalt International Energy, for their support and permission for this publication. The authors also acknowledge Petrobras for pioneering the understanding of the pre-salt lacustrine play. Special thanks to Art Saller and Cobalt for providing isotope data from the Kwanza Basin and the extremely useful discussions on all aspects of the carbonate geology. In addition, Leslie Neal, Scott Lepley, Dan Finucane and Jeff Pietras from BP Brazil assets provided invaluable interpretations and data. We acknowledge ION and TGS Nopec for kindly providing the regional seismic lines. Early versions of the paper benefited greatly by revisions from Mark Thompson and Anna Matthews. Early work and interpretations from the Campos Basin from Steven Dorobek were invaluable in the integration of data from both conjugates. We thank Paul Wright for providing the groundwork and solid fundamentals of many aspects of lacustrine systems. A special thank you goes to Dave Johnson and Andrew Kimber, who helped us to draft the figures.

References

ALTENHOFEN, S.D. 2013. *Caracterização petrográfica de depósitos carbonáticos lacustres do Grupo Lagoa Feia, Bacia de Campos, Brasil*. Monografia (Trabalho de Conclusão de Curso). Universidade Federal do Rio Grande do Sul, Porto Alegre.

ANDREWS, J.E. 2006. Palaeoclimatic records from stable isotopes in riverine tufas: synthesis and review. *Earth-Science Reviews*, **75**, 85–104.

ASLANIAN, D., MOULIN, M. *ET AL.* 2009. Brazilian and African passive margins of the Central Segment of the South Atlantic Ocean: kinematic constraints. *Tectonophysics*, **468**, 98–112.

BASKIN, R.L., WRIGHT, V.P., DRISCOLL, N., GRAHAM, K. & HEPNER, G. 2012. Microbialite bioherms in Great Salt Lake: Influence of active tectonics and anthropogenic effects. AAPG Search and Discovery Article 90153. American Association of Petroleum Geologists, Tulsa, OK.

BOHACS, K.M., CARROLL, A.R., NEDE, J.E. & MANKIROWICZ, P.J. 2000. Lake-basin type, source potential, and hydrocarbon character: an integrated sequence-stratigraphic-geochemical framework. *In*: GIERLOWSKI-KORDESCH, E.H. & KELTS, K.R. (eds) *Lake Basins through Space and Time*. AAPG Studies in Geology, **46**, 3–33.

BOWEN, G.J. & WILKINSON, B. 2002. Spatial distribution of $\delta^{18}O$ in meteoric precipitation. *Geology*, **30**, 315–318.

BRACCINI, E., DENISON, J.R., SCHEEVEL, P., JERONIMO, P., ORSOLINI, P. & BARLETTA, V. 1997. A revised chrono-litho-stratigraphic framework for the pre-salt (Lower Cretaceous) in Cabinda, Angola. *Bulletin des Centres de Recherches Exploration-Production Elf-Aquitaine*, **21**, 125–151.

BUENO, G.V. 2004. Diacronismo de eventos no rifte Sul-Atlantico. *Boletim de Geosciencias da Petrobras (Rio de Janeiro)*, **12**, 203–229.

BURNE, R.V., MOORE, L.S., CHRISTY, A.G., TROITZSCH, U., KING, P.L., CARNERUP, A.M. & JOSEPH HAMILTON, P. 2014. Stevensite in the modern thrombolites of Lake Clifton, Western Australia: a missing link in microbialite mineralization? *Geology*, **42**, 575–578.

BURWOOD, R. 1999. Angola: source rock control for Lower Congo Coastal and Kwanza Basin petroleum systems. *In*: CAMERON, N., BATE, R.H. & CLURE, V. (eds) *The Oil and Gas Habitats of the South Atlantic*. Geological Society, London, Special Publications, **153**, 181–194, http://doi.org/10.1144/GSL.SP.1999.153.01.12

CARROLL, A.R. & BOHACS, K.M. 1999. Stratigraphic classification of ancient lakes balancing tectonic and climatic controls. *Geology*, **27**, 99–102.

CARVALHO, M.D., PRACA, U.M., SILVA-TELLES, A.C., JAHNERT, R.J. & DIAS, J.L. 2000. Bioclastic carbonate lacustrine facies models in the Campos Basin (Lower Cretaceous), Brazil. *In*: GIERLOWSKI-KORDESCH, E.H. & KELTS, K.R. (eds) *Lake Basins through Space and Time*. AAPG Studies in Geology, **46**, 245–256.

CASANOVA, J. 1986. East African Rift stromatolites. *In*: FROSTICK, L.E., RENAUT, R.W., REID, I. & TIERCELIN, J.J. (eds) *Sedimentation in the African Rifts*. Geological Society, London, Special Publications, **25**, 201–210, http://doi.org/10.1144/GSL.SP.1986.025.01.17

CAZIER, E.C., BARGAS, C. *ET AL.* 2014. Petroleum geology of the Cameia Field, Deepwater Pre-Salt Kwanza Basin, Angola, West Africa. AAPG Search and Discovery Article 20275, presented at the International Conference & Exhibition, 14–17 September, Istanbul, Turkey, http://www.searchanddiscovery.com/pdfz/documents/2014/20275cazier/ndx_cazier.pdf.html

CHABOUREAU, A.C., GUILLOCHEAU, F., ROBIN, C., ROHAIS, S., MOULIN, M. & ASLANIAN, D. 2013. Paleogeographic evolution of the central segment of the South Atlantic during Early Cretaceous times: paleotopographic and geodynamic implications. *Tectonophysics*, **24**, 191–223.

COHEN, A.S. & THOUIN, C. 1987. Nearshore carbonate deposits in Lake Tanganyika. *Geology*, **155**, 414–418.

CONTRUCCI, I., MATIAS, L. & MOULIN, M. 2004. Deep structure of the West African continental margin (Congo, Zaire, Angola)., between 5°S and 8°S, from reflection/refraction seismics and gravity data. *Geophysical Journal International*, **158**, 529–553.

CROSBY, A.G., WHITE, N.J., EDWARDS, G.R.H., THOMPSON, M., CORFIELD, R. & MACKAY, L. 2011. Evolution of deep-water rifted margins: testing depth-dependent extensional models. *Tectonics*, **30**, TC1004.

DAVIS, M. & KUSZNIR, N.J. 2004. Depth-dependent lithospheric stretching at rifted continental margins. *In*: KARNER, G.D. (ed.) *Proceedings of NSF Rifted Margins Theoretical Institute*. Columbia University Press, New York, 92–136.

DAVISON, I. 2007. Geology and tectonics of the South Atlantic Brazilian salt basins. *In*: REIS, A.C., BUTLER, R.W.H. & GRAHAM, R.H. (eds) *Deformation of the Continental Crust: the Legacy of Mike Coward*. Geological Society, London, Special Publications, **272**,

345–359, http://doi.org/10.1144/GSL.SP.2007.272.01.18

DELLA PORTA, G. & BARILARO, F. 2011. Non-marine carbonate precipitates: a review based on recent and ancient studies. AAPG Search and Discovery Article 30217, presented at the AAPG International Conference and Exhibition, 23–26 October, Milan, Italy.

DINGLE, R.V. 1999. Walvis Ridge barrier: its influence on palaeoenvironments and source rock generation deduced from ostracod distributions in the early South Atlantic Ocean. *In*: CAMERON, N.R., BATE, R.H. & CLURE, V.S. (eds) *The Oil and Gas Habitats of the South Atlantic*. Geological Society, London, Special Publications, **153**, 293–302, http://doi.org/10.1144/GSL.SP.1999.153.01.18

DRISCOLL, N.W. & KARNER, G.D. 1998. Lower crustal extension across the Northern Carnarvon basin, Australia: evidence for an eastward dipping detachment. *Journal of Geophysical Research-Solid Earth*, **103**, 4975–4991.

DUPRE, S., BERTOTTI, G. & CLOETINGH, S.A.P.L. 2007. Tectonic history along the South Gabon Basin: anomalous early post-rift subsidence. *Marine and Petroleum Geology*, **24**, 151–172.

EICHENSEER, H.T., WALGENWITZ, F.R. & BIONDI, P.J. 1999. Stratigraphic control on facies and diagenesis of dolomitized oolitic siliciclastic ramp sequences Pinda Group, Albian, offshore Angola. *AAPG Bulletin*, **83**, 1729–1758.

FRIEDMAN, I. & O'NEIL, J.R. 1977. Compilation of stable isotope fractionation factors of geochemical interest. *In*: FLEISHER, M. (ed.) *Data of Geochemistry*. 6th edn. US Geological Survey Professional Paper, 440-KK.

GROSDIDIER, E., BRACCINI, E., DUPONT, G. & MORON, J.M. 1996. Biozonation du Cretace inferieur non marin des bassins du Gabon et du Congo. *Geologie de l'Afrique et de l'Atlantique Sud: Actes Colloques Angers*, **1994**, 67–82.

GUARDADO, L.R., SPADINI, A.R., BRANDAO, J.S.L. & MELLO, M.R. 2000. Petroleum system of the Campos Basin. *In*: MELLO, M.R. & KATZ, B.J. (eds) *Petroleum Systems of South Atlantic Margins*. AAPG Memoir, **73**, 317–324.

HARDIE, L.A. & EUGSTER, H.P. 1980. Evaporation of seawater: calculated mineral sequences. *Science*, **208**, 498–500.

HARRIS, N.B., SORRIAUX, P. & TOOMEY, D.F. 1994. Geology of the Lower Cretaceous Viodo Carbonate, Congo Basin: a lacustrine carbonate in the South Atlantic rift. *In*: LOMANDO, A.J., SCHREIBER, B.C. & HARRIS, P.M. (eds) *Lacustrine Reservoirs and Depositional Systems*. SEPM Core Workshop, **19**, 143–172.

HARRIS, P.M. 2000. Toca Carbonate, Congo Basin: response to an evolving rift lake. *In*: MELLO, M.R. & KATZ, B.J. (eds) *Petroleum Systems of South Atlantic Margins*. AAPG Memoir, **73**, 341–360.

HARRIS, P.M., ELLIS, J. & PURKIS, J.S. 2013. Assessing the extent of carbonate deposition in early rift settings. *AAPG Bulletin*, **97**, 27–60.

HICKS, M., HARGRAVE, J. & SCHOLZ, C. 2012. Macro-, meso-, and microstructures of microbialites from Lake Turkana, East African rift: paleoenvironment and diagenesis. AAPG Search and Discovery Article 90142, presented at the AAPG Annual Convention and Exhibition, 22–25 April 2012, Long Beach, California.

HILLAIRE-MARCEL, C. & CASANOVA, J. 1987. Isotopic hydrology and palaeohydrology of the Magadi (Kenya)-Natron (Tanzania) basin during the late Quaternary. *Palaeogeography Palaeoclimatology Palaeoecology*, **58**, 155–181.

HORSCHUTZ, P.M.C. & SCUTA, M.S. 1992. Facies-perfis e mapeamento de qualidade do reservatorio de coquinas da Formacao Lagoa Feia do Campo de Pampo. *Boletim Geociencias Petrobras*, **6**, 45–58.

JACKSON, M. & HUDEC, M. 2009. Interplay of basement tectonics, salt tectonics, and sedimentation in the Kwanza Basin, Angola. AAPG Search and Discovery Article 30091, presented at the AAPG International Conference and Exhibition, Cape Town, South Africa, 26–29 October.

JAHNERT, R.J. & COLLINS, L.B. 2012. Characteristics, distribution and morphogenesis of subtidal microbial systems in Shark Bay, Australia. *Marine Geology*, **303-306**, 115–136.

JAMES, N.P. & KENDALL, A.C. 1992. Introduction to carbonate and evaporite facies models. *In*: WALKER, R.G. & JAMES, N.P. (eds) *GT 1: Facies Models. Response to Sea Level Change*. Geological Association of Canada, St John's, 265–275.

JONES, B.F. 1986. Clay minerals diagenesis in lacustrine sediments. *US Geological Survey Bulletin*, **1578**, 291–300.

KARNER, G.D. & DRISCOLL, N.W. 1999a. Style, timing, and distribution of tectonic deformation across the Exmouth Plateau, northwest Australia, determined from stratal architecture and kinematic basic modeling. *In*: MACNIOCAILL, C. & RYAN, P.D. (eds) *Continental Tectonics*. Geological Society, London, Special Publications, **164**, 287–323, http://doi.org/10.1144/GSL.SP.1999.164.01.14

KARNER, G.D. & DRISCOLL, N.W. 1999b. Tectonic and stratigraphic development of the West African and eastern Brazilian margins; insights from quantitative basin modelling. *In*: CAMERON, N.R., BATE, R.H. & CLURE, V.S. (eds) *The Oil and Gas Habitats of the South Atlantic*. Geological Society, London, Special Publications, **153**, 11–40, http://doi.org/10.1144/GSL.SP.1999.153.01.02

KARNER, G.D. & GAMBOA, L.A.P. 2007. Timing and origin of the South Atlantic pre-salt sag basins and their capping evaporates. *In*: SCHREIBER, B.C., LUGLI, S. & BAÇBEL, M. (eds) *Evaporites Through Space and Time*. Geological Society, London, Special Publications, **285**, 15–35, http://doi.org/10.1144/SP285.2

KARNER, G.D., DRISCOLL, N.W. & BARKER, D.H.N. 2003. Syn-rift regional subsidence across the West African continental margin: the role of lower plate ductile extension. *In*: ARTHUR, T., MACGREGOR, D.S. & CAMERON, N.R. (eds) *Petroleum Geology of Africa: New Themes and Developing Technologies*. Geological Society, London, Special Publications, **207**, 105–129, http://doi.org/10.1144/GSL.SP.2003.207.6

KAZMIERCZAK, J., KEMPE, S., KREMER, B., LÓPEZ-GARCÍA, P., DAVID MOREIRA, D. & ROSALUZ TAVERA, R. 2011. Hydrochemistry and microbialites of the alkaline crater Lake Alchichica, Mexico. *Facies*, **57**, 543–570.

KUSZNIR, N.J. & KARNER, G.D. 2007. Continental lithospheric thinning and breakup in response to upwelling divergent mantle flow: application to the Woodlark, Newfoundland and Iberia margins. *In*: KARNER, G.D., MANATSCHAL, G. & PINHEIRO, L.M. (eds) *Imaging, Mapping and Modelling Continental Lithosphere Extension and Breakup*. Geological Society, London, Special Publications, **282**, 389–419, http://doi.org/10.1144/SP282.16

LENTINI, M.R., FRASER, S.I., SUMMER, H.S. & DAVIES, R.J. 2010. Geodynamics of the South Atlantic conjugate margins: implications for hydrocarbon potential. *Petroleum Geoscience*, **16**, 217–229, http://doi.org/10.1144/1354-079309-909

LEPLEY, S. & PICCOLI, L. 2013. Utilizing chemostratigraphic techniques to improve lacustrine carbonate reservoir characterization. AAPG Search and Discovery Article 90166, presented at the AAPG International Conference & Exhibition, 8–11 September, Cartagena, Colombia.

MARRETT, R. & ALLMENDINGER, R.W. 1992. Amount of extension on 'small' faults: an example from the Viking graben. *Geology*, **20**, 47–50.

MARTON, L.G., TARI, G.C. & LEHMANN, C.T. 2000. Evolution of the Angolan passive margin, West Africa, with emphasis on post-salt structural styles. *In*: MOHRIAK, W.U. & TALWANI, M. (eds) *Atlantic Rifts and Continental Margins*. Geophysical Monograph, American Geophysical Union, **115**, 129–149.

McHARGUE, T.R. 1990. Stratigraphic development of proto-South Atlantic rifting in Cabinda, Angola – a petroliferous lake basin. *In*: KATZ, B.J. (ed.) *Lacustrine Basin Exploration Case Studies and Modern Analogs*. AAPG Memoir, **50**, 307–326.

MELLO, M.R., AZAMBUJA, N., MOHRIAK, W.U., CATTO, A.J. & FRANCOLIN, J.B. 2011. Promising giant new hydrocarbon frontier: the Namibian continental margin. *GeoExpro*, **8**, 64–69.

MEREDITH, D.J. & EGAN, S.S. 2002. The geological and geodynamic evolution of the eastern Black Sea basin: insights from 2-D and 3-D tectonic modelling. *Tectonophysics*, **350**, 157–179.

MONTARON, B. & TAPPONNIER, P. 2010. A quantitative model for salt deposition in actively spreading basin. AAPG Search and Discovery Article 30117, presented at the AAPG International Conference and Exhibition 15–18 November, Rio de Janeiro, Brazil.

MOREIRA, J.L.P., MADEIRA, C.V., GIL, J.A. & MACHADO, M.A.P. 2007. Bacia de Santos. *Boletim de Geociencias da Petrobras, Rio de Janeiro*, **15**, 531–549.

MOULIN, M., ASLANIAN, D. & UNTERNEHR, P. 2010. A new starting point for the south and equatorial Atlantic ocean. *Earth-Science Reviews*, **98**, 1–37.

MUNIZ, M. 2013. *Tectono-stratigraphic evolution of the Barremian–Aptian continental rift carbonates in southern Campos Basin, Brazil*. PhD thesis, Royal Holloway, University of London.

PETERSOHN, E. & ABELHA, M. 2013. *Geological Assessment of Libras Prospectus – Brasil Rounds*. ANP, Brazil, http://www.brasil-rounds.gov.br/arquivos/Seminarios_P1/Apresentacoes/partilha1_tecnico_ambiental_ingles.pdf

PLATT, N.H. & WRIGHT, V.P. 1991. Lacustrine carbonates: facies models, facies distributions and hydrocarbon aspects. *In*: ANADON, P, CABRERA, L.L. & KELTS, K. (eds) *Lacustrine Facies Analysis*. International Association of Sedimentologists, Special Publications, **13**, 57–74.

POROPAT, S.F. & COLIN, J.P. 2012. Early Cretaceous Ostracode biostratigraphy of eastern Brazil and western Africa: an overview. *Gondwana Research*, **22**, 772–798.

QUIRK, D.G., SCHODT, N., LASSEN, B., INGS, S.J., HSU, D., HIRSCH, K.K. & VON NICOLAI, C. 2012. Salt tectonics on passive margins: examples from Santos, Campos and Kwanza basins. *In*: ALSOP, G.I., ARCHER, S.G., HARTLEY, A.J., GRANT, N.T. & HODGKINSON, R. (eds) *Salt Tectonics, Sediments and Prospectivity*. Geological Society, London, Special Publications, **363**, 207–244, http://doi.org/10.1144/SP363.10

QUIRK, D.G., HERTLE, M. *ET AL.* 2013. Rifting, subsidence and continental break-up above a mantle plume in the central South Atlantic. *In*: MOHRIAK, W.U., DANFORTH, A., POST, P.J., BROWN, D.E., TARI, G.C., NEMČOK, M. & SINHA, S.T. (eds) *Conjugate Divergent Margins*. Geological Society, London, Special Publications, **369**, 185–214, http://doi.org/10.1144/SP369.20

RENAUT, R.W., OWEN, R.B., JONES, B., TIERCELIN, J.J., TARITS, C., EGO, J.K. & KONHAUSER, K.O. 2013. Impact of lake-level changes on the formation of thermogene travertine in continental rifts: evidence from Lake Bogoria, Kenya Rift Valley. *Sedimentology*, **60**, 428–468.

RESTON, T.J. 2007. The extension discrepancy at North Atlantic non-volcanic rifted margins: depth-dependent stretching or unrecognised faulting? *Geology*, **35**, 367–370.

RESTON, T.J. 2009. The extension discrepancy and synrift subsidence deficit at rifted margins. *Petroleum Geoscience*, **15**, 217–237, http://doi.org/10.1144/1354-079309-845

RESTON, T.J. 2010. The opening of the central segment of the South Atlantic: symmetry and the extension discrepancy. *Petroleum Geoscience*, **16**, 199–206, http://doi.org/10.1144/1354-079309-907

SARG, F.J., SURIAMIN, TÄNAVSUU-MILKEVICIENE, K. & HUMPHREY, J.D. 2013. Lithofacies, stable isotopic composition, and stratigraphic evolution of microbial and associated carbonates, Green River Formation (Eocene), Piceance Basin, Colorado. *AAPG Bulletin*, **97**, 1937–1966.

SCOTESE, C.R. 1997. *Continental Drift*. 7th edn. Paleomap Project, Arlington.

SIBUET, J.C. 1992. Formation of non-volcanic passive margins: a composite model applies to the conjugate Galicia and southeastern Flemish Cap margins. *Geophysical Research Letters*, **19**, 769–772.

SOUZA-EGIPSY, V., WIERZCHOS, J., ASCASO, C. & KENNETH, H. 2005. Mg–silica precipitation in fossilization mechanisms of sand tufa endolithic microbial community, Mono Lake (California). *Chemical Geology*, **217**, 77–87.

STARK, D.M. 1991. *Well Evaluation Conference Angola 1991*. Schlumberger, Paris.

TALBOT, M.R. 1990. A review of the palaeohydrological interpretation of carbon and oxygen isotopic ratios in primary lacustrine carbonates. *Chemical Geology*, **80**, 261–279.

TERRA, G.J.S., SPADINI, A.R. *ET AL.* 2010. Carbonate rock classification applied to Brazilian sedimentary basins. *Boletin Geociencias Petrobras*, **18**, 9–29.

THOMPSON, D.L., STILWELL, J.D. & HALL, M. 2015. Lacustrine carbonate reservoirs from Early Cretaceous rift lakes of Western Gondwana: pre-salt coquinas of Brazil and West Africa. *Gondwana Research*, **28**, 26–51.

TOSCA, N.J. & WRIGHT, V.P. 2014. The formation and diagenesis of Mg-clay minerals in lacustrine carbonate reservoirs. AAPG Search and Discovery Article 51002, presented at the AAPG Annual Convention and Exhibition, 6–9 April, Houston, Texas, USA, http://www.searchanddiscovery.com/documents/2014/51002tosca/ndx_tosca.pdf

TUCKER, M.E. & WRIGHT, V.P. 1990. *Carbonate Sedimentology*. Blackwell Scientific, Oxford.

UNTERNEHR, P., PÉRON-PINVIDIC, G., MANATSCHAL, G. & SUTRA, E. 2010. Hyper-extended crust in the South Atlantic: in search of a model. *Petroleum Geoscience*, **16**, 207–215, http://doi.org/10.1144/1354-079309-904

WARREN, J.K. 2010. Evaporites through time: tectonic, climatic and eustatic controls in marine and nonmarine deposits. *Earth-Science Reviews*, **98**, 217–268.

WATCHARANANTAKUL, R. & MORLEY, C.K. 2000. Syn-rift and post-rift modelling of the Pattani Basin, Thailand: evidence for a ramp-flat detachment. *Marine and Petroleum Geology*, **17**, 937–958.

WRIGHT, P.V. 2012. Lacustrine carbonates in rift settings: the interaction of volcanic and microbial processes on carbonate deposition. *In*: GARLAND, J., NEILSON, J.E., LAUBACH, S.E. & WHIDDEN, K.J. (eds) *Advances in Carbonate Exploration and Reservoir Analysis*. Geological Society, London, Special Publications, **370**, 39–47, http://doi.org/10.1144/SP370.2

Pre-rift and synrift exhumation, post-rift subsidence and exhumation of the onshore Namibe Margin of Angola revealed from apatite fission track analysis

PAUL F. GREEN[1]* & VLADIMIR MACHADO[2]

[1]*Geotrack International Pty Ltd, West Brunswick, Victoria, Australia*

[2]*Sonangol Pesquisa & Produção, Luanda, Angola*

Corresponding author (e-mail: mail@geotrack.com.au)

Abstract: A series of cooling events in the development of the Namibe margin of Angola is defined by apatite fission track analysis data from samples of outcropping Cretaceous sandstones and crystalline Precambrian basement. Regional exhumation in the Late Carboniferous–Early Permian and Jurassic preceded Early Cretaceous rifting. Further episodes of uplift and erosion affected the margin in the Early and Late Cretaceous before it was buried by up to 2 km of post-break-up section, which was subsequently removed during Cenozoic uplift and erosion beginning between 35 and 20 Ma. This timing is consistent with published analyses of river profiles suggesting that uplift of the margin began at *c.* 30 Ma. Estimates of between 1.5 and 2 km of section removed during Cenozoic exhumation are consistent with burial depths estimated from the diagenesis of former evaporite horizons now at outcrop. These results add further support to a growing body of evidence showing that the evolution of 'passive' margins is anything but passive. The key episodes of exhumation defined in this study are broadly synchronous with events identified in many areas of the West Africa margin from Namibia to Equatorial Guinea and are regarded as representing a continent-scale response to stresses related to tectonic plate movements.

Understanding the evolution of elevated passive continental margins is important in the exploration for hydrocarbons in adjacent offshore basins for a number of reasons, including the prediction of the development of reservoir horizons as a result of uplift and erosion (MacGregor 2013). Less widely appreciated, perhaps, is the common observation that the episodes of uplift and erosion that affect onshore margins also often extend into the adjacent offshore basins (Green *et al.* 2013*a*). As a result of their regional extent, these events commonly result in low-angle unconformities that are often misinterpreted as representing periods of stability or non-deposition, whereas the application of palaeothermal methods (e.g. apatite fission track analysis (AFTA®) or vitrinite reflectance) reveals significant heating (due to burial) and cooling (due to exhumation) during the corresponding time intervals (Green & Duddy 2012; Green *et al.* 2013*a*). Exhumation in offshore basins can have serious consequences for exploration (Corcoran & Doré 2005), including regional tilting leading to the remigration of hydrocarbon accumulations and the effects of uplift leading to seal breach and a loss of charge.

Discussion of onshore rifted margins has traditionally been presented within a framework of margins that have been permanently uplifted since before break-up (e.g. Weissel & Karner 1989) and this view has dominated geomorphological and thermochronological studies of continental margins for many years (e.g. Ollier & Marker 1985; Gilchrist & Summerfield 1991, 1994; van der Beek *et al.* 1994, 1995, 2002; Gallagher & Brown 1997, 1999*a*, *b*; Ollier & Pain 1997; Gallagher *et al.* 1998; Bishop 2007). However, a growing body of evidence now suggests that the evolution of 'passive' margins is anything but passive (Paton 2012; Green *et al.* 2013*a*). In contrast with simple models in which offshore basins continuously subside while onshore margins are permanently uplifted, recent studies of a number of margins have suggested that regions now represented by onshore elevated margins have undergone post-rift subsidence and burial following break-up and have been uplifted more recently (e.g. Machado 2007, 2012; Japsen *et al.* 2012*a*, *b*; Green *et al.* 2013*a*, *b*). In many regions, offshore basins have experienced uplift and erosion at the same time as the now onshore region, with major implications for the timing of hydrocarbon generation and the magnitude of maturity levels, in addition to the other consequences of exhumation discussed above. Therefore a detailed investigation of burial and uplift histories in passive margin basins forms an important component of exploration for hydrocarbons.

In this study, AFTA data in samples of outcropping Cretaceous sandstones and crystalline

From: SABATO CERALDI, T., HODGKINSON, R. A. & BACKE, G. (eds) 2017. *Petroleum Geoscience of the West Africa Margin*. Geological Society, London, Special Publications, **438**, 99–118.
First published online December 23, 2015, http://doi.org/10.1144/SP438.2

Precambrian basement from the Namibe margin of Angola were used to assess the tectonic development of this relatively little-studied margin.

Geological background

Early Cretaceous rifting resulted in the segmentation of the South Atlantic margin of Africa defined by numerous oceanic fracture zones, the Namibe margin representing the southern portion of the Angolan margin south of the Benguela Fault Zone. Little detail has been published on the regional geology of the Namibe margin, but Moulin *et al.* (2013) and Quirk *et al.* (2013) have provided insights into the development of the margin, while Guiraud *et al.* (2010) provided a detailed description of the Benguela margin immediately to the north. According to Quirk *et al.* (2013), rifting began in the Valanginian (138 Ma), with break-up at the Barremian–Aptian boundary (123 Ma). Gindre-Chanu *et al.* (2015) have described Aptian evaporites exposed at outcrop on the Namibe margin; these display conspicuous effects of diagenetic alteration under mesogenetic conditions as a result of the effects of deeper burial by up to 1.5–2 km prior to subsequent exhumation. These evaporites are overlain by post-break-up fluvial to marginal marine upper Aptian to Albian carbonate and clastic sediments, overlain unconformably by Campanian to Maastrichtian marine limestones and volcanic rocks. The marine nature of the post-salt section demonstrates post-break-up subsidence of this margin. The inland geology is dominated by Precambrian granitic basement capped unconformably by the epicratonic sediments (arkose and quartzite) of the Precambrian Chela Group (Pereira *et al.* 2011), forming a prominent escarpment at an elevation >2 km. Inboard of the main basin-bounding fault, the Campanian–Maastrichtian units lie directly on the basement, covering a prominent Late Cretaceous planation surface.

The geomorphology of the region is dominated by the Bié dome, which formed during the Miocene (Roberts & White 2010) and reaches elevations >2.5 km above sea-level; it is truncated towards the Atlantic margin, forming a conspicuous escarpment separating the inland plateau from the coastal plain (Fig. 1). Roberts & White (2010) reported that river profiles draining the inland plateau are immature, suggesting recent uplift, and their numerical analysis of these profiles suggested the early initiation of uplift at *c.* 30 Ma, with a more pronounced phase in the last 10 myr.

Published thermochronology studies for Namibia (Gallagher & Brown 1999*a*, *b*; Raab *et al.* 2002, 2005; Brown *et al.* 2014) have provided extensive evidence of Early and Late Cretaceous cooling episodes. Cooling in these studies was interpreted solely in terms of exhumation, with up to 5 km of section removed from the coastal locations. However, it seems likely that a range of other processes (e.g. igneous intrusions, hydrothermal activity, increased heat flow) may have contributed to the heating prior to the Cretaceous cooling recorded in these studies (Green *et al.* 2013*a*) and that the true amounts of section removed may have been smaller, although Rouby *et al.* (2009) suggested that the estimated amounts of denudation based on constant heat flow were consistent with the thickness of sediments preserved offshore.

Most of these studies did not consider the nature of the rocks that were removed during Cretaceous exhumation. Many samples from the basement outcrops along the northern Namibian coast did not cool below *c.* 110°C until the Late Cretaceous (*c.* 80 Ma). As many of these samples were collected adjacent to Early Cretaceous Etendeka volcanic rocks, which were erupted on to a land surface on which Late Jurassic–Early Cretaceous sands were deposited, the Late Cretaceous palaeotemperatures >110°C must have been related, at least in part, to burial by larger thicknesses of Etendeka basalts and/or post-Etendeka sedimentary units. Thus these data show that the margin subsided after Early Cretaceous rifting and was buried prior to the onset of Late Cretaceous exhumation. This scenario was adopted by Dauteuil *et al.* (2013) in their modelling study of the development of the Namibia margin.

Previous studies in Namibia have tended to downplay Cenozoic exhumation, but Laughland *et al.* (2008) reported evidence for the Cenozoic removal of 1100 m of section in the offshore Lüderitz Basin. To the north, in the Inner Kwanza Basin and adjacent basement rocks of northern Angola, Machado (2007) reported further evidence of the Mesozoic exhumation of basement, Early Cenozoic reburial and Late Cenozoic exhumation based on AFTA and vitrinite reflectance data. These data were interpreted in terms of the erosional removal of 1000 m of section during the Miocene in the Inner Kwanza Basin, compared with 900 m estimated using fluid inclusion microthermometry (Machado 2007). These values have been confirmed by the analysis of stacking velocity data, indicating 500–1000 m of removed section along the Congo and northern Angolan continental shelves (Walford & White 2005). Based on AFTA data in sedimentary units of the onshore Kwanza Basin and underlying basement samples, Jackson *et al.* (2005) reported evidence for three phases of cooling, representing exhumation that began in the intervals *c.* 150, 100–70 and 20–10 Ma (although they suggested that some Miocene cooling may have been due to processes other than exhumation).

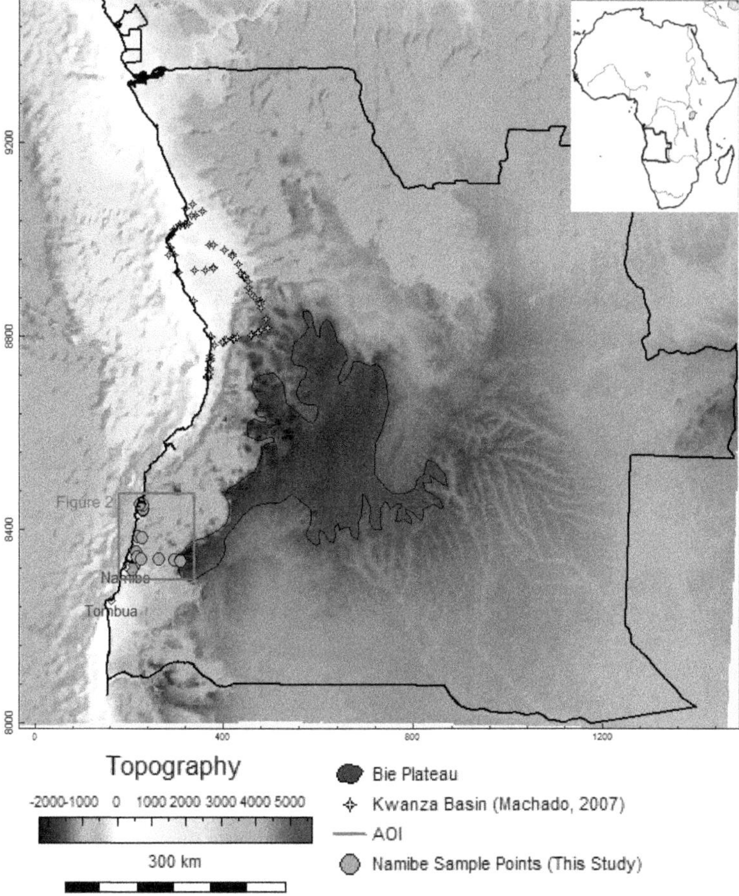

Fig. 1. Topography of Angola, showing the area discussed in this paper (outlined rectangle, shown in more detail in Fig. 2) and the locations of samples, including the samples analysed by Machado (2007). The Bié dome is the dominant feature of the region and reaches elevations >2 km above sea-level. AOI, area of interest.

Further north, in Gabon, Walgenwitz *et al.* (1992) reported evidence for Early Cretaceous (*c.* 110 Ma), Late Cretaceous (84–76 Ma), Eocene (*c.* 45 Ma) and Miocene (25–21 Ma) phases of cooling from apatite fission track (AFT) data, each of which was interpreted in terms of phases of exhumation. These four phases of exhumation are remarkably similar to those identified from AFTA data in the Rio Muni well and onshore outcrop samples from Equatorial Guinea, to the north of Gabon, by Turner *et al.* (2008) beginning in the intervals 110–95, 85–70, 45–35 and 15–10 Ma. Only the most recent of these events is discordant between the Gabon and Equatorial Guinea studies.

Thus from Namibia in the south to Equatorial Guinea in the north, AFT studies from the West Africa margin have consistently defined four major periods of cooling. Although not all episodes are identified at every location, the results reviewed here suggest that the effects of the four episodes are likely to have been felt over much, if not all, of the margin.

Sample details

The sampled rocks were from the Namibe sedimentary basin and its corresponding basement/ hinterland. The sample locations (Fig. 2) spanned the region from the coast to the inland Great Escarpment of Angola at elevations from close to sea-level up to *c.* 1600 m above sea-level (asl). The rock samples consisted of different lithologies: granite–granodiorite and quartzites from the basement (oldest rocks); sandstones and calcareous sandstones from the sedimentary basin succession ranging in age from Aptian to Maastrichtian; and one sample of basanite reported to contain apatite phenocrysts, but which did not yield apatite suitable for analysis

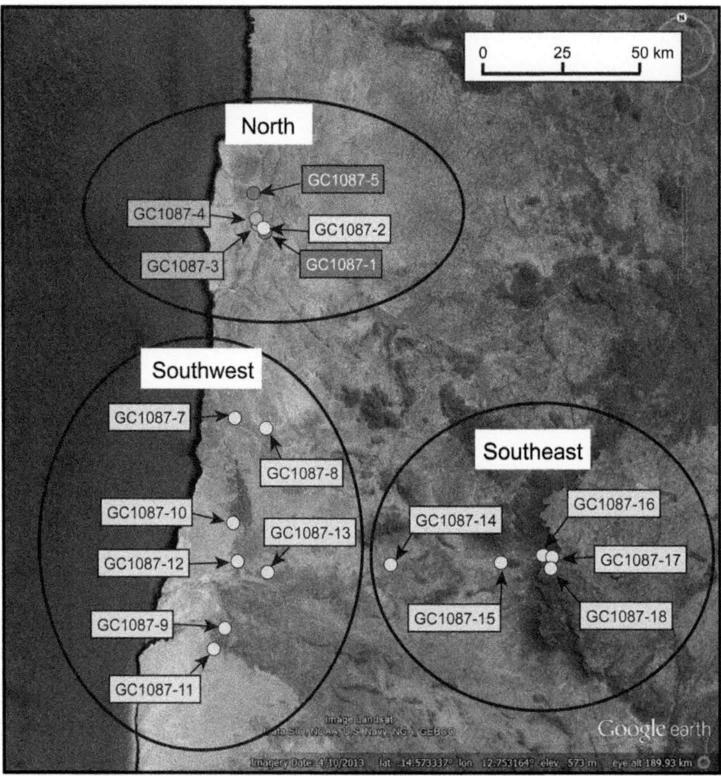

Fig. 2. Locations of samples analysed in this study, superimposed on a Google Earth image of the region. Samples of Cretaceous sandstones are shown with a green background (see online version for colours); basement samples have a white background.

(semi-quantitative X-ray analysis showed the phenocryst mineral to be a K, Ca, Al-silicate mineral, possibly a zeolite.). Of the remaining 17 samples analysed, 12 provided excellent yields of apatite, with lesser yields in five samples. In all the samples that yielded apatite, the data quality was sufficiently high to ensure that reliable thermal history constraints (within the stated confidence limits) were obtained in all 17 samples.

Results

Basic AFTA parameters (fission track age and mean track length) in the 17 samples that yielded apatite are summarized in Table 1, where details of the sample locations and stratigraphic age are also provided. The measured AFT ages varied between 137.1 ± 18.8 and 294.2 ± 30.2 Ma and the mean track lengths were generally between 11 and 13 μm. Fission track ages and mean track lengths are plotted in Figure 3, where the fission track age measured in each sample is contrasted with the respective depositional age. Although the AFT ages do not indicate the timing of a specific event

(Green *et al.* 2013*a*), values consistently younger than 300 Ma measured in samples of Precambrian basement immediately show that the AFTA data in these samples are dominated by Phanerozoic events and preserve little or no information on the earlier (Precambrian) history. All the samples of Mesozoic sandstones give AFT ages older than the depositional age, showing that the apatites in these samples retain the fission tracks formed before they were deposited in the Mesozoic sandstones. The mean track lengths in these samples are significantly less than those expected if they had remained at surface temperatures since deposition (i.e. the default thermal history; Green & Duddy 2012). This could be due either to a dominance of shorter tracks formed prior to deposition or to post-depositional heating; it is important to distinguish the effects of pre- and post-depositional heating in these samples before any firm conclusions are made regarding the post-depositional history of these samples.

The mean track lengths in the 17 outcrop samples are plotted against the corresponding AFT age in Figure 4. Also shown in this figure are previously published AFT data from Namibia (Raab *et al.* 2002, 2005). The different datasets define a consistent trend, suggesting that a common overall style of thermal history is applicable over the region.

The AFT ages in the 17 samples from the Namibe margin are plotted against elevation in Figure 5. The lack of any coherent relationship between these two parameters suggests that elevation alone does not exert a first-order control on thermal history in this region. AFT ages are mapped in Figure 6, which highlights a trend towards older ages in the NW and east and younger ages in the SW.

Thermal history interpretation of AFTA data

Background

AFTA is based on the analysis of radiation damage features ('fission tracks') in apatite. These tracks are produced continuously through time by the spontaneous fission of ^{238}U atoms present as impurities. Tracks form with lengths within a finite range and, as the host rock is heated, the tracks are shortened at a rate that depends on the temperature as a result of the progressive repair of the radiation damage. In a sample heated to a palaeothermal maximum and then cooled, those tracks formed prior to the onset of cooling are all shortened to a length determined by the maximum palaeotemperature and are then 'frozen' at that length when cooling begins. Tracks formed after cooling began are longer as a result of the lower prevailing temperatures. The proportion of short to long tracks is therefore diagnostic of the time when cooling began in relation to the total time over which the tracks have been retained. In samples that are heated to a sufficiently high temperature (typically *c.* 110°C, but dependent on the apatite composition), all the tracks formed prior to cooling are totally annealed (i.e. all damage is repaired) and the sample only begins to retain tracks once the sample cools below the critical temperature once more. Only a minimum estimate of the maximum palaeotemperature is possible in such samples.

Tracks are revealed on etching by their intersections with a polished surface and the areal density of tracks coupled with the uranium content provides an indication of the time over which the tracks have accumulated based on the laws of radioactive decay. Measurement of the uranium content and calibration against age standards allows the conversion of the measured track density to a fission track age. However, because of the reduced track lengths and the corresponding reduction in the probability of tracks intersecting the polished surface, the measured age does not represent the actual time over which tracks have been retained. Instead, the measured age is determined by the balance between the production of tracks by spontaneous fission and the reduced probability of revelation as a result of the reduction in track lengths (i.e. the thermal history).

Thermal history constraints are extracted from AFTA data by forward modelling the parameters (fission track age and the distribution of track lengths) expected from a range of viable thermal history scenarios, comparing these with the measured data and iterating towards a best-fit history with corresponding 95% confidence limits. In samples that have been heated to a maximum palaeotemperature and then cooled, the data are dominated by the palaeothermal maximum and the prior history is overprinted. For this reason, it is not possible to constrain the entire history and interpretation is instead focused on defining the maximum palaeotemperature and the time at which cooling from that maximum began. If a sample is reheated following the initial cooling, it is often possible to define a later event from the shortening of tracks formed during this part of the history, provided the peak palaeotemperature is lower than the earlier maximum. In some cases it is possible to define three discrete events. The AFTA technique has been reviewed in detail by Green & Duddy (2012) and Green *et al.* (2013*a*).

Interpretation of data from this study

Details of the thermal history constraints extracted from the AFTA data in each sample analysed for this study, using the outlined approach, are summarized in Table 1. We reiterate that we do not attempt to constrain the whole history. Instead, these thermal history constraints are presented in terms of the maximum palaeotemperature and the timing of cooling in one or more discrete episodes because these aspects of the history dominate the data. For each sample, the range of palaeotemperatures and timing of cooling quoted in Table 1 represent 95% confidence intervals.

The AFTA data in all samples of Mesozoic sedimentary units can be explained in terms of a single episode of heating and cooling after deposition. In these samples, the maximum post-depositional palaeotemperature and the onset of cooling are defined primarily from the form of the track length distribution. Many of these samples also provide definition of two pre-depositional episodes (Table 1). Several samples contain two distinct age populations, signifying derivation from provenance regions characterized by different thermal histories. As summarized in Table 1, these different age populations have been separated and interpreted separately. With the AFTA data in each sample capable of defining at least two palaeothermal episodes, up to four episodes during the

Table 1. *AFTA data, sample details and associated thermal history interpretations; Namibe margin of Angola*

Sample No.	Latitude and longitude	Stratigraphic division[1]	Stratigraphic age[1] (Ma)	Elevation (m asl)	ρ_D^2 (10^6 tracks cm^{-2})	ρ_s^2 (10^6 tracks cm^{-2})	ρ_i^2 (10^6 tracks/ cm^2)	Fission track age[3] (Ma)	$P(\chi^2)^4$ (%) (no. of grains)	Mean track length[5] (μm)	Standard deviation[6] (μm)	Thermal history constraints[7]
GC1087-1	−14.107700° S 12.503020° E	Pre-salt sandstone	130–112	200	1.375 (2189)	1.081 (845)	1.869 (1461)	155.9 ± 19.7	<1 (20)	12.1 ± 0.2 (91)	1.56	*>105°C; >165 Ma 100–105°C; 165–95 Ma 65–80°C; 40–0 Ma*
GC1087-2	−14.107700° S 12.503020° E	Basement	540		1.365 (2189)	0.857 (789)	1.415 (1303)	155.8 ± 11.8	<1 (20)	12.6 ± 0.2 (103)	1.75	*>105°C; >195 Ma 100–105°C; 195–105 Ma 60–80°C; 85–20 Ma*
GC1087-3	−14.105290° S 12.499410° E	Albian sand	112–100	178	1.354 (2189)	1.495 (1115)	1.576 (1175)	231.5 ± 18.9	<1 (20)	12.4 ± 0.1 (100)	1.42	
GC1087-3 (older ages)								299.0 ± 18.0	43 (11)	12.4 ± 0.3 (21)	1.47	*>100°C; 320–270 Ma 80–95°C; 280–140 Ma*
GC1087-3 (young ages)								174.9 ± 15.1	13 (9)	12.4 ± 0.2 (77)	1.38	*>100°C; 210–90 Ma 90–95°C; 155–60 Ma 60–75°C; 47–0 Ma*
GC1087-4	−14.095930° S 12.493620° E	Albian sand	112–100	164	1.344 (2189)	2.245 (1892)	1.811 (1526)	294.2 ± 30.2	<1 (20)	12.4 ± 0.2 (104)	1.54	
GC1087-4 (older ages)								284.1 ± 16.2	12 (14)	12.2 ± 0.2 (79)	1.50	*>100°C; 390–285 Ma 80–100°C; 320–700 Ma*
GC1087-4 (young ages)								107.5 ± 14.0	34 (3)	13.9 ± 0.5 (7)	1.42	*>100°C; 165–85 Ma 60–80°C; 900 Ma*
GC1087-5	−14.012960° S 12.506810° E	Pre-salt sandstone	130–112	86	1.333 (2189)	1.658 (1523)	1.740 (1599)	242.1 ± 20.0	<1 (20)	12.4 ± 0.1 (105)	1.33	
GC1087-5 (oldest ages)								389.3 ± 29.5	87 (5)	12.7 ± 0.3 (16)	1.28	*>100°C; 560–375 Ma*
GC1087-5 (intermediate ages)								233.9 ± 11.0	80 (12)	12.3 ± 0.1 (79)	1.25	*>100°C; 320–240 Ma 75–90°C; 265–70 Ma*
GC1087-5 (young ages)								129.8 ± 11.4	43 (3)	12.2 ± 0.3 (10)	0.95	*>100°C; 200–130 Ma 60–75°C; 45–0 Ma*
GC1087-6	−13.994810° S 12.450930° E	Maastrichtian basanite	71–66	53								No apatite
GC1087-7	−14.606310° S 12.394680° E	Campanian– Maastrichtian	84–66	397	1.323 (2189)	1.723 (590)	2.839 (972)	147.9 ± 18.3	<1 (20)	12.7 ± 0.1 (102)	1.50	
GC1087-7 (older ages)								255.6 ± 23.4	26 (6)	12.7 ± 0.2 (5)	1.52	*>100°C; 360–220 Ma*
GC1087-7 (young ages)								113.9 ± 10.1	10 (13)	12.7 ± 0.2 (97)	1.47	*>100°C; 150–100 Ma 60–75°C; 37–0 Ma*
GC1087-8	−14.635010° S 12.480970° E	Basement	540	453	1.570 (2421)	0.615 (180)	0.745 (218)	251.6 ± 33.7	4 (22)	10.9 ± 0.3 (108)	2.62	*>105°C; >210 Ma 95–105°C; 210–75 Ma 50–85°C; 60–0 Ma*
GC1087-9	−15.160110° S 12.308640 E	Basement	540	149	1.566 (2421)	0.920 (136)	2.522 (373)	113.6 ± 22.3	<1 (11)	11.5 ± 0.3 (45)	2.29	*>105°C; >140 Ma 95–105°C; 140–50 Ma 70–55°C; 45–5 Ma*

Sample	Location	Stratigraphy	Stratigraphic age (Ma)	No. grains	ρ_D	ρ_s	ρ_i	Central age (Ma)	$P(\chi^2)$	Mean track length (µm)	Std dev	Thermal history
GC1087-10	−14.885070° S 12.657140° E	Campanian–Maastrichtian	84–66	270	1.562 (2421)	1.867 (686)	4.152 (1526)	140.2 ± 14.4	<1 (20)	13.2 ± 0.1 (104)	1.36	
GC1087-10 (older ages)								184.2 ± 15.4	11 (9)	13.4 ± 0.3 (21)	1.40	*>100°C; 225–165Ma*
GC1087-10 (young ages)								88.4 ± 7.5	87 (9)	12.8 ± 0.3 (36)	1.57	*>100°C; 125–70 Ma* *50–75°C; 40–0 Ma*
GC1087-11	−15.214590° S 12.273750° E	Campanian–Maastrichtian	84–66	149	1.558 (2421)	1.299 (631)	3.135 (1523)	137.1 ± 18.8	<1 (20)	12.6 ± 0.2 (106)	1.76	
GC1087-11 (older ages)								278.0 ± 26.9	19 (6)	12.1 ± 0.4 (24)	1.70	*>100°C; 650–200 Ma* *80–100°C; 325–60 Ma*
GC1087-11 (young ages)								87.2 ± 5.7	47 (14)	12.7 ± 0.2 (82)	1.77	*>120°C; 120–85 Ma* *80–100°C; 95–40 Ma* *40–75°C; 40–0 Ma*
GC1087-12	−14.985590° S 12.369270° E	Basement	540	277	1.553 (2421)	0.860 (152)	0.927 (164)	180.2 ± 61.7	8 (6)	10.9 ± 0.5 (33)	2.83	>105°C; >200 Ma 95–105°C; 200–40 Ma 30–95°C; 100–0 Ma
GC1087-13	−15.019820° S 12.444320° E	Basement	540	425	1.549 (2421)	2.485 (1323)	3.599 (1916)	206.8 ± 9.4	11 (20)	12.2 ± 0.2 (111)	2.20	>100°C; c. 300 Ma? 95–100°C; 205–130 Ma 60–80°C; 75–15 Ma
GC1087-14	−15.031180° S 12.797950° E	Basement	540	416	1.545 (2421)	1.085 (396)	1.342 (490)	242.9 ± 21.8	12 (20)	11.9 ± 0.2 (108)	2.31	>105°C; >300 Ma 100–105°C; 330–160 Ma 80–90°C; 110–40 Ma
GC1087-15	−15.055180° S 13.108490° E	Basement	540	504	1.541 (2421)	0.999 (418)	1.054 (441)	285.9 ± 25.9	6 (20)	11.5 ± 0.2 (108)	2.12	>95°C; c. 300 Ma? 85–95°C; 230–70 Ma 50–80°C; 60–0 Ma
GC1087-16	−15.047600° S 13.226860° E	Basement	540	979	1.537 (2421)	0.280 (215)	0.335 (257)	247.7 ± 23.9	53 (15)	10.7 ± 0.3 (63)	2.26	>100°C; >215 Ma 85–100°C; 215–25 Ma 45–80°C; 35–0 Ma
GC1087-17	−15.052600° S 13.236430° E	Basement	540	1249	1.533 (2421)	4.309 (442)	5.801 (595)	219.9 ± 15.1	83 (11)	11.9 ± 0.3 (50)	1.78	>100°C; c. 300 Ma 85–100°C; 250–80 Ma 50–80°C; 80–0 Ma
GC1087-18	−15.076790° S 13.233070° E	Basement	540	1628	1.528 (2421)	1.828 (268)	2.155 (316)	260.0 ± 36.6	<1 (14)	12.5 ± 0.2 (76)	1.68	>105°C; 370–250 Ma 80–90°C; 235–110 Ma 45–70°C; 100–0 Ma

[1] All numerical values for stratigraphic ages were assigned following Gradstein *et al.* (2012).

[2] ρ_s, spontaneous track density; ρ_i, induced track density; ρ_D, glass dosimeter track density. Numbers in parentheses show the number of tracks counted in determining all track densities.

[3] Central age (Galbraith 2005) used for samples containing a significant spread in single grain ages ($P(\chi^2) < 5\%$), otherwise the pooled age is quoted. All ages were calculated using the zeta calibration approach of Hurford & Green (1983), using zeta values for CN5 glass of 380.4 ± 5.7 (samples GC1087-1 to GC1087-7) or 392.9 ± 7.4 (samples GC1087-8 to GC1087-18). All errors quoted at ± 1σ. All analytical details are as described by Green (1986), with the exception that thermal neutron irradiations for this study showed a significant flux gradient and the appropriate value of ρ_D for each sample was determined by linear interpolation through the stack of grain mounts. Cl contents were determined by electron microprobe.

[4] Probability that all single grain ages belong to a single population (Galbraith 2005).

[5] Numbers in parentheses show the number of track lengths measured. All errors quoted at ± 1 sigma.

[6] Standard deviation of the track length distribution.

[7] Thermal history solutions derived from AFTA data based on assumed heating and cooling rates of 1 and 10°C Ma^{-1}, respectively. Quoted ranges correspond to correspond to ± 95% confidence limits on the maximum/peak palaeotemperature and the onset of cooling in discrete episodes of heating and cooling. Conditions shown in italics represent pre-depositional cooling events in samples of sedimentary rock. Where samples contain discrete age populations, solutions are provided for each population wherever possible. Isolated grains in these samples may not fall into either population and do not feature in the interpretations.

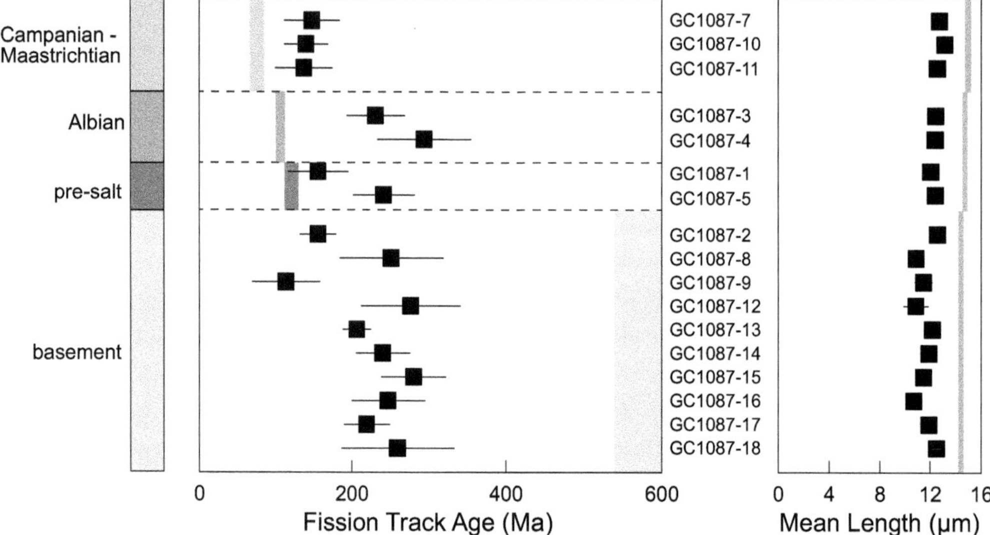

Fig. 3. Apatite fission track ages compared with depositional ages and mean track lengths compared with values if samples had never been heated above present day temperatures (i.e. the default thermal history; Green & Duddy 2012).

pre-depositional history may be obtained from the multiple age populations in a single sample. Alternatively, different age populations in the same sample may define a common palaeothermal event, but of a different magnitude. Sample GC1087-5 contains three discrete age populations, although in this case, with a smaller proportion of grains in each population, only a single pre-depositional episode can be defined for two of the populations.

The AFTA data in most of the basement samples require three discrete episodes of post-depositional heating and cooling to explain all facets of the

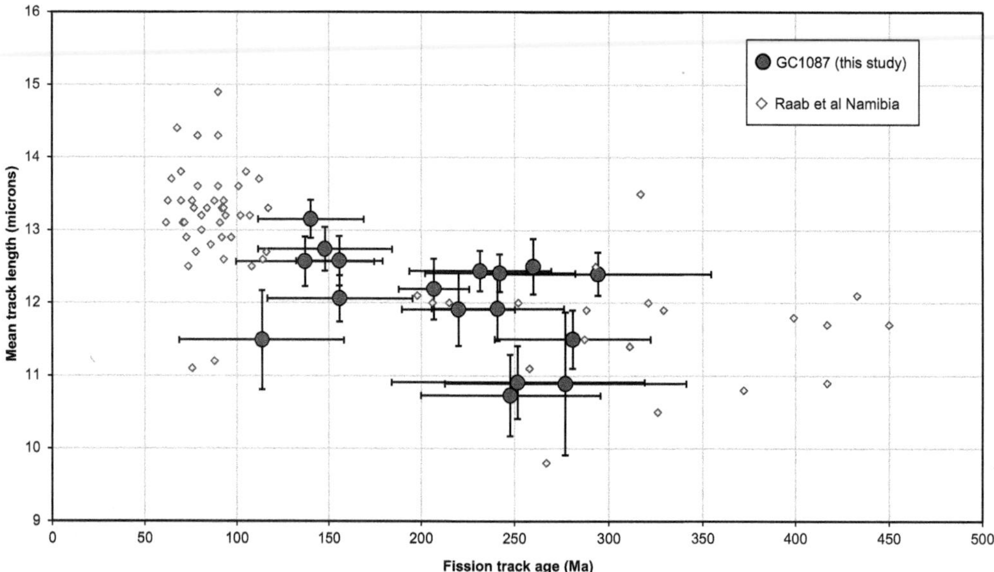

Fig. 4. Mean track length data plotted against fission track age for apatites analysed in this study together with data in outcrop samples from Namibia (Raab *et al.* 2002, 2005). Data from this study follow the same general trend as the Namibia data, suggesting a broadly similar thermal history.

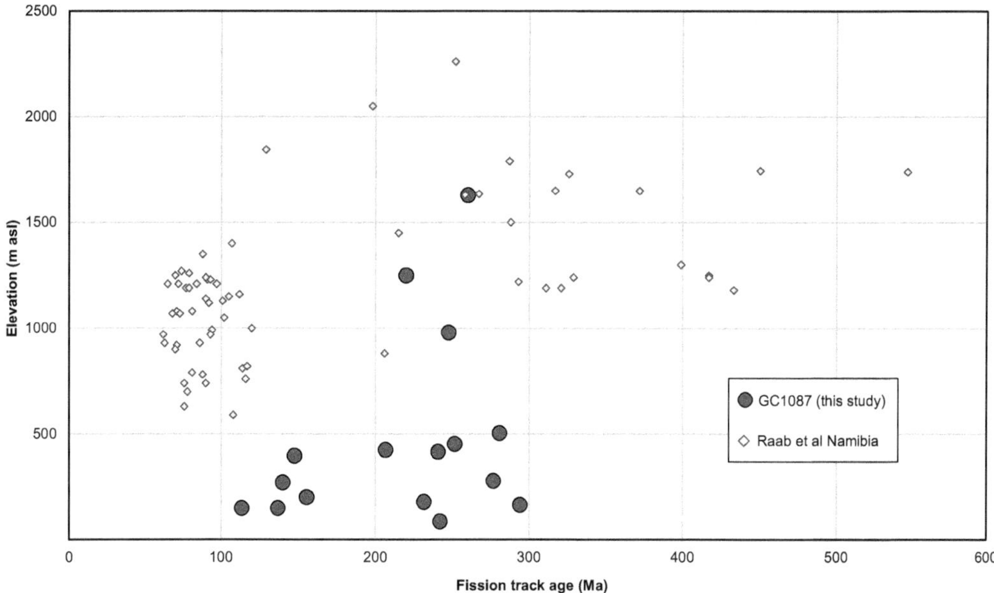

Fig. 5. Apatite fission track ages in samples analysed in this study plotted against elevation above sea-level. Data in outcrop samples from Namibia (Raab *et al.* 2002, 2005) are also shown. Results from this study show no significant relationship with elevation.

data. The earliest episode is typically defined by the AFT age reduction, while the later events are defined from the form of the track length distribution. In some cases, precise constraints on one of the events cannot be defined in detail, usually because the earlier history is masked by later events.

Timing constraints derived from the AFTA data in each sample (summarized in Table 1) are compared in Figure 7, with both pre-depositional and post-depositional episodes shown for the Mesozoic sandstone samples. The samples in Figure 7 are divided into three regions (indicated in Fig. 2), with the results in samples from each region plotted in stratigraphic order. In general, these results display a high degree of consistency between different samples and, on this basis, together with the consistent trend shown by the data in Figure 4, seems reasonable to assume that the AFTA data in all samples within the study region represent the effects of regionally synchronous cooling episodes. We have therefore sought to identify the minimum number of common episodes that can satisfy the timing constraints on cooling identified from AFTA data in all samples.

On this basis, a comparison of the timing constraints in individual samples in Figure 7 shows that almost all samples show evidence (including during the pre-depositional history for earlier episodes in Mesozoic sandstone samples) of cooling

that began in one of the following intervals (illustrated by the vertical columns in Fig. 7):

320–285 Ma	Late Carboniferous–Early Permian
195–165 Ma	Jurassic
35–20 Ma	Latest Eocene–Early Miocene

Note that the Carboniferous–Early Permian and Jurassic episodes, which dominate the data from the basement samples, are also very strongly expressed in the pre-depositional histories of the samples from Mesozoic sandstones.

An earlier episode is also recognized, in which cooling began in the interval:

560–375 Ma   Infra-Cambrian–Late Devonian

This event is only recognized in the pre-depositional history of two Mesozoic sandstone samples, in which the effects of later events have not been sufficient to overprint the early history.

In addition, most samples also show evidence of Cretaceous cooling, but the onset of cooling identified from AFTA in individual samples shows some possible differences across the region. Samples of Early Cretaceous sedimentary units in the northern region show consistent evidence in their

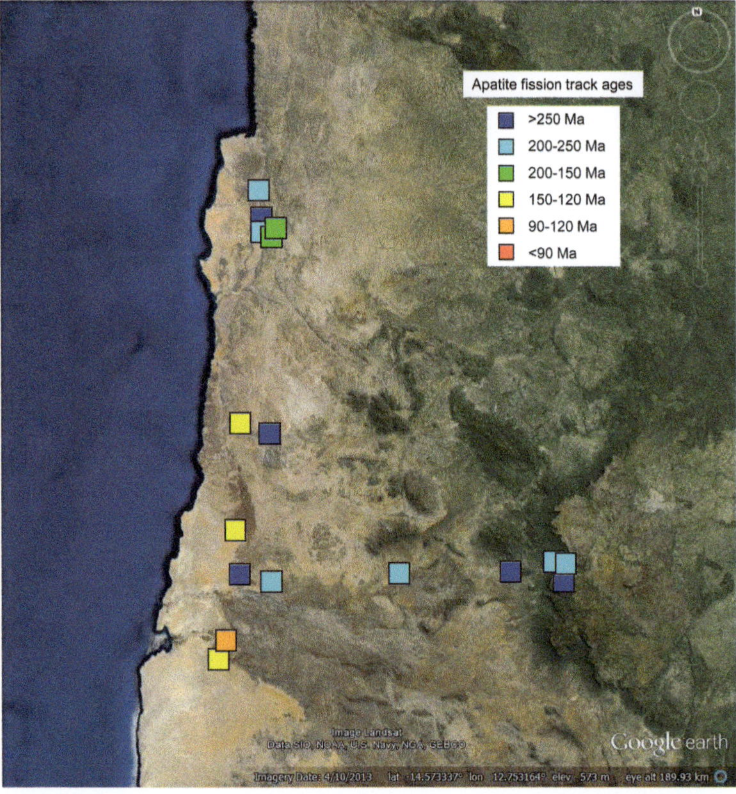

Fig. 6. Apatite fission track ages in samples analysed for this study. Youngest ages are measured in samples in the SW of the study area.

pre-depositional history of cooling, which began some time later than 155 Ma (red bars in Fig. 7, combining results from samples GC1087-1, GC1087-3 and GC1087-4). As the oldest of these units are attributed to the interval 130–112 Ma (pre-salt; Table 1), cooling must have begun at the latest prior to 112 Ma, showing that cooling must have begun at some time in the interval:

155–112 Ma   Latest Jurassic–Aptian

In contrast, samples of apatites from Late Cretaceous sedimentary units in the southwestern region show evidence of pre-depositional cooling beginning in the interval 120–100 Ma (combining constraints from AFTA data in samples GC1087-7, GC1087-10 and GC1087-11). Sample GC1087-11 also shows evidence of a later pre-depositional event that began in the interval between 95 and 66 Ma (taking the younger limit from the youngest possible depositional age; Table 1). Therefore samples in this region define two discrete cooling episodes in sediment provenance regions, beginning in the intervals:

120–100 Ma   Aptian–Albian
95–66 Ma   Late Cretaceous

The timing of the Aptian–Albian cooling episode overlaps to some extent with the 'Latest Jurassic–Aptian' episode identified in the northern region, but is younger than the older limit of the depositional age of the sedimentary units in the north. Thus it is possible at the extremes of the allowed ranges that the Aptian–Albian episode in the SE and the Latest Jurassic–Aptian episode in the north represent a common episode of cooling that began in the interval 120–112 Ma. However, more complex interpretations involving diachronous cooling across the region or a series of discrete cooling episodes are also possible.

Basement sample GC1087-9 shows cooling that began in the interval 140–50 Ma. As this sample is within a region of Campanian–Maastrichtian

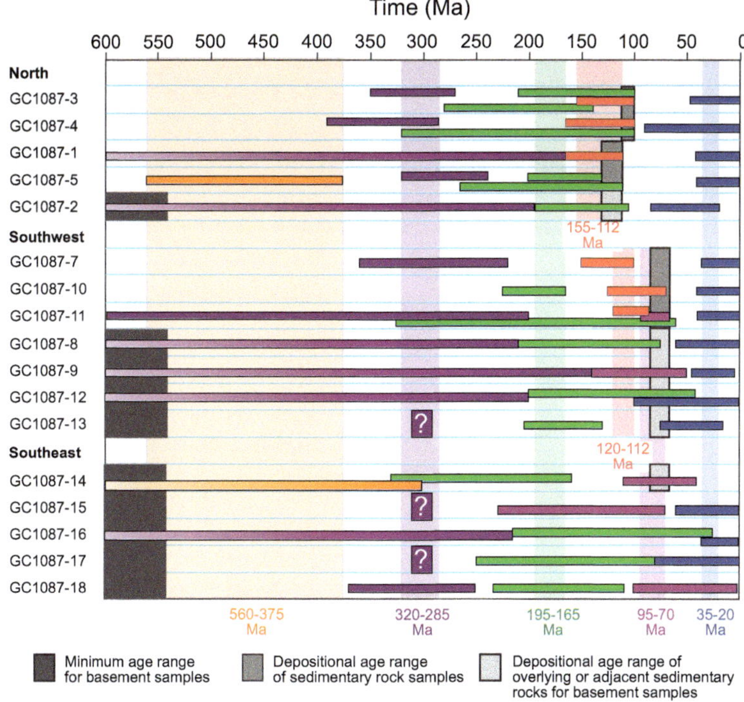

Fig. 7. Onset of cooling determined from AFTA in individual samples from this study. Cooling events are attributed to regional cooling episodes by colour. Multiple cooling episodes are identified in individual samples as described in the text. In some samples, multiple age populations provide different solutions.

sedimentary cover, we interpret this cooling to have occurred prior to the deposition of this sedimentary cover, i.e. prior to 66 Ma at the very latest. Similarly, AFTA data in basement sample GC1087-14 in the SW region define cooling from 80 to 90°C beginning between 110 and 40 Ma. This sample was collected from close to the Campanian–Maastrichtian land surface and again we interpret cooling as prior to the deposition of the sedimentary cover, i.e. prior to 66 Ma at the youngest limit. Sample GC1087-15 from a location *c.* 30 km east of sample GC1087-14, also cooled from 85 to 95°C in the interval 230–70 Ma. Given the similarity in palaeotemperatures in these two samples, collected from similar elevations, it seems reasonable to identify cooling in both samples, together with that in sample GC1087-9, in terms of a common event, in which case cooling must have begun between 110 and 70 Ma. We therefore also attribute these cooling episodes to the Late Cretaceous episode (Fig. 7).

Finally, the onset of cooling (100–0 Ma) in the most recent episode identified in sample GC1087-18 overlaps both this Late Cretaceous episode and the Latest Eocene–Early Miocene episode and could, in principle, represent either episode. Samples GC1087-15 to GC1087-18 were collected over an elevation range of >1100 m and comparing the thermal history solutions in these samples in Table 1 shows little difference in the peak palaeotemperatures from which the samples cooled in the most recent event. If these palaeotemperatures were attributed to a common event, this would imply a very low palaeogeothermal gradient in this episode. However, it seems more likely that the most recent cooling identified in sample GC1087-18 represents an earlier episode than that recognized in the 'deeper' samples; this interpretation is preferred here. On this basis, the most recent cooling episode in sample GC1087-18 is correlated with the Late Cretaceous cooling episode identified in other samples.

In summary, the results of this study define at least six periods of cooling, one of which may represent two separate phases:

560–375 Ma	Infra-Cambrian–Late Devonian
320–285 Ma	Late Carboniferous–Early Permian
195–165 Ma	Jurassic
120–112 Ma	Aptian (or 155–112 and 120–100 Ma)
95–70 Ma	Late Cretaceous
35–20 Ma	Latest Eocene–Early Miocene

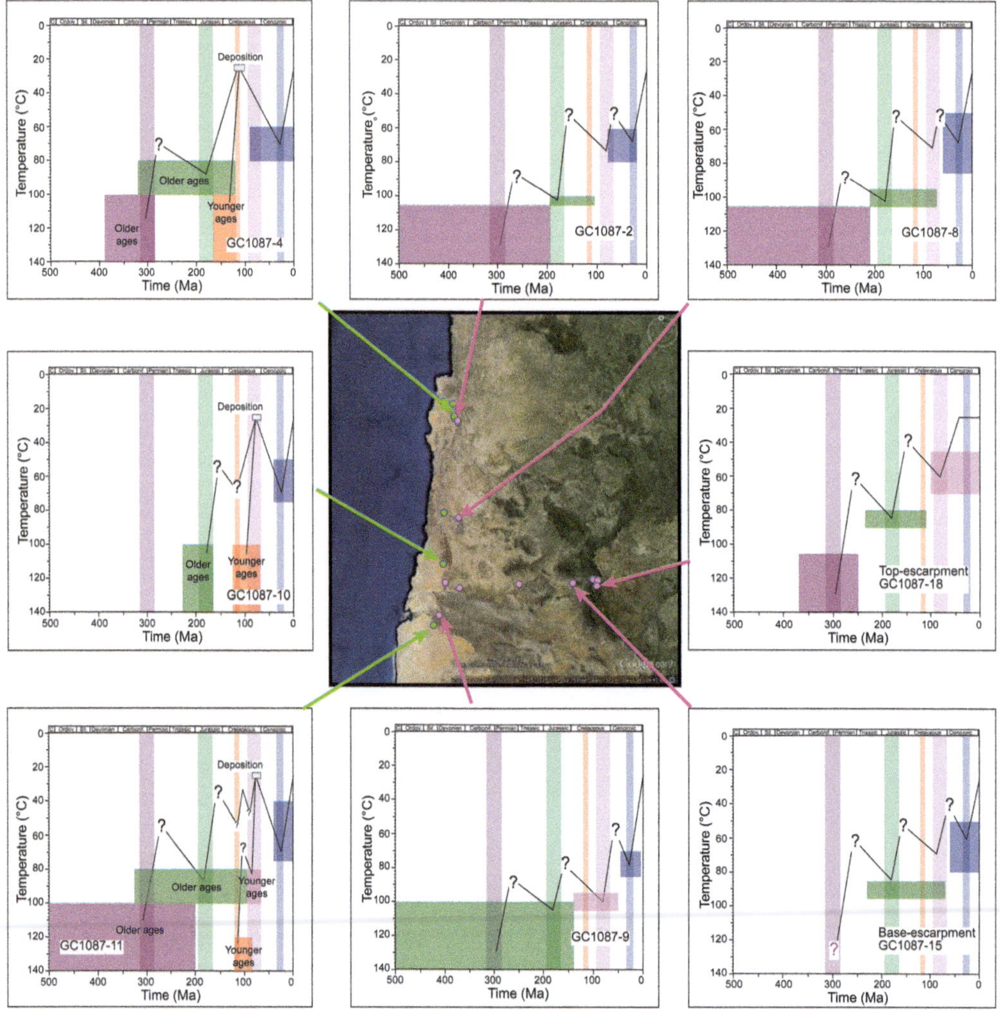

Fig. 8. Schematic representation of thermal histories derived from AFTA in selected samples around the study region. Boxes represent thermal history solutions derived from the data and listed in Table 1, coded by colour to the regional cooling episodes shown in Figure 7.

Whatever the precise details of the Early Cretaceous cooling history, data from almost all samples of the sedimentary rocks show evidence of cooling in sedimentary provenance regions shortly before deposition, implying fairly rapid cooling. This is interpreted as the exhumation of sediment provenance regions prior to the deposition of these sedimentary units.

Evidence for a final stage of cooling from temperatures generally around 60–80°C is seen in all samples (Table 1). For the sedimentary rock samples, this episode is the only post-depositional event recognized from the AFTA data. In Figure 7, this event is attributed to a cooling episode that began in the interval 35–20 Ma (Latest Eocene to Early Miocene) in all samples except sample GC1087-18, in which the most recent cooling event is attributed to the Late Cretaceous cooling episode.

Thermal history solutions derived from the AFTA data in each outcrop sample are summarized in Table 1, while Figure 8 provides a schematic illustration of the thermal history interpretation of AFTA data in selected samples analysed for this study.

Integration with regional geology

The results presented in this paper show that sedimentary units deposited along the Namibe margin in the Early and Late Cretaceous were derived from provenance terrains that had undergone major cooling shortly before the deposition of these sediments. These cooling episodes most likely resulted from the uplift and erosion of basin margins during rifting and break-up. However, only one of the analysed basement samples from coastal locations close to the margin shows any evidence of Cretaceous cooling, suggesting that Early and Late Cretaceous denudation was less pronounced in this region than elsewhere. Provenance regions for the Cretaceous sediments, which underwent greater levels of denudation, were therefore probably outside the immediate study region.

Basement samples from coastal locations were dominated by pre-rift events, as well as the Cenozoic episode identified in almost all samples. Jurassic cooling, which began between 195 and 165 Ma, presumably indicates kilometre-scale denudation, suggesting a phase of pre-rift doming as a precursor to the development of the Atlantic margin. An earlier cooling episode that began between 320 and 285 Ma (Late Carboniferous–Early Permian) coincided with a major regional hiatus between the Cape and Karoo supergroups in southern Africa, which Tankard *et al.* (2009) assigned to the interval 330–302 Ma. Effects of both Jurassic and Late Carboniferous–Early Permian episodes have been identified across many areas of southern Africa (Green *et al.* 2009, 2011, 2013*b*) and are regarded as major regional periods of exhumation in which kilometre-scale thicknesses of section were removed. The Late Carboniferous–Early Permian cooling episode resulted in the development of a regional erosion surface displaying the effects of glacial erosion characteristic of the period (Green *et al.* 2009).

Late Cretaceous cooling, which began between 95 and 70 Ma, is identified in basement samples from inland locations (samples GC1087-14 to GC1087-18, Fig. 7) as well as one basement sample closer to the coast (GC1087-9). This cooling episode is contemporaneous with a prominent Late Cretaceous unconformity, represented onshore by the sub-Campanian planation surface on which Campanian to Maastrichtian sedimentary units, including marine limestones, were deposited. The margin clearly underwent a major degree of denudation prior to the deposition of these Campanian to Maastrichtian units on the resulting erosion surface (evidence of cooling during this denudation episode in most of the samples closer to the coast was probably masked by the effects of Cenozoic palaeotemperatures). Cooling within a similar Late Cretaceous time interval has also been recognized over much of southern Africa and is well documented in Namibia (Raab *et al.* 2002, 2005; Brown *et al.* 2014). Late Cretaceous cooling has been interpreted

solely in terms of exhumation in these previous studies, although an increase in Late Cretaceous palaeotemperatures towards the coast in this study suggests the possibility of increased heat flow around the margin. However, kilometre-scale

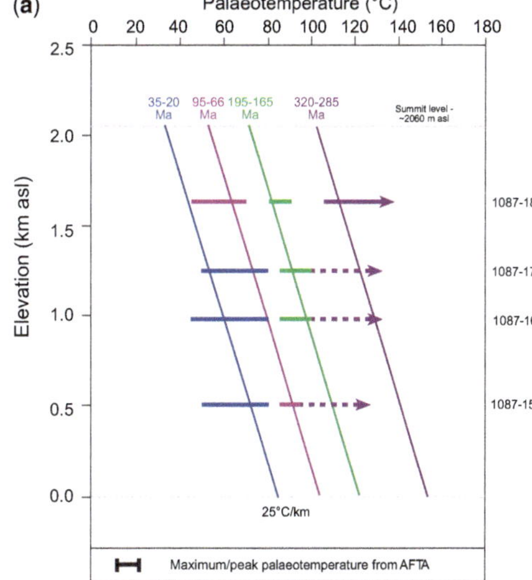

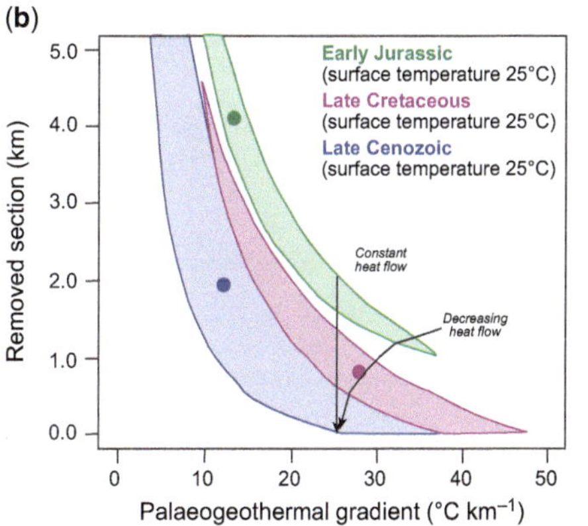

Fig. 9. (**a**) Palaeotemperature constraints in four samples collected from the escarpment in the east of the study area plotted against elevation. Data in four episodes are consistent with palaeogeothermal gradients around 25°C km^{-1}, as shown by the profiles, although a wide range of alternative interpretations is possible. (**b**) Constraints on the ranges of palaeogeothermal gradients and corresponding values of additional section required above the present day summit level at *c.* 2060 m required to explain the palaeotemperatures defined from AFTA in individual samples. For palaeogeothermal gradients around 25–30°C km^{-1}, no additional burial is required above the rim of the escarpment at the onset of Cenozoic cooling. Different possible trajectories for constant and variable heat flow conditions are indicated.

                            P. F. GREEN & V. MACHADO

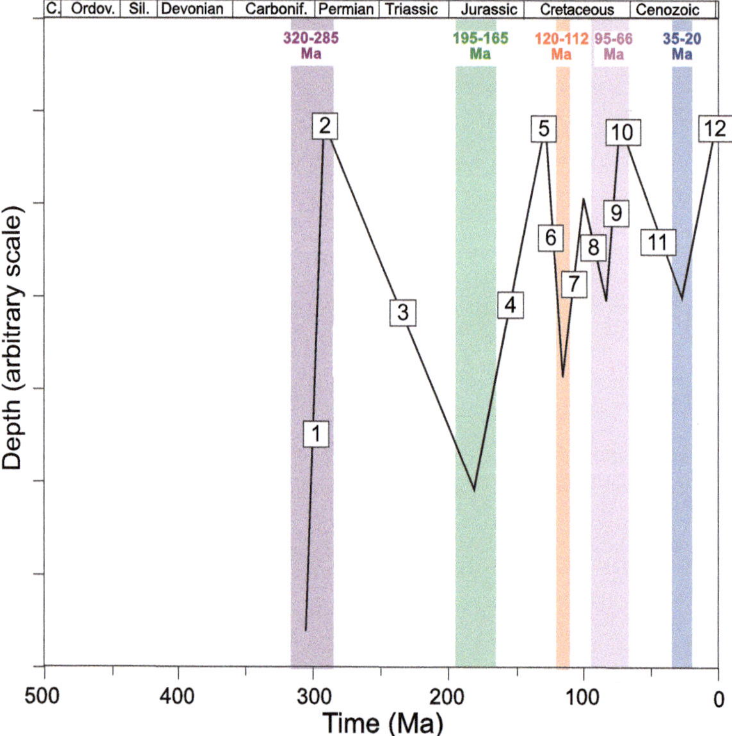

1: Regional exhumation leading to development of ...
2: Late Carboniferous–Early Permian glacial landscape.
3: Reburial by Karoo sediments.
4: Post-Karoo exhumation leading to development of ..
5: Early Cretaceous landscape on which Etendeka lavas were erupted.
6: Burial by Early Cretaceous syn-rift section.
7: Early Cretaceous exhumation (more pronounced outside study region)
8: Further mid-Cretaceous reburial.
9: Late Cretaceous exhumation leading to ..
10: Planation surface on which Campanian–Maastrichtian section was deposited.
11: Deeper burial through Late Cretaceous and Early Cenozoic.
12: Cenozoic exhumation leading ultimately to development of present-day landscape.

Fig. 10. Summary of the evolution of the Namibe margin of Angola based on the results in this paper. The depth scale is arbitrary and the diagram is meant to indicate the relative sense of vertical movements through time rather than the absolute magnitude, but movements are on a scale of kilometres.

denudation of the margin during the Late Cretaceous is required to explain the AFTA data for any reasonable values of palaeogeothermal gradient. As noted earlier, Late Cretaceous cooling has also been identified in other areas of the West Africa margin from northern Angola to Equatorial Guinea, and this event clearly affected much, perhaps all, of the continental margin.

A final phase of cooling related to uplift and erosion began during the Late Cenozoic (between 35 and 20 Ma) and most likely resulted in the development of the present-day topography. As noted earlier, the analysis of river profiles around the Bié dome (Roberts & White 2010) has suggested major uplift of the margin beginning at *c.* 30 Ma and the results from AFTA are consistent with this conclusion. Geomorphological data, together with river longitudinal profile analyses, have previously been interpreted to suggest that significant Cenozoic erosion has affected the Benguela and Kwanza margins, whereas Cenozoic uplift was less significant along the Namibe basin. Integrating this with our results suggests that the Miocene acceleration in uplift defined by Roberts & White (2010) was possibly stronger in the north, whereas exhumation accompanying the initial uplift at *c.* 30 Ma may have been stronger in the south around the Namibe margin.

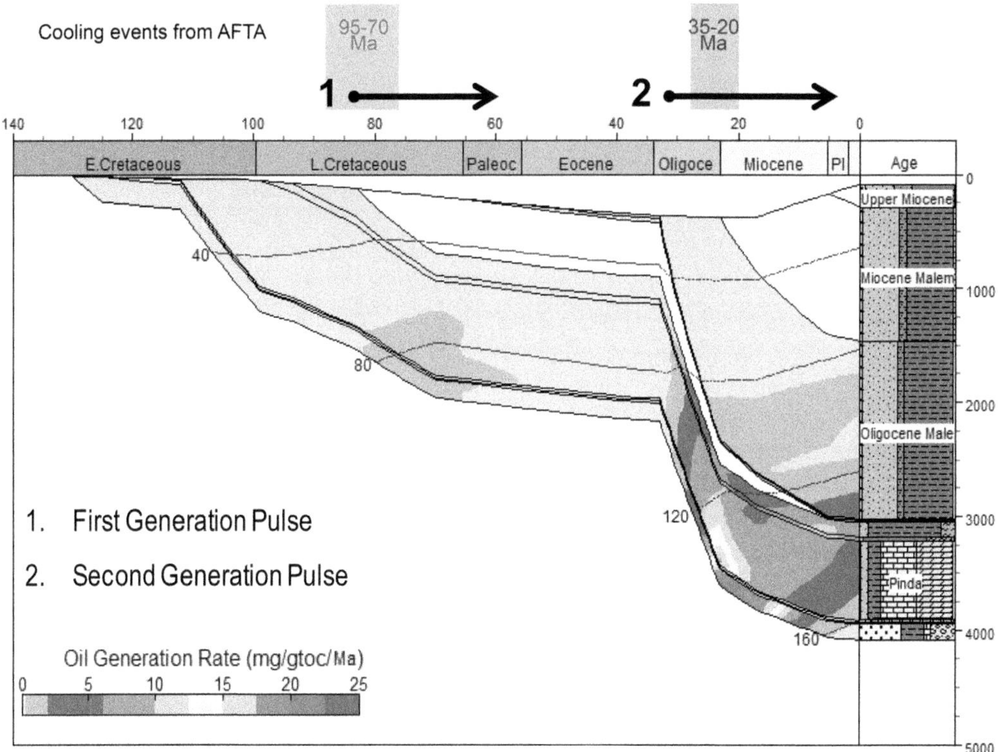

Fig. 11. Basin modelling results of a typical Kwanza Basin section showing: (1) first generation pulse for potential Cretaceous source rocks; and (2) burial-induced second pulse of hydrocarbon generation and expulsion involving Cretaceous and Tertiary source rocks (Machado 2012). Also shown are the timings of the two post-break-up periods of exhumation of the onshore Namibe margin identified from AFTA in this study. The link between onshore exhumation and offshore burial is clear.

Figure 9 shows palaeotemperatures plotted against elevation above sea-level in the five samples from the easternmost locations at the escarpment for four palaeothermal episodes, with notional palaeotemperature profiles indicated for each episode. Samples close to the summit of the inland escarpment show no evidence of Cenozoic cooling and this region may not have been covered by sediment prior to the onset of uplift, or was perhaps covered by only a thin veneer, the thermal effects of which cannot be resolved in the AFTA data. Also shown in Figure 9 are the ranges of allowed values of palaeogeothermal gradient and the amounts of additional section (above the summit level of 2060 m asl subsequently removed during uplift and erosion) for the Jurassic, Late Cretaceous and Late Cenozoic episodes. For palaeogeothermal gradients of *c*. 25°C km^{-1} or higher, no additional section above present day summit levels is required for the Cenozoic episode, supporting our earlier comments. For palaeogeothermal gradients of 25–30°C km^{-1}, Late Cretaceous

palaeotemperatures require the presence of around 1 km of additional section prior to the onset of exhumation, whereas Jurassic palaeotemperatures require higher values nearer to 2 km.

The combination of events identified here defines an episodic style of development for the Namibe margin of Angola (Fig. 10). At least two major pre-rift periods of exhumation affected the region prior to Valanginian–Barremian rifting (Quirk *et al.* 2013), while during the rift period adjacent sections of the margin were uplifted and eroded, shedding sediments into the developing marginal basins. Following break-up at the Barremian–Aptian boundary (Quirk *et al.* 2013), the margin underwent further erosion (although denudation at this time was less pronounced in our study area), leading to a prominent planation surface on which Cenomanian sediments were deposited. The margin then subsided and was buried by Late Cretaceous to mid-Cenozoic sediments and was subsequently uplifted and eroded in the Late Cenozoic, beginning between 35 and 20 Ma. Most

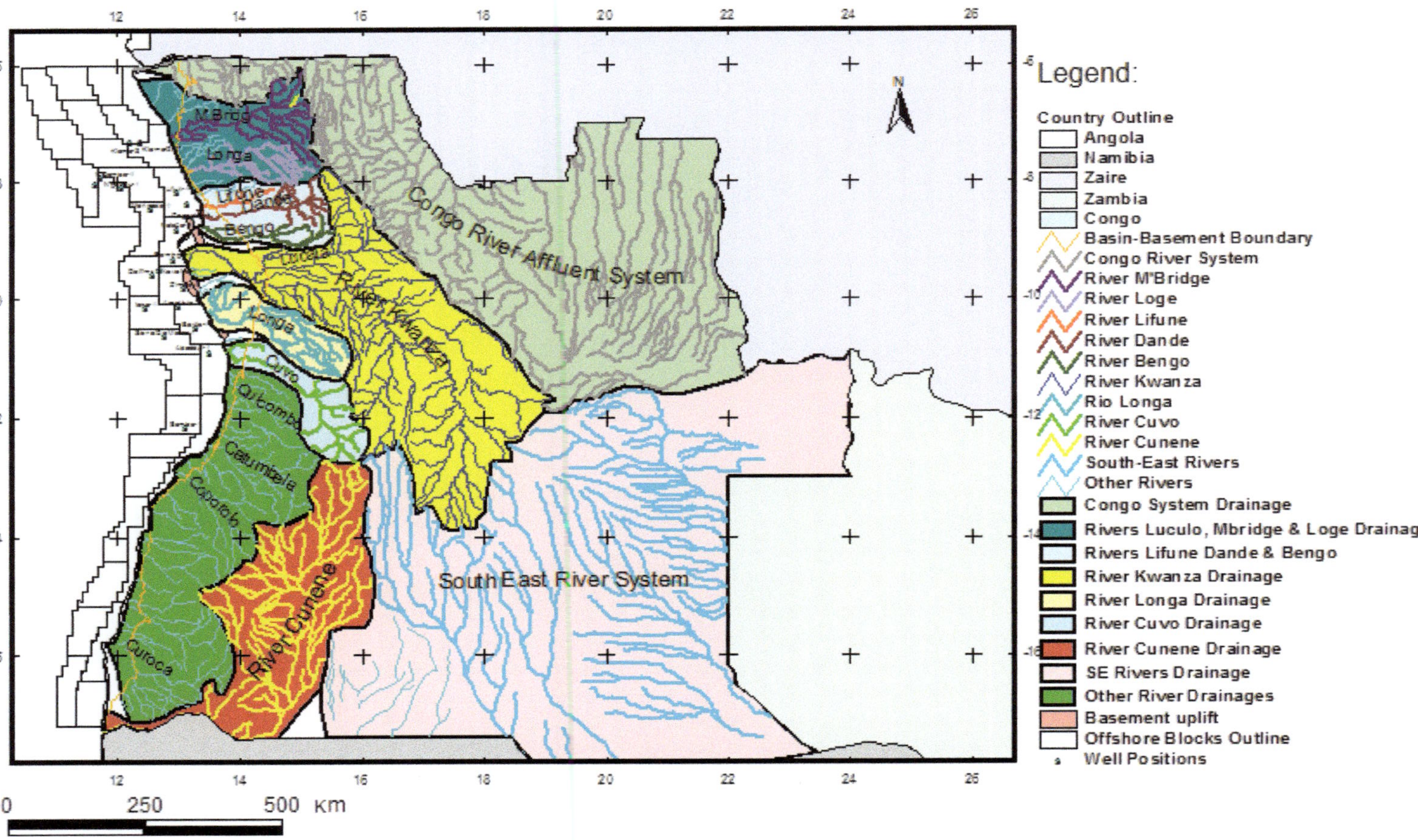

Fig. 12. Mapping of the main drainage basins of Angola showing key river systems.

samples from the coastal plain cooled from palaeo-temperatures around 60–70°C at this time, equivalent to burial by around 1.5–2 km of section subsequently removed during Late Cenozoic exhumation. This amount of burial is highly consistent with that indicated by the diagenesis of evaporites in the coastal region (Gindre-Chanu *et al.* 2015). The minimal Cenozoic cover required at the rim of the escarpment, at an elevation of *c.* 2 km, also supports the suggestion that up to 2 km of section was removed from the coastal plain during uplift of the Bié dome beginning at *c.* 30 Ma.

A key component of this interpretation is that the results require that the margin subsided and was buried following break-up and was uplifted and eroded much later, leading to the development of the present-day topography. Post-break-up burial and subsequent uplift is emphasized by the presence of Late Cretaceous sediments containing marine fossils now found at elevations of around 400 m asl. Similar observations have been made at other passive continental margins (e.g. Japsen *et al.* 2005, 2006, 2009, 2014; Green *et al.* 2011, 2013*a*), suggesting that this style of development is a general feature of passive margin development. Existing models, which commonly involve permanently uplifted margins following break-up (Ollier & Marker 1985; Weissel & Karner 1989; Gilchrist & Summerfield 1991, 1994; van der Beek *et al.* 1994, 1995, 2002; Gallagher & Brown 1997, 1999*a*, *b*; Ollier & Pain 1997; Gallagher *et al.* 1998; Bishop 2007), do not accurately reproduce this observed behaviour and require significant revision.

Implications for hydrocarbon exploration

As demonstrated in a previous study in the Kwanza Basin (Machado 2007, 2012), the most recent (Cenozoic) erosion event of the Namibe hinterland may have had a significant influence on hydrocarbon generation, expulsion and migration in the offshore basin. Time–temperature paths from AFTA indicate an Oligocene–Miocene timing of rapid cooling (?uplift and erosion) for the Inner Kwanza Basin and hinterland. The eroded material has been shown to have been deposited in the Outer Kwanza Basin. Basin modelling studies in the Outer Kwanza Basin calibrated by AFTA and vitrinite reflectance data resulted in a consistent burial-induced temperature increase in the Oligocene–Miocene that prompted a second phase of hydrocarbon generation and expulsion during the Neogene (Fig. 11) (Machado 2007, 2012). Given the similar overall history defined for the Namibe margin in this study, similar implications probably apply to the offshore basin in this sector of the Angolan margin.

Source to sink relationship

In the Kwanza Basin, there is a very clear correlation between modelled time–temperature paths from the onshore basement and the offshore sedimentary record in the Outer Kwanza Basin (Machado 2007). Phases of increased erosion onshore generally match with increased sedimentation offshore (Fig. 11). This observation is supported by biostratigraphic data and basin modelling, which indicate enhanced Oligocene to Miocene sedimentation after a Late Cretaceous to Eocene period of sedimentary quiescence.

The similarity between the data from the Namibe and Kwanza basins onshore again suggests similarities in offshore basin development in the two regions. The pronounced Cenozoic erosion in northern Namibia and Angola is generally confirmed in the offshore sedimentary record of the basins north of the Walvis Ridge (Namibe, Benguela, Kwanza and Congo). However, direct source to sink links are difficult to identify as a result of many uncertainties in the onshore to offshore correlation.

By analogy, the very pronounced Late Cretaceous erosional event in central Namibia is documented by an increased sediment influx in the Namibe, Walvis and Lüderitz basins offshore (Raab *et al.* 2005 and references cited therein). However, it is worth noting that the river drainage systems for the Benguela and Namibe basins are not so well developed as those of the Kwanza and Congo basins. Most of Angola's rivers originate in the elevated areas of the Bié/Central Plateau (Angola shield), which is one of the largest river dispersal heads in Africa.

In the SW of Angola a series of rivers, including the Longa, Cuvo, Quicombo and Catumbela, flow westwards from the Bié Plateau into the Atlantic Ocean. Two of the major rivers of Angola (the Kwanza and Cunene) take a more indirect route to the Atlantic Ocean. The fact that the SW rivers have short courses and display poorly developed dendritic patterns suggests that they are relatively young compared with the Kwanza and Cunene rivers, thus giving us an insight into their antecedence (Fig. 12). Ultimately, these short rivers with limited drainage are able to transport limited amounts of sediment, which is bypassed into the offshore Benguela and Namibe basins.

Conclusions

The Namibe margin of southern Angola has undergone complex development involving a number of episodes of exhumation and intervening reburial.

(1) Late Carboniferous–Early Permian exhumation, which began between 320 and 285 Ma,

was part of a continental-scale erosional event leading to the development of a major erosional surface preserving glacial features characteristic of the time.

(2) Jurassic exhumation, which began between 195 and 165 Ma, appears to be a precursor to the development of the South Atlantic rift.

(3) Early Cretaceous denudation, which began between 120 and 112 Ma (or possibly two episodes which began between 155 and 112 Ma and 120 and 112 Ma) in sediment provenance regions, is interpreted in terms of exhumation in relation to the developing Atlantic rift.

(4) Late Cretaceous denudation, which began between 95 and 70 Ma, defined both in the sediment provenance regions and in the basement samples within the study area, post-dates rifting and break-up and led to the development of a regional planation surface on which Campanian–Maastrichtian sediments and volcanic rocks were deposited.

(5) Following further deposition leading to the burial of the Late Cretaceous planation surface by up to 2 km, Late Cenozoic denudation, which began between 35 and 20 Ma, led to the development of the present day landscape, involving uplift of the Bié dome to its present altitude of >2 km above the present-day sea-level, and subsequent dissection.

These key episodes of denudation are identified in many areas of the West Africa margin, from Namibia to Equatorial Guinea, and are regarded as representing a continent-scale response to stresses related to tectonic plate movements.

We gratefully acknowledge Sonangol for permission to publish this work and both Sonangol and Statoil Angola for their support during fieldwork. We thank Heike Gröger of Statoil ASA, Stavanger, Norway for useful discussions and Roger Swart for sample collection and assistance in the field.

References

BISHOP, P. 2007. Long-term landscape evolution: linking tectonics and surface processes. *Earth Surface Processes and Landforms*, **32**, 329–365.

BROWN, R., SUMMERFIELD, M., GLEADOW, A.J.W., GALLAGHER, K., CARTER, A., BEUCHER, R. & WILDMAN, M. 2014. Intracontinental deformation in southern Africa during the Late Cretaceous. *Journal of African Earth Sciences*, **100**, 20–41.

CORCORAN, D.V. & DORÉ, A.G. 2005. A review of techniques for the estimation of magnitude and timing of exhumation in offshore basins. *Earth-Science Reviews*, **72**, 129–168.

DAUTEUIL, O., DESCHAMPS, F., BOURGEOIS, O., MOCQUET, A. & GUILLOCHEAU, F. 2013. Post-breakup evolution and palaeotopography of the North Namibian Margin during the Meso-Cenozoic. *Tectonophysics*, **589**, 103–115.

GALBRAITH, R.F. 2005. *Statistics for Fission Track Analysis*. Chapman & Hall/CRC, London.

GALLAGHER, K. & BROWN, R. 1997. The onshore record of passive margin evolution. *Journal of the Geological Society, London*, **154**, 451–457, http://doi.org/10.1144/gsjgs.154.3.0451

GALLAGHER, K. & BROWN, R.W. 1999a. The Mesozoic denudation history of the Atlantic margins of southern Africa and southeast Brazil and the relationship to offshore sedimentation. *In*: CAMERON, N., BATE, R. & CLURE, V. (eds) *The Oil and Gas Habitats of the South Atlantic*. Geological Society, London, Special Publications, **153**, 41–53, http://doi.org/10.1144/GSL.SP.1999.153.01.03

GALLAGHER, K. & BROWN, R.W. 1999b. Denudation and uplift at passive margins: the record on the Atlantic Margin of southern Africa. *Philosophical Transactions of the Royal Society, London*, **357**, 835–859.

GALLAGHER, K., BROWN, R. & JOHNSON, C. 1998. Fission track analysis and its applications to geological problems. *Annual Review of Earth and Planetary Science*, **26**, 519–572.

GILCHRIST, A.R. & SUMMERFIELD, M.A. 1991. Denudation, isostasy and landscape evolution. *Earth Surface Processes and Landforms*, **16**, 555–562.

GILCHRIST, A.R. & SUMMERFIELD, M.A. 1994. Tectonic models of passive margin evolution and their implications for theories of long-term landscape development. *In*: KIRKBY, M.J. (ed.) *Process Models and Theoretical Geomorphology*. British Geomorphological Research Group Symposia Series, **8**. Wiley, 55–84.

GINDRE-CHANU, L., WARREN, J.K. ET AL. 2015. Diagenetic evolution of Aptian evaporites in the Namibe Basin (south-west Angola). *Sedimentology*, **62**, 204–233.

GRADSTEIN, F.M., OGG, J.G., SCHMIDTZ, M.D. & OGG, G.M. 2012. *A Geologic Time Scale 2012*. Elsevier BV.

GREEN, P.F. 1986. On the thermo-tectonic evolution of Northern England: evidence from fission track analysis. *Geological Magazine*, **123**, 493–506.

GREEN, P.F. & DUDDY, I.R. 2012. Thermal history reconstruction in sedimentary basins using apatite fission-track analysis and related techniques. *In*: HARRIS, N.D. & PETERS, K. (eds) *Analyzing the Thermal History of Sedimentary Basins: Methods and Case Histories*. SEPM Special Publications, **103**, 65–104.

GREEN, P.F., SWART, R., JACOB, J., WARD, J. & BLUCK, B. 2009. Thermochronology and landscape development in Southern Africa. Extended abstract presented at the PESGB/HGS Africa Meeting, 9–10 September, London.

GREEN, P.F., DUDDY, I.R. & MALAN, J. 2011. AFTA data show that the southern margin of Africa was buried by Early Cretaceous sediments prior to the onset of regional exhumation. Extended abstract presented at the PESGB/HGS Africa Meeting, 7–8 September, London.

GREEN, P.F., LIDMAR-BERSGTRÖM, K., JAPSEN, P.J., BONOW, J.M. & CHALMERS, J.A. 2013a. *Stratigraphic Landscape Analysis, Thermochronology and the*

Episodic Development of Elevated, Passive Continental Margins. Geological Survey of Denmark and Greenland Bulletin, **30**.

GREEN, P.F., DUDDY, I.R., JAPSEN, P., BONOW, J. & CHALMERS, J.A. 2013*b*. The tectonic development of Africa's elevated passive continental margins and implications for exploration. Extended abstract presented at the PESGB/HGS Africa Meeting, 11–12 September, London.

GUIRAUD, M., BUTA-NETO, A. & QUESNE, D. 2010. Segmentation and differential post-rift uplift at the Angola margin as recorded by the transform-rifted Benguela and oblique-to-orthogonal-rifted Kwanza basins. *Marine & Petroleum Geology*, **27**, 1040–1068.

HURFORD, A.J. & GREEN, P.F. 1983. The zeta age calibration of fission-track dating. *Chemical Geology (Isotope Geosciences Section)*, **1**, 285–317.

JACKSON, M.P.A., HUDEC, M.R. & HEGARTY, K.A. 2005. The great West African Tertiary coastal uplift: fact or fiction? A perspective from the Angolan divergent margin. *Tectonics*, **24**, TC6014, http://doi.org/10.1029/2005TC001836

JAPSEN, P., GREEN, P.F. & CHALMERS, J.A. 2005. Separation of Palaeogene and Neogene uplift on Nuussuaq, West Greenland. *Journal of the Geological Society, London*, **162**, 299–314, http://doi.org/10.1144/0016-764904-038

JAPSEN, P., BONOW, J.M., GREEN, P.F., CHALMERS, J.A. & LIDMAR-BERGSTRÖM, K. 2006. Elevated passive continental margins: long-term highs or Neogene uplifts? New evidence from West Greenland. *Earth and Planetary Science Letters*, **248**, 315–324.

JAPSEN, P., BONOW, J.M., GREEN, P.F., CHALMERS, J.A. & LIDMAR-BERGSTRÖM, K. 2009. Formation, uplift and dissection of planation surfaces at passive continental margins – a new approach. *Earth Surface Processes and Landforms*, **34**, 683–699.

JAPSEN, P., BONOW, J.M. ET AL. 2012*a*. Episodic burial and exhumation in NE Brazil after opening of the South Atlantic. *Bulletin of the Geological Society of America*, **124**, 800–816.

JAPSEN, P., CHALMERS, J.A., GREEN, P.F. & BONOW, J.M. 2012*b*. Elevated passive continental margins: not rift shoulders but expressions of episodic post-rift burial and exhumation. *Global & Planetary Change*, **90–91**, 73–86.

JAPSEN, P., GREEN, P.F., BONOW,. J.M., NIELSEN, T.F.D. & CHALMERS, J.A. 2014. From volcanic plains to glaciated peaks: burial, uplift and exhumation history of southern East Greenland after opening of the NE Atlantic. *Global and Planetary Change*, **116**, 91–114.

LAUGHLAND, M., JUDGE, N. & GREEN, P.F. 2008. Uplift of the Namibian passive margin and its potential implications from hydrocarbon generation in the Lüderitz Basin. Abstract presented at the AAPG meeting, 26–29 October, Cape Town, South Africa.

MACGREGOR, D.S. 2013. Late Cretaceous–Cenozoic sediment and turbidite reservoir supply to South Atlantic margins. *In*: MOHRIAK, W.U., DANFORTH, A., POST, P.J., BROWN, D.E., TARI, G.C., NEMČOK, M. & SINHA, S.T. (eds) *Conjugate Divergent Margins*. Geological Society, London, Special Publications, **369**, 109–128, http://doi.org/10.1144/SP369.7

MACHADO, V. 2007. *Sand provenance, diagenesis and hydrocarbon charge history of the Kwanza Basin, Angola*. PhD thesis, University of Aberdeen.

MACHADO, V. 2012. AFTA approach to constrain tectono-thermal events in the Kwanza Basin: implications for sedimentation and basin evolution. Paper presented at the SPE Angola section general meeting, 15 February, Luanda, Angola.

MOULIN, M., ASLANIAN, D., RABINEAU, M., PATRIAT, M. & MATIAS, L. 2013. Kinematic keys of the Santos–Namibe basins. *In*: MOHRIAK, W.U., DANFORTH, A., POST, P.J., BROWN, D.E., TARI, G.C., NEMČOK, M. & SINHA, S.T. (eds) *Conjugate Divergent Margins*. Geological Society, London, Special Publications, **369**, 91–107, http://doi.org/10.1144/SP369.3

OLLIER, C.D. & MARKER, M.E. 1985. The Great Escarpment of southern Africa. *Zeitschrift für Geomorphologie, Neue Folge, Supplementbände*, **54**, 37–56.

OLLIER, C.D. & PAIN, C.F. 1997. Equating the basal unconformity with the palaeoplain: a model for passive margins. *Geomorphology*, **19**, 1–15.

PATON, D.A. 2012. Post-rift deformation of the North East and South Atlantic margins: are passive margins really passive? *In*: BUSBY, C.A. & AZOR, A. (eds) *Tectonics of Sedimentary Basins: Recent Advances*. Wiley-Blackwell, Chichester, 249–269.

PEREIRA, E., TASSINARI, C.C.G., RODRIGUES, J.F. & VAN-DÚNEM, M.V. 2011. New data on the deposition age of the volcano-sedimentary Chela Group and its Eburnean basement: implications to post-Eburnean crustal evolution of the SW of Angola. *Comunicações Geológicas*, **98**, 29–40.

QUIRK, D., HERTLE, M. ET AL. 2013. Rifting, subsidence and continental break-up above a mantle plume in the central South Atlantic. *In*: MOHRIAK, W.U., DANFORTH, A., POST, P.J., BROWN, D.E., TARI, G.C., NEMČOK, M. & SINHA, S.T. (eds) *Conjugate Divergent Margins*. Geological Society, London, Special Publications, **369**, 185–214, http://doi.org/10.1144/SP369.20

RAAB, M.J., BROWN, R.W., GALLAGHER, K., CARTER, A. & WEBER, K. 2002. Late Cretaceous reactivation of major crustal shear zones in northern Namibia: constraints from apatite fission track analysis. *Tectonophysics*, **349**, 75–92.

RAAB, M.J., BROWN, R.W., GALLAGHER, K., WEBER, K. & GLEADOW, A.J.W. 2005. Denudational and thermal history of the Early Cretaceous Brandberg and Okenyenya igneous complexes on Namibia's passive margin. *Tectonics*, **24**, TC3006, http://doi.org/10.1029/2004TC001688

ROBERTS, G.G. & WHITE, N. 2010. Estimating uplift rate histories from river profiles using African examples. *Journal of Geophysical Research*, **115**, B02406, http://doi.org/10.1029/2009JB006692

ROUBY, D., BONNET, S. ET AL. 2009. Sediment supply to the Orange sedimentary system over the last 150 My: an evaluation from sedimentation/denudation balance. *Marine and Petroleum Geology*, **26**, 782–794.

TANKARD, A., WELSINK, H., AUKES, P., NEWTON, R. & STETTLER, E. 2009. Tectonic evolution of the Cape and Karoo basins of South Africa. *Marine and Petroleum Geology*, **26**, 1379–1412.

TURNER, J.P., GREEN, P.F., HOLFORD, S.P. & LAWRENCE, S.R. 2008. Thermal history of the Rio-Muni (West Africa) – NE Brazil margins during continental breakup. *Earth and Planetary Science Letters*, **270**, 354–367.

VAN DER BEEK, P., CLOETINGH, S. & ANDRIESSEN, P. 1994. Mechanisms of extensional basin formation and vertical motions at rift flanks: constraints from tectonic modelling and fission-track thermochronology. *Earth and Planetary Science Letters*, **121**, 417–433.

VAN DER BEEK, P., ANDRIESSEN, P. & CLOETINGH, S. 1995. Morphotectonic evolution of rifted continental margins: inferences from a coupled tectonic–surface processes model and fission track thermochronology. *Tectonics*, **14**, 406–421.

VAN DER BEEK, P., SUMMERFIELD, M.A., BRAUN, J., BROWN, R.W. & FLEMING, A. 2002. Modelling postbreakup landscape development and denudational history across the southeast African (Drakensberg Escarpment) margin. *Journal of Geophysical Research*, **107**, 2351, http://doi.org/10.1029/2001JB000744

WALFORD, H.J. & WHITE, N.J. 2005. Constraining uplift and denudation of west African continental margin by inversion of stacking velocity data. *Journal of Geophysical Research*, **110**, B04403, http://doi.org/10.1029/2003JB002893

WALGENWITZ, F., RICHERT, J.P. & CHARPEMTIER, P. 1992. Southwest African plate margin: thermal history and geodynamical implications. *In*: POAG, C.W. & DE GRACIANSKY, P.C. (eds) *Geological Evolution of Atlantic Continental Rises*. Springer, 20–45.

WEISSEL, J.K. & KARNER, G.D. 1989. Flexural uplift of rift flanks due to mechanical unloading of the lithosphere during extension. *Journal of Geophysical Research*, **94**, 13919–13950.

An unusual Proterozoic petroleum play in Western Africa: the Atar Group carbonates (Taoudeni Basin, Mauritania)

ANTONIO MARTÍN-MONGE[1]*, ROGER BAUDINO[1], LUIS M. GAIRIFO-FERREIRA[1], RAFAEL TOCCO[1], MARCELLO BADALÌ[1], MARÍA OCHOA[1], SIGIT HARYONO[1], SOFÍA SORIANO[2], NIDAL EL HAFIZ[1], JULIO HERNÁN-GÓMEZ[1], BEATRIZ CHACÓN[1], IGNACIO BRISSON[1,3], GIOVANNI GRAMMATICO[1,4], ROBERTO VARADÉ[1] & HUSSEIN ABDALLAH[1]

[1]*Repsol Exploración SA, calle de Méndez Álvaro, 44, 28045 Madrid, Spain*

[2]*Repsol Services Company, 2001 Timberloch Place #3000, Spring, TX 77380, USA*

[3]*Present address: YPF SA, Macacha Güemes 515, 1425 Buenos Aires, Argentina*

[4]*Present address: Dana Petroleum, 62 Huntly Street, Aberdeen AB10 1RS, UK*

**Corresponding author (e-mail: martin.monge.antonio@repsol.com)*

Abstract: The Taoudeni Basin is the largest sedimentary basin in Africa. This intracratonic basin, which forms the sedimentary cover of the West African Craton, records episodic sedimentation since the Proterozoic. It is a largely unexplored basin, with only eight exploration wells drilled to date, although its petroleum prospectivity was established in 1974 when gas and liquids were tested in the Abolag-1 well. This paper focuses on the identification and assessment of potential source rocks and reservoirs in the Taoudeni Basin in Mauritania and aims to establish a geochemical genetic link between the occurrence of petroleum and potential sources. Various disciplines and techniques were integrated to reach a better understanding of the petroleum systems potentially existing in this basin. We also define and describe the elements and processes of the Atar-Atar (!) petroleum system. This study is an example of the exploration of an unusual and high-risk play in some of the oldest sedimentary rocks on Earth, with a long, polyphase structural evolution, a complex thermal history (where burial might not be the only controlling factor), a massively diagenetized carbonate reservoir, and large uncertainties on the timing of generation, trap formation and preservation.

The Taoudeni Basin, one of the main structural units of the West African Craton (WAC), is the largest sedimentary basin in Africa and has an area of roughly two million square kilometres (Fig. 1). It is well developed in Mauritania and Mali and extends marginally into Algeria, Burkina Faso and Senegal. The Taoudeni Basin records a discontinuous history of sedimentation spanning from the Mesoproterozoic to the present day. Although its hydrocarbon prospectivity was established early during its exploration, the Taoudeni Basin remains largely under-explored and still stands as one of the last frontiers in the petroleum exploration of northwestern Africa. The basin is characterized by a long and multiphase tectonic evolution, a complex thermal history with relatively young magmatism, and potentially problematic carbonate reservoirs with no modern analogue and affected by a prolonged diagenetic history. However, the existence of active petroleum systems holds promise in the quest for economic accumulations.

Numerous petroleum systems involving sourcing from Precambrian rocks, Precambrian reservoirs or in which both source and reservoir are of Precambrian age have been defined globally (e.g. Craig *et al.* 2013). Significant commercial petroleum production exists from petroleum systems sourced from Proterozoic source rocks, e.g. in the South Oman Salt Basin (Grosjean *et al.* 2009), the Lena–Tunguska province of Russia (Ulmishek 2001) and the Sichuan Basin of China (Korsch *et al.* 1991). Unravelling the complexities of the elements and processes that define the petroleum system becomes particularly challenging when dealing with these 'older' and 'deeper' plays in poorly explored areas. Proterozoic primitive life was dominated by microbial and algal organisms, resulting in source rocks and reservoirs that cannot easily be compared with those from the Phanerozoic.

The purpose of this paper is two-fold. The stratigraphy, structure and evolution of the Taoudeni Basin are addressed first, providing a synthesis

From: SABATO CERALDI, T., HODGKINSON, R. A. & BACKE, G. (eds) 2017. *Petroleum Geoscience of the West Africa Margin*. Geological Society, London, Special Publications, **438**, 119–157.
First published online January 7, 2016, http://doi.org/10.1144/SP438.5

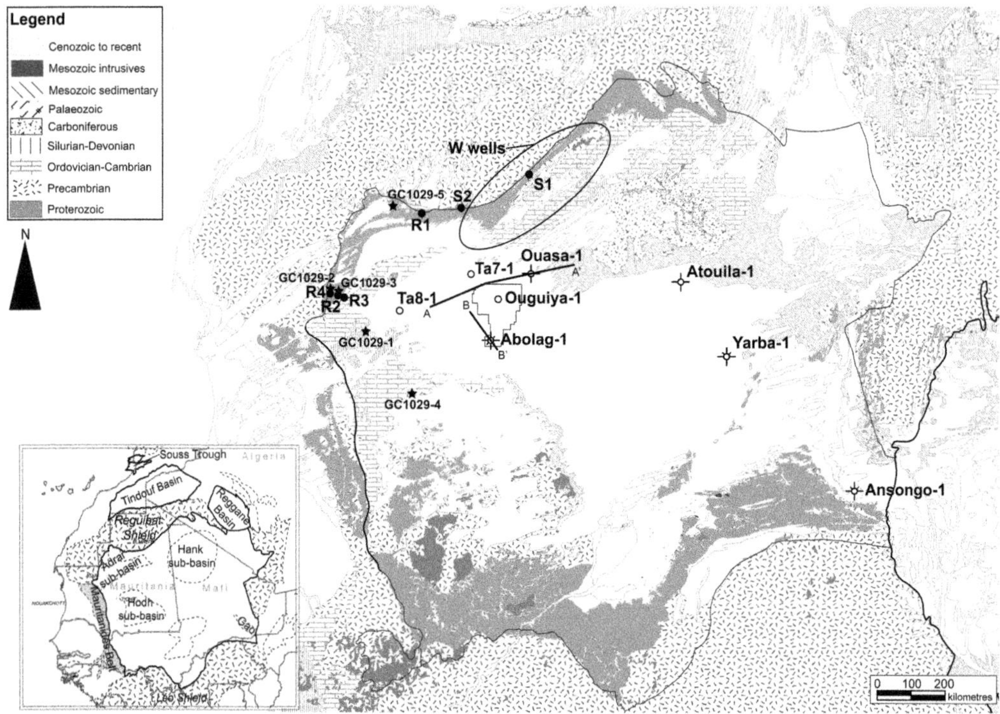

Fig. 1. Geological sketch map of the Taoudeni Basin. Map inset shows the location of the bounding Precambrian shields, Panafrican orogens and neighbouring sedimentary basins in addition to the main sub-basins of the Taoudeni Basin. Outcrops of the Mesozoic intrusive rocks are shown. Shallow wells are indicated by circles along the northern margin of the basin. The precise location of the W wells discussed in the text is proprietary of Wintershall Holdings GmbH. Sampling locations for the apatite fission track analysis study are indicated by stars. Line A–A′ is the trace of the composite seismic line shown in Figure 3. Line B–B′ is the trace of the seismic line shown in Figure 20.

of its geology. We then discuss and update its petroleum prospectivity, integrating published data with new research and interpretations.

Exploration history

The Taoudeni Basin is a largely unexplored frontier basin. Only eight exploration wildcat wells have been drilled to date (Fig. 1) and about 37 000 linear kilometres of two-dimensional seismic data have been acquired over the entire basin. However, the petroleum prospectivity was established early during the exploration of the basin. Between 1972 and 1985, exploration activities were carried out in both Mali and Mauritania, involving several international operators who conducted several geophysical acquisition campaigns and drilled five exploration wells. The first well drilled in the basin was Abolag-1 in Mauritania (drilled by Texaco in 1974), which proved the existence of a working petroleum system in the Taoudeni Basin and provided invaluable information regarding the geology of

the subsurface portion of the basin. The well tested gas and condensate from the upper part of the Proterozoic carbonate–siliciclastic sequence (Atar Group) and gas shows were detected while drilling throughout the intersected sedimentary column of Mesoproterozoic to Devonian age. The Ouasa-1 well followed the same year. Operated by Agip, the well did not reach the Proterozoic carbonates tested in Abolag-1 and did not record any significant petroleum shows. In 1975, both Texaco and Agip relinquished their licences in Mauritania.

In Mali, the Ansongo-1 well was drilled by the Société Nationale Elf Aquitaine in 1978. It targeted a late Cretaceous sequence in the Gao graben in southern Mali. The well was dry, but reported fluorescence and oil impregnations in cores; it proved the geological analogy of this Cretaceous depocentre with other western and central African Mesozoic rifts. The Yarba-1 well, drilled by the same operator in 1981 in the Araouane permit of Mali, above the Achkaikar high, reached the Proterozoic carbonate–siliciclastic sequence, where oil and gas shows were reported. The Atouila-1 well, drilled by Esso in 1985, was located at an intermediate position between the Proterozoic outcrops to the north and the centre of the Hank sub-basin. The primary objective was to test the Saharan Palaeozoic play in the northern part of the Taoudeni Basin in Mali. The well was abandoned with minor gas shows.

Petroleum exploration remained dormant in the area for the following three decades. Interest from international companies has recently increased and a new phase of exploration is underway. The Malian and Mauritanian governments have granted a large number of exploration contracts during the last decade. In Mali, no drilling has yet resulted from these new efforts, although numerous companies have acquired new gravimetric, magnetic and seismic data. In onshore Mauritania, intense exploration is also underway involving various international operators. Since 2007, Repsol and its partners have been engaged in an intense exploration programme involving wide-ranging surface geological studies, shallow coring and the acquisition of geophysical data (airborne gravimetric and magnetic surveys and various seismic campaigns). In 2014, as a result of this exploration work, the Ouguiya-1 well was drilled. Two other new wildcat wells were also drilled by Total in 2007 and 2013 in the northern part of the Mauritanian sector of the Taoudeni Basin.

Stratigraphy, structure and evolution of the Taoudeni Basin

Stratigraphy

Intracratonic sags are characterized by long-lived, but slow, subsidence, with numerous breaks in sedimentation, the presence of multiple and often shifting depocentres, and limited evidence of synsedimentary tectonic activity. The sedimentary environments existing in these basins were generally continental, but in many cases a large portion of the sedimentation also took place in shallow epicontinental seas (Allen & Allen 2005).

The Taoudeni Basin presents all these characteristics. It records sedimentation since the Mesoproterozoic; however, the basin-fill thickness rarely exceeds 5–6 km and is often much lower. The lateral continuity of the facies is remarkable (Bertrand-Sarfati et al. 1991) and the low subsidence rates have resulted in very low stratal depositional angles. Deformation affected relatively narrow bands along the basin margins, limiting the infilling exposure to the edges.

The extensive fieldwork carried out during the 1970s and 1980s provided invaluable information for unravelling the geology of the Adrar (Trompette

1973; Deynoux 1978), Hank (Villemur 1967), Hodh (Deynoux 1978) and Tagant (Dia 1984) regions. Other researchers revisited this basin in subsequent years, improving our knowledge of its tectonostratigraphic framework (Moussine-Pouchkine & Bertrand-Sarfati 1997) and easing correlations between depocentres (Moussine-Pouchkine et al. 1988; Bertrand-Sarfati et al. 1991). The recent geological mapping undertaken by the Bureau de Recherches Géologiques et Minières (BRGM) (Lahondère et al. 2003, 2005), including work commanded by Repsol and partners during 2007 and 2008, and the British Geological Survey (Pitfield et al. 2004) have resulted in harmonized basin-scale stratigraphic models. A profusion of subordinate units that are not easy to correlate have been defined historically inside the basin. Figure 2 shows a generalized stratigraphic column, including nomenclature equivalences for the Adrar, Hank and Hodh depocentres of the Taoudeni Basin. In this paper we adopt the BRGM revised nomenclature (Lahondère et al. 2005) and use the Adrar region lithostratigraphic names to refer to a series of age-equivalent units across depocentres.

Trompette (1973) divided the sedimentary succession exposed in the Adrar region into three unnamed major stratigraphic units, which were formally redefined as the Hodh, Adrar and Dahr supergroups by Pitfield et al. (2004). The bases of the Adrar and Dahr supergroups are bounded by major basin-wide unconformities (D4 and D6 on Fig. 2), both overlain by glaciogenic deposits. Within each ensemble, further unconformities of varying extent are used to subdivide these units into groups (Fig. 2).

At the base, the Hodh Supergroup comprises the Char, Atar and Assabet el Hassiane groups. The Char Group consists of a predominantly siliciclastic succession, with subordinate carbonates, deposited in a shallow-marine environment. The Atar Group is composed of alternating fine-grained terrigenous deposits and stromatolite-bearing shallow-marine carbonate intervals, with subordinate evaporites. The lithostratigraphic succession is further subdivided into eight formations according to the recent stratigraphic revision proposed by the BRGM (Fig. 2). The overlying Assabet el Hassiane Group is made up of terrigenous alluvial plain to outer shelf sediments. Lithologically, it contains shale, very fine-grained to conglomeratic sandstones and sparse limestones. Historically, the age of Hodh Supergroup strata was only poorly constrained by Rb–Sr analyses performed on sedimentary glauconite and illite (Clauer 1981; Clauer et al. 1982). According to these radiometric data, the age of the Char and Atar groups ranged from Tonian to mid-Cryogenian (1000–770 Ma). These ages were also supported by palynological studies (e.g. Amard

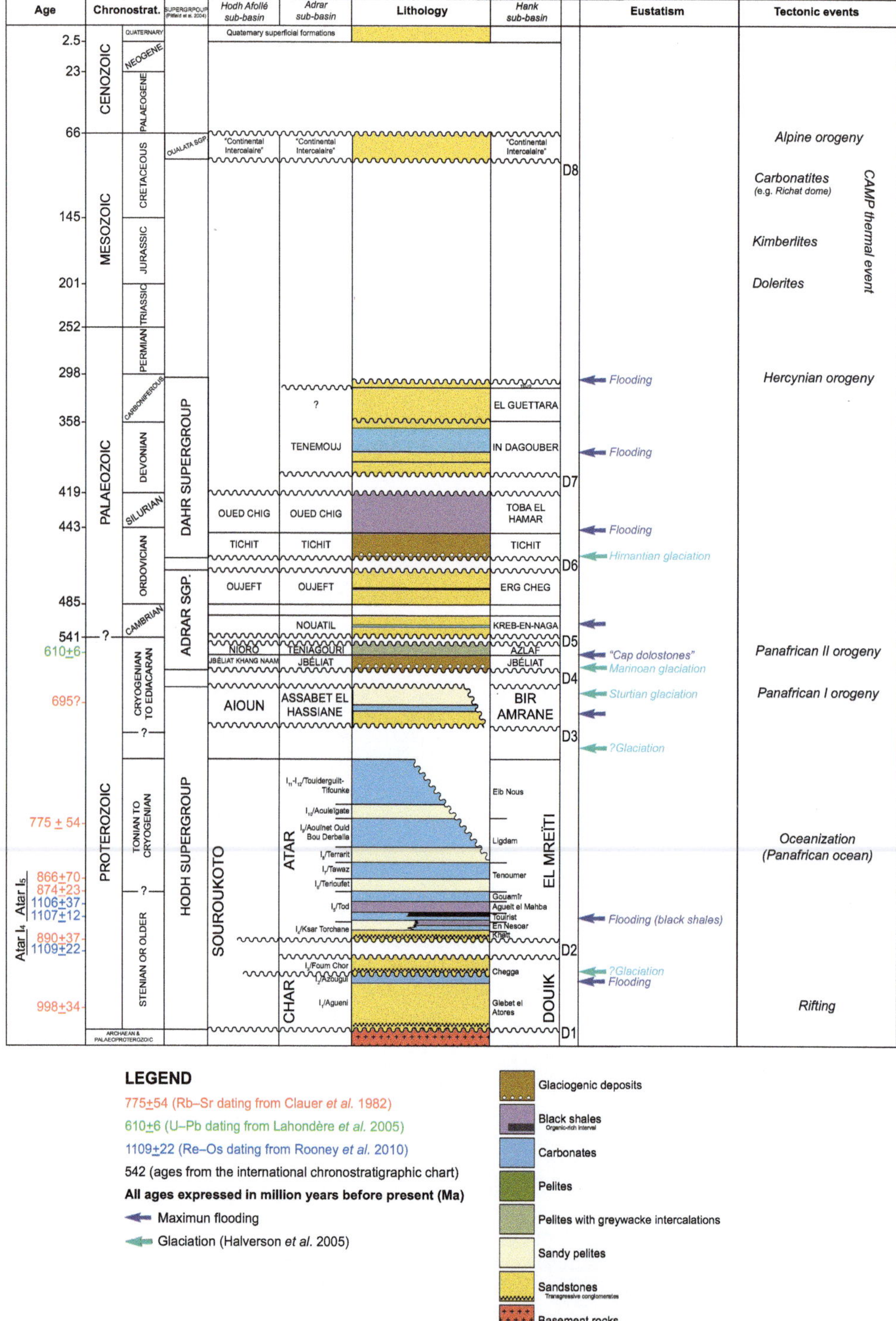

Fig. 2. Generalized stratigraphy of the Taoudeni Basin in Mauritania. Lithostratigraphic name equivalences are provided for the Adrar, Hank and Hodh depocentres. The major discontinuities are numbered and reference to global glacial and tectonic events is provided. Modified after Lahondère *et al.* (2003, 2005).

1986; Lottaroli *et al.* 2009), although some workers have cast doubts on the actual biostratigraphic distributions of the microbial and algal forms used (e.g. Kah *et al.* 2012). Teal & Kah (2005) and Shields *et al.* (2007) presented, respectively, carbon and strontium isotope chemostratigraphic data that were consistent with a Mesoproterozoic age. Confirmation of this older age arrived with new Re–Os absolute age constraints presented by Rooney *et al.* (2010) and Rooney (2011) for organic-rich sediments of the Ksar Torchane (Atar I_4) and Tod (Atar I_5) formations of the basal part of the Atar Group. They indicate that sedimentation of the Atar Group started before 1100 Ma (Fig. 2).

The Adrar Supergroup of late Neoproterozoic to Ordovician age consists of the Jbéliat, Téniagouri, Nouatil and Oujeft groups. The unconformity at the base of the Adrar Supergroup is related to incision during a glacial lowstand (D4; Fig. 2), although it also displays a slight angular character. The lower part of the supergroup represents a typical Ediacaran glaciogenic succession and includes the tillites and fluvial deposits of the Jbéliat Group and the barytic dolostones (cap dolostones) and silicified mudstones of the Téniagouri Group (Shields *et al.* 2007). The overlying Nouatil Group succession is composed of red beds and other continental siliciclastic deposits, shallow-marine stromatolitic and laminated dolostones and, ultimately, siliciclastics and minor evaporites deposited in shallow lake and floodplain environments. The lower part of the Oujeft Group is dominated by fluvial sandstones. Marine transgression marks the upper part of the Oujeft Group, which is composed of a sandstone-dominated succession deposited in a shallow-marine shelf.

The latest Ordovician to Devonian Dahr Supergroup consists of the Tichit, Oued Chig and Tenemouj groups. The unconformity at the base of the Dahr Supergroup is also of glacio-eustatic origin (D6; Fig. 2). The Tichit Group crops out discontinuously, occupying sinuous, channel-like structures. It is composed of a quartzitic sandstone-dominated succession deposited in glacial, proglacial and periglacial environments. The Oued Chig Group contains a lower sandstone-dominated transgressive succession that was deposited in shallow-marine environments. Continued marine transgression is associated with the finer grained deposits in the upper part of the Oued Chig Group. A lowstand period marks the top of the Oued Chig Group, which locally shows evidence of emergence and weathering (D7; Fig. 2). The Devonian Tenemouj Group, a succession of shallow-marine siliciclastic sediments, marks a new period of marine transgression in the basin.

The Mesozoic to Cenozoic 'Continental Intercalaire', redefined by Pitfield *et al.* (2004) as the Oualata Supergroup, overlies unconformably the Palaeozoic succession of the Taoudeni Basin in western Mauritania and Mali. This relatively thin succession is entirely continental and consists of fluviatile, lacustrine and playa-type deposits.

Structure and evolution of the Taoudeni Basin

Magnetic surveys have revealed the overall structure of the WAC, showing several major depocentres (Bayer & Lesquer 1978; Toft *et al.* 1992). The Taoudeni Basin has been subdivided into eight sub-basins (Villeneuve 2005 and references cited therein), but only three contain significant amounts of Proterozoic and Palaeozoic sediments: the Adrar sub-basin to the NW, the Hank sub-basin to the NE and the Hodh sub-basin in the central western part of the basin (Fig. 1). The depth of these depocentres does not exceed 5–6 km. They are separated by basement highs bounded by major structural faults that experienced important tectonism during the Proterozoic and were later reactivated in response to changing tectonic stresses. The Khatt high separates the Adrar and Hodh sub-basins, and the Abolag-Ouasa high separates the former from the Hank sub-basin. Finally, the Achkaikar high separates the Hank sub-basin from the failed Proterozoic rift of the Gourma aulacogen, located to the SE in Mali (Fig. 1).

The Taoudeni Basin is floored in Mauritania by the Archaean and Palaeoproterozoic plutonic and metamorphic terranes of the Reguibat Shield, which constitutes the northern segment of the WAC (Fig. 1). The Taoudeni Basin occupies a relatively internal position within the WAC. This explains the general sag geometry of the basin, with low subsidence rates and condensed sections, and the weak deformation experienced by the sediments deposited during its prolonged history. The structural and stratigraphic frameworks of the main depocentres of the Taoudeni Basin are fairly simple. Although some of the basin-wide unconformities are of glacio-eustatic origin (D4 and D6 on Fig. 2), the tectonic evolution of the Mauritanides mobile belt, located along the western margin of the basin in Mauritania and Senegal (Fig. 1), is the major control on the basin evolution as a whole (Waters & Schofield 2004; Villeneuve 2005).

Subsidence and sedimentation began during the late Mesoproterozoic. Rapid thickness changes associated with faults within the Char Group suggest a tectonic control on sedimentation. This early tectonic instability has been interpreted as a rifting period, speculated to have developed during the first stages of the break-up of Rodinia in the Neoproterozoic (Pitfield *et al.* 2004). However, the new age constraints for the Atar Group (Rooney *et al.* 2010) have highlighted the uncertainties of current palaeogeographical reconstructions for the Precambrian and emphasized the need to revisit the role of the WAC in the Rodinia supercontinent. The Khatt and Abolag-Ouasa highs that separate the Adrar, Hank and Hodh sub-basins were probably delineated during this early tectonism. The Char Group sediments, only exposed in the northwestern margin of the basin, are mostly developed over the Archaean basement of the Reguibat Shield, which was subjected to earlier deformation and flexuring than the Palaeoproterozoic terranes. In fact, the shear zone that juxtaposes the Archaean and Palaeoproterozoic terranes, which resulted from the early Palaeoproterozoic Eburnean orogeny (Schofield & Gillespie 2007), is interpreted to run along the Abolag-Ouasa basement high.

A basin-wide unconformity of unknown duration separates the Char and Atar groups (D2 on Fig. 2). The Atar Group records a period of regionally widespread sedimentation in shallow-marine environments, fluctuating between low-energy siliciclastics and carbonates. It is unclear whether this cyclic sedimentation represents repeated sea-level changes or a varying influx of sediments of possible climatic origin (Pitfield *et al.* 2004; Kah *et al.* 2012). No tectonic control is evident on the sedimentation of the Atar Group in Mauritania, but synsedimentary extensional deformation has been described in the Algerian Taoudeni (Moussine-Pouchkine & Bertrand-Sarfati 1997). The widespread carbonate platform that occupied the northern part of the Taoudeni Basin coincided with rifting to the south. The Gourma aulacogen is interpreted as a pericratonic failed rift arm that accumulated up to 8 km of sediments during the Meso- and Neoproterozoic, and finally closed during the Panafrican tectonic events without achieving oceanization (Bayer & Lesquer 1978; Black *et al.* 1979; Bronner *et al.* 1980; Villeneuve 2005; Álvaro & Vizcaïno 2012).

The unconformity that separates the Atar and Assabet el Hassiane groups (D3 on Fig. 2) is interpreted to represent the onset of compression along the margins of the WAC during the Panafrican orogeny (Panafrican I tectonic event; Villeneuve & Cornée 1994). This angular unconformity is markedly diachronous on the different margins of the Taoudeni Basin (Villeneuve & Cornée 1994; Deynoux *et al.* 2006). The sedimentation above the unconformity is dominated by clastic materials mainly sourced from the east. The stratigraphic architecture shows typical foreland basin deposition, with higher subsidence and sediments thickening towards the west (Moussine-Pouchkine & Bertrand-Sarfati 1997).

The upper part of the Assabet el Hassiane Group and the overlying Jbéliat Group are separated by a major unconformity of glacial erosional origin (D4; Fig. 2). It represents the first period of generalized emergence of the basin and has been correlated with the Marinoan ice sheets that covered most of North Africa towards the end of the Neoproterozoic era. The Jbéliat Group records the Neoproterozoic deglaciation prior to the major marine transgression that followed (Shields *et al.* 2007). The Téniagouri Group basal fine-grained sediments represent this transgressive episode, which marked the beginning of a new period of extensional tectonism in the Taoudeni Basin with deepening and strong subsidence, particularly in southern Mauritania. The age of this transgression is Ediacaran; it is well-constrained and coincides with plutonic activity in the Mauritanides (Lahondère *et al.* 2005). The subsequent rapid filling of the basin with the sandstones and shales of the Téniagouri Group is interrupted by another major unconformity that marks a new period of emergence in the Taoudeni and Tindouf basins (D5; Fig. 2). This event has been tied to a second phase of Panafrican compression along the margins of the WAC (Panafrican II tectonic event; Dallmeyer & Villeneuve 1987; Dallmeyer & Lécorché 1989; Culver & Hunt 1991; Villeneuve & Cornée 1994), which is also markedly diachronous (Pitfield *et al.* 2004).

The early Cambrian major transgression is largely eustatic, although an important tectonic component has been identified locally (Waters & Schofield 2004). The Cambrian Nouatil Group of the Taoudeni Basin is composed of terrigenous sediments deposited in a range of depositional environments (Pitfield *et al.* 2004). The Cambrian to early Ordovician Oujeft Group lies conformably above the Nouatil Group and evolves from fluviatile to marine depositional settings in the upper part of the unit. The erosional unconformity separating this unit from the overlying Tichit Group is again the result of a glacio-eustatic sea-level fall (D6; Fig. 2) and indicates a new period of emergence associated with the end-Ordovician glaciations (Pitfield *et al.* 2004). The large-scale channel-fill glaciogenic successions of the Tichit Group, described in the Adrar, Hank and Hodh sub-basins (Ghienne & Deynoux 1998), are overlain by the Silurian marine transgressive deposits of the Oued Chig Group, marking a new period of generalized subsidence. The marine flooding took place over an irregular palaeotopography inherited from previous glacial erosion and led to the development of two important Silurian depocentres in the Adrar and Hank sub-basins. The Souss Basin, in the northern margin of the Reguibat Shield, preserves a thick early Cambrian to Ordovician marine succession, with associated volcanic activity, interpreted to represent rifting along the margins of the WAC. By contrast, the early Palaeozoic continental to marine terrigenous successions of the Taoudeni Basin are comparatively condensed. Waters & Schofield (2004) interpreted these contrasting successions as reflecting the relative positions of these basins within the WAC, showing the effects of local intraplate tectonics rather than global eustatic control.

The Tenemouj Group of Devonian age, well developed in the Adrar region, is composed of shallow-marine clastic sediments deposited during a new marine transgression that followed a marine lowstand during which the erosional unconformity at the base of the group developed (D7; Fig. 2). The transgressive period was followed by tilting and compression affecting the entire basin during the Hercynian orogeny (Pitfield *et al.* 2004). The Hercynian orogeny resulted from the collision between Gondwana and North America and produced a generalized emergence at the end of the Palaeozoic. The Hercynian deformation is associated with broadly north–south-trending reverse faults and fold axes in the Tagant-Assaba region of the Mauritanides belt (Villeneuve 2005). The western part of the Taoudeni and Tindouf basins became weakly folded forelands of the intensely deformed bands along their occidental margins, tightly folded and thrust towards the east and NE.

Exposures of post-Carboniferous sedimentary rocks are very rare in northern Mauritania. During the Mesozoic, most of central and western Africa experienced widespread extension and magmatism related to the early phases of the Pangaea destabilization. These polyphased events led to the opening of the central Atlantic Ocean and to the development of the Mesozoic basins of the West and Central African Rift System. Two small elongated basins developed during the early Cretaceous towards the southern margin of the Taoudeni Basin: the Gao and Nara graben. The Gao graben, considered to be a part of the West and Central African Rift System, is an asymmetrical pericratonic half-graben containing up to 6 km of Cretaceous sediments. The Nara graben, originally one of the troughs of the interpreted Gourma triple junction (Bronner *et al.* 1980; Álvaro & Vizcaïno 2012), was reactivated during these extensional episodes, but subsided much less. Its structure is distinctive with respect to other western and central African rifts. In the northern Taoudeni Basin, this extensional period is recorded as a generalized epeirogenic uplift of the bounding shields, which became eroded and acted as the source of the sediments that were deposited over most of the northern Taoudeni Basin. The Oualata Supergroup ('Continental Intercalaire') lies unconformably over the late Palaeozoic sediments (D8; Fig. 2) and forms an extensive and relatively thin siliciclastic unit of late Cretaceous age deposited after a prolonged period of erosion.

On seismic sections, the structure of the Taoudeni Basin appears smooth (Fig. 3). The deformation of the Mauritanides fold–thrust belt is only clearly visible on the westernmost seismic lines. Away from the basin margins, the timing of deformation is generally difficult to predict. Neither complex deformation nor major truncation can be observed. The regional deformation and fracturing seem to be related to uplift during the Hercynian and Mesozoic events, with little evidence of Panafrican tectonism.

The Mesozoic intrusive activity recorded in the Taoudeni Basin is usually related to the first stages of the central Atlantic continental rifting (Sebai *et al.* 1991; Verati *et al.* 2005). It is conspicuously represented across northern and western Africa (in Algeria, Burkina Faso, Ivory Coast, Ghana, Guinea, Liberia, Mali, Mauritania, Morocco and Senegal) and forms part of the Central Atlantic Magmatic Province (CAMP), the world's largest igneous province, extending over North and South America, Europe and Africa (Ernst & Buchan 2002). According to plentiful geochronological data, the CAMP magmatism peaks around *c.* 199–200 Ma (Deckart *et al.* 1997; Marzoli *et al.* 1999, 2004; Knight *et al.* 2004; Verati *et al.* 2005, 2007; Nomade *et al.* 2007), predating by about 10 myr the initial stages of the opening of the central Atlantic Ocean. The early Jurassic magmatic activity recorded in the Taoudeni Basin is of regional extent and yields a well-constrained Hettangian–Sinemurian age (Dosso *et al.* 1979; Sebai *et al.* 1991; Pitfield *et al.* 2004; Verati *et al.* 2005), in good agreement with the timing of the CAMP magmatism elsewhere.

The origin of the CAMP remains unclear. Although some researchers attribute it to a mantle plume (e.g. Janney & Castillo 2001; Ernst & Buchan 2002), others favour a shallower mantle origin (e.g. McHone 2000; Verati *et al.* 2005; Coltice *et al.* 2007, 2009). Coltice *et al.* (2007, 2009), following Anderson (1982), hypothesized that the mantle magmatic activity associated with the break-up of Pangaea could be related to shallow mantle heating primarily induced by the aggregation of the supercontinent. Pangaea would have acted as a thermal insulator, enlarging the wavelength of mantle convective flow and naturally leading to significant subcontinental warming.

In the Adrar, Hank and Hodh sub-basins, as well as in the structural highs between them, the intense phase of Mesozoic magmatism led to the intrusion of doleritic dykes, sills, laccoliths and chimneys that cut through the basement and the Proterozoic and Palaeozoic sedimentary sequences. These intrusions were preferentially emplaced within clayey intervals and show thicknesses that are proportional to those of the host fine-grained interval, similar to what is observed in other sedimentary basins in

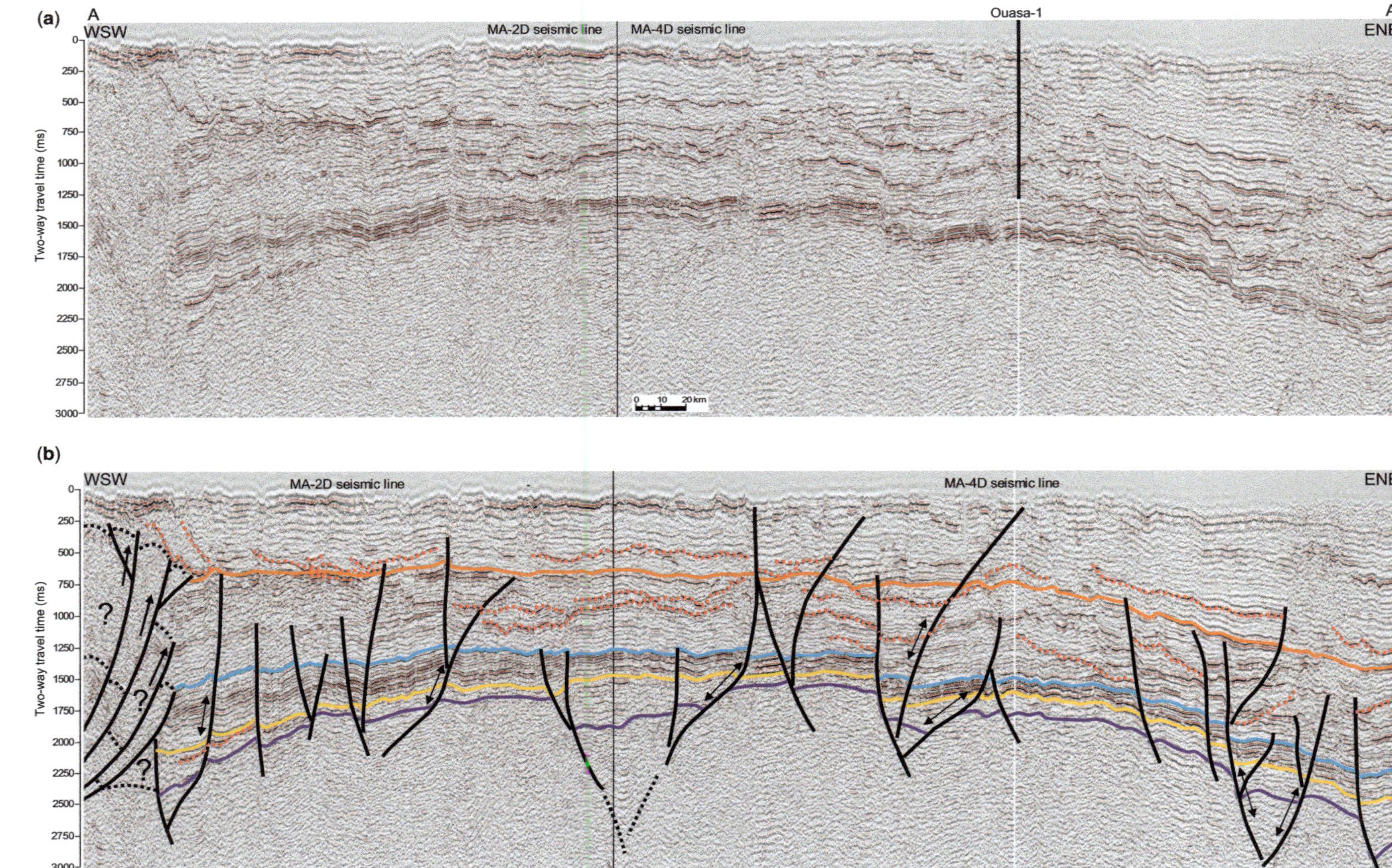
(a)
A
WSW
A'
ENE
MA-2D seismic line
MA-4D seismic line
Ouasa-1
Two-way travel time (ms)
0
250
500
750
1000
1250
1500
1750
2000
2250
2500
2750
3000
0 10 20 km

(b)
WSW
MA-2D seismic line
MA-4D seismic line
ENE
Two-way travel time (ms)
0
250
500
750
1000
1250
1500
1750
2000
2250
2500
2750
3000
?
?
?
?

North Africa (e.g. Murzuq, Reggane). In outcrop, they have been observed intruding fine-grained intervals of the Atar Group and its lateral equivalents, particularly the Tod (Atar I_5) and Aouleïgate (Atar I_6) formations, the Assabet el Hassiane Group and several other intervals of the Neoproterozoic and Palaeozoic sequences. The section drilled at Abolag-1 found numerous doleritic intervals within the Proterozoic, Devonian and Silurian sections (see Fig. 5). In fact, about 200 m of doleritic sills were intersected at this well out of 2942 m drilled. A similar situation was encountered in Ouasa-1, although this well did not reach the Atar Group. Intrusive bodies are also easy to recognize in the subsurface across the whole area through seismic data (e.g. Fig. 3). It is worth mentioning that differences in the frequency of occurrence of intrusions can be appreciated both vertically and areally on seismic data.

Apatite fission track analysis (AFTA) was performed on five sandstone samples collected from outcrops and shallow cores along the western and northwestern margins of the basin (see Fig. 1 for sampling locations). The AFTA approach is based on the kinetics of the annealing of radiation damage features (fission tracks) on detrital apatite grains recovered from sedimentary rocks and was conducted for this study by Geotrack International Pty Ltd. A detailed explanation of the basis of the technique and the methodology used is given by Bray *et al.* (1992) and references cited therein.

The selected sandstone samples, with depositional ages ranging from the ?Mesoproterozoic to the lower Palaeozoic, yielded excellent apatite populations and therefore the analyses are considered to be highly reliable. The mean measured fission track lengths in all the analysed samples range from 12.5 to 13 μm and the measured track ages vary from *c.* 100 to *c.* 400 Ma (Fig. 4a). The

measured track lengths are significantly shorter than the lengths resulting from modelling a prolonged residence at their present day temperature and the measured ages are much lower than the depositional ages (Fig. 4a). This indicates that the host units from which the samples were collected must have been much hotter than present day temperatures at some time after deposition. A complex thermal history can be interpreted from these data, with at least five cooling episodes recorded at a regional scale (Fig. 4b). These five cooling episodes can be related to three main events in the evolution of the basin (Fig. 2): the Hercynian uplift; cooling after Mesozoic magmatism; and Cenozoic uplift, probably related to Alpine movements. The Hercynian and Alpine events can be interpreted in all the samples. The Mesozoic thermal episode appears to be more localized. This event is nevertheless recorded in samples from opposite sides of the sampling area and the fact that it is not associated with increased burial, nor is it related to actual intrusions, probably indicates a subregional- or regional-scale process. This is consistent with the large-scale record of CAMP magmatism and with the implications of the model of Coltice *et al.* (2007, 2009). Similar studies carried out on other North African sedimentary basins have also identified a major late Triassic to early Jurassic thermal episode, which has been interpreted to reflect a regional deep crustal thermal perturbation (e.g. Logan & Duddy 1998; Boote *et al.* 2008).

In summary, the Taoudeni Basin records episodic sedimentation from the Mesoproterozoic to the Palaeozoic over the northern part of the WAC. The overall sedimentation rates are low, reflecting the internal position of the basin within the craton. A series of major regional unconformities has been identified within the sedimentary succession and serves as a useful correlation tool over the vast

Fig. 3. Composite seismic line (MA-2D/MA-4D) through the Abolag-Ouasa High (**a**) without and (**b**) with structural interpretation. See Figure 1 for location. Interpreted horizons include top-basement (violet), top Char Group (yellow), top Atar Group (blue) and top Assabet el Hassiane Group (orange). The black solid lines are faults. Notice how the Char Group was deposited in graben and half-graben with great variations in thickness during a rifting tectonic phase. The Atar Group shows less variation, but the existence of highs during sedimentation is demonstrated by thinning towards the centre of this section. Regarding deformation, the fault interpretation suggests a transtensional dominant regime during the Char Group deposition, with graben-limiting faults tending to join at depth. Numerous normal faults can also be observed affecting the Atar Group; several of these tensional faults were reactivated during the following compressive phases and the more evident are marked with double-headed arrows. On the western end of the section, the Mauritanides deformation front is clearly visible with reverse faults reaching the surface on occasion (black arrows indicate relative movement). No stratigraphic interpretation was possible in this area and some continuous reflectors are drawn as dotted black lines. It is difficult to assign the compressive and transpressive deformation to a single phase, but the fact that many faults are sealed by Mesozoic deposits suggests that most of the present day structuration was achieved during the Hercynian phase and that Alpine deformation had little effect in the centre of the basin. The red dotted lines outline dolerite intrusions, which were relatively easy to interpret due to their high-amplitude seismic response and their obliquity to the sedimentary strata. Many others exist, but are more difficult to recognize because their thickness is below seismic resolution and they are parallel to the strata. It is interesting to note how the presence of intrusions seems to be associated with deep faults, which might have acted as conduits to the magma.

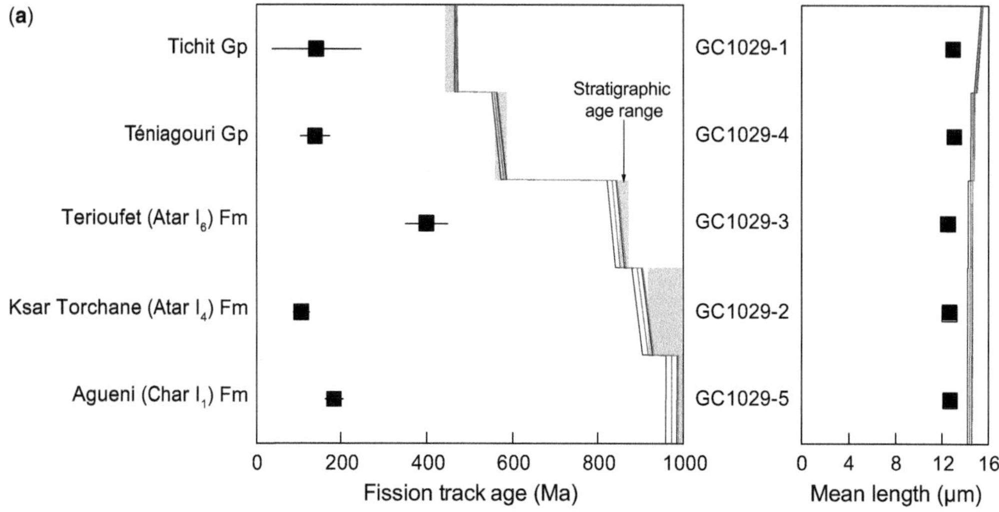

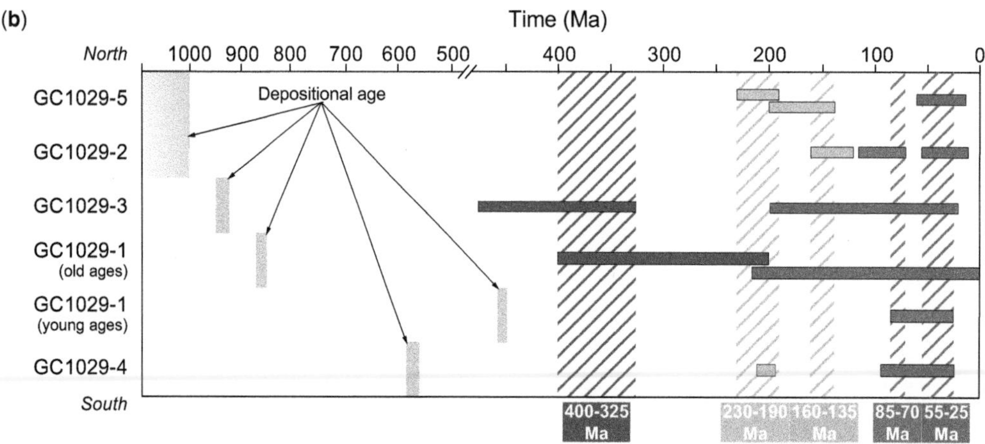

Fig. 4. (**a**) Apatite fission track ages and mean track lengths of five outcrop and shallow well samples from the Taoudeni Basin in Mauritania. See Figure 1 for sample locations. Fission track ages are compared with depositional ages. The lines show the ranges of the fission track ages and the mean track length predicted from modelling a prolonged residence at their present day temperature for apatites containing 0.0–0.1, 0.4–0.5, 0.9–1.0 and 1.5–1.6 wt% Cl. All samples display fission track ages that are significantly younger than the depositional age, suggesting that these samples had been much hotter at some time after deposition. (**b**) The integration of the results from all samples suggests that at least five cooling episodes affected the region. The fission track data of sample GC1029-1 appear to define two distinct groups of ages.

area occupied by this intracratonic basin. These unconformities record the complex interplay of the global eustatic cycles with major tectonic events that affected the WAC margins. Post-Hercynian subsidence in the basin is very low, with thin Mesozoic and Cenozoic deposits covering the northern part of the basin. The Mesozoic magmatic and extensional events are thought to have had a significant impact on the thermal evolution of the basin, as suggested by the abundance of intrusions. The integration of these observations into geological

and petroleum system models is a critical step towards the assessment of the prospectivity of the Taoudeni Basin.

Source rock potential

Petroleum occurrences and indications

Hydrocarbon occurrences and indications are numerous across the Taoudeni Basin. The Abolag-1 well tested hydrocarbons on the lower part of the

Touiderguilt (Atar I_{11}/I_{12}) Formation (interval 2523–2536 mMD; Fig. 5). The produced fluids were methane-dominated (>80 vol%) and flowed at a rate of approximately 0.48×10^6 SCF per day. Although the data at our disposal regarding the exact composition of the produced petroleum are incomplete, we can nevertheless exclude the possibility that the tested interval was either a dry gas or a black oil reservoir. Considering the methane-rich composition of the fluids and the production of liquid hydrocarbons at the separator during the well test, the tested interval most probably corresponds to a wet gas or a retrograde condensate reservoir. Gas was also recorded in significant amounts while drilling at several intervals, including the basal sandstones of the Assabet el Hassiane Group and several intervals within the Atar Group (possibly also within the Char Group; Fig. 5).

Petroleum shows, mostly gas, were also recorded in other exploration wells. The Ouasa-1 well did not reach the Atar Group, but found minor gas shows within the Assabet el Hassiane Group clastic sediments. The Atouila-1 and Yarba-1 wells in Mali also found gas shows. The former tested in the Taoudeni Basin the potential of the North African Palaeozoic play, well established in the Saharan platform of Algeria, Libya and Tunisia, and the source of substantial ongoing petroleum production in the region. Atouila-1 was dry, but had minor gas indications in the Carboniferous and Silurian sections, although mostly associated with shaly intervals. In Yarba-1, gas shows were encountered on the Proterozoic carbonate–siliciclastic sequence (the Atar Group lateral equivalent). More recently, the Ouguiya-1 well has also encountered significant gas peaks in several intervals within the Assabet el Hassiane, Atar and Char groups (Baudino *et al.* 2014). A direct light yellow fluorescence was observed on sidewall cores recovered from the basal sandstone interval of the Assabet el Hassiane Group.

Our recent work revealed other hydrocarbon indications of great relevance. In particular, some of the shallow stratigraphic wells drilled along the northern rim of the Taoudeni Basin in Mauritania have encountered oil stains and impregnations within the Atar and Char groups. The detailed geochemical characterization of these shows is described later in this paper. Residual solid hydrocarbons ('bitumen') are pervasive in outcrops and shallow well samples at various stratigraphic levels within the Char, Atar and the lower part of the Assabet el Hassiane groups. Bitumen is also significantly abundant in the Proterozoic carbonate sequence intersected by the Abolag-1, Ouguiya-1 and Yarba-1 exploration wells. We interpret the ubiquitous presence as a clear indication that significant quantities of liquid petroleum circulated through the petroleum system at some stage.

Source rock identification

Several potential source intervals have been identified within the Proterozoic Char, Atar and Assabet el Hassiane groups. Some black shale intervals have been described in outcrops of the Azougui (Char I_2) Formation and assessed for source rock potential. The surface samples were completely weathered and therefore no evaluation of the original source potential has been possible. Thin organic-rich black shale intervals have also been described within the Assabet el Hassiane Group, interbedded with stromatolite-bearing carbonate levels in the Atar area, although the source rock potential remains to be fully evaluated.

The most interesting levels in terms of source potential lie within the fine-grained intervals of the Atar Group. Dark-coloured siltstones and shales are found at several levels within the carbonate–siliciclastic alternations, namely within the Ksar Torchane (Atar I_4), Tod (Atar I_5), Terioufet (Atar I_6), Aouleïgate (Atar I_{10}) and Touiderguilt (Atar I_{11}/I_{12}) formations. We will focus solely on the Ksar Torchane (Atar I_4) and Tod (Atar I_5) formations, which have been found to be the most promising. The upper part of the Ksar Torchane (Atar I_4) Formation records the onset of carbonate deposition in the Taoudeni Basin and is composed of intercalated stromatolite-bearing carbonates and black shales. The three distinct stromatolitic reefs that characterize the overlying Tod (Atar I_5) Formation (Bertrand-Sarfati & Moussine-Pouchkine 1985; Kah *et al.* 2012) are separated by intervals of fine-grained siliciclastic sediments. These fine-grained units, containing finely laminated, black shales, paper-sheet-like, and minor siltstones and calcarenites, show a notable lateral continuity along the northern edge of the Taoudeni Basin (Kah *et al.* 2012).

The discussion that follows is primarily based on the analysis of source rock samples and organic extracts from the Ksar Torchane (Atar I_4) and Tod (Atar I_5) formations, intersected and sampled by the shallow wells drilled by Repsol and partners along the northern margin of the Taoudeni Basin in Mauritania (wells R1, R2, R3 and R4; Fig. 1) as well as in some outcrop samples and cutting samples from Abolag-1. Within the framework of a data exchange agreement with Wintershall Holding GmbH, we had access to samples from another four shallow wells drilled in the northeastern part of the Taoudeni Basin in Mauritania (referred to in this work as wells W1, W2, W3 and W4; the locations of these wells are proprietary of Wintershall), which were subsequently analysed by Repsol. Oil stains and impregnations in clastic intervals intersected by the R4 and W3 shallow wells were also identified, sampled and analysed. Our

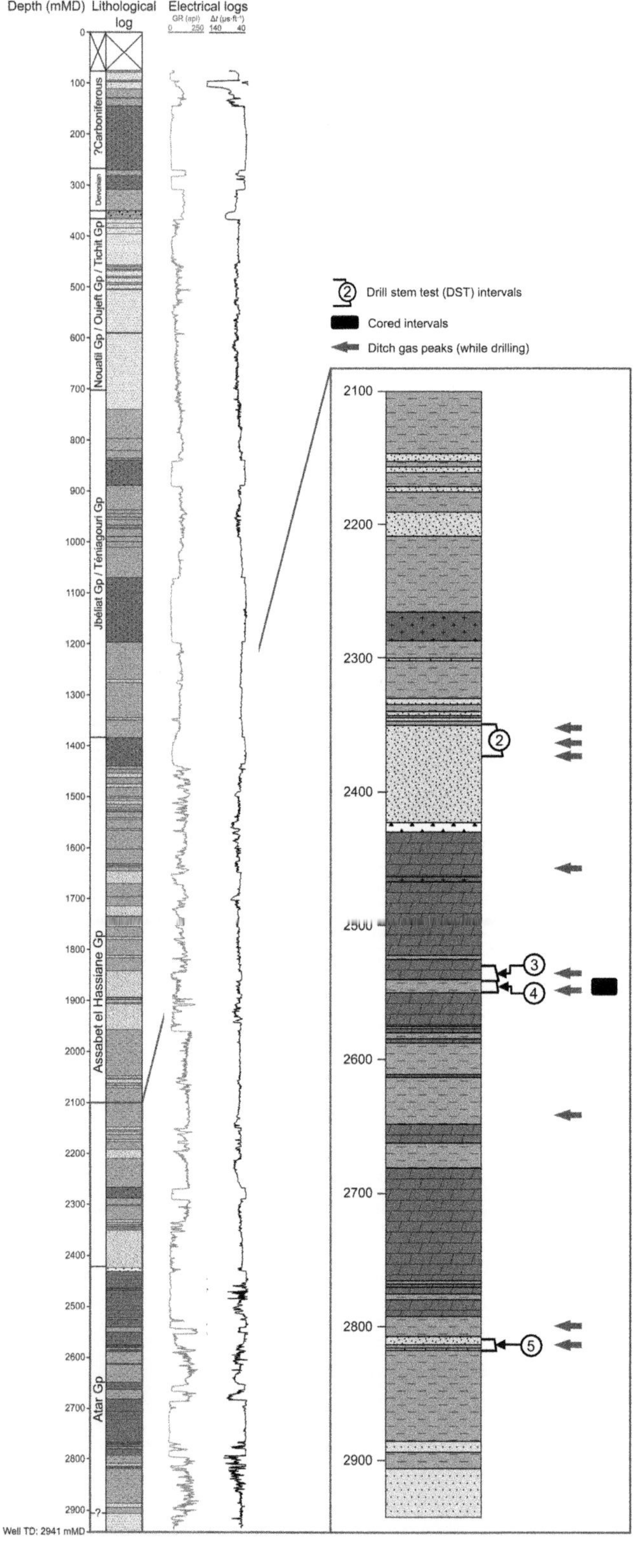

Depth (mMD)
Lithological log
Electrical logs
GR (api)
Δt (µs·ft⁻¹)
Drill stem test (DST) intervals
Cored intervals
Ditch gas peaks (while drilling)
?Carboniferous
Devonian
Tichit Gp
Oujeft Gp
Nouatil Gp
Téniagouri Gp
Jbéliat Gp
Assabet el Hassiane Gp
Atar Gp
Well TD: 2941 mMD

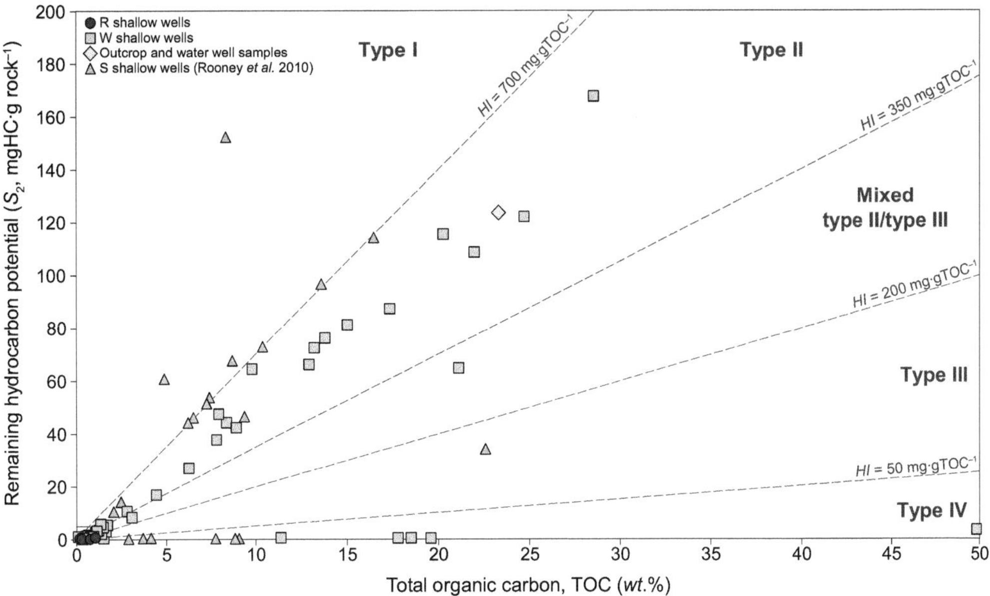

Fig. 6. Remaining petroleum-generative potential (as Rock-Eval S_2) versus organic richness (TOC content) for source rock samples from shallow wells and outcrops in the Ksar Torchane (Atar I_4) and Tod (Atar I_5) formations. Shallow core data are from the R and W wells and from Rooney *et al.* (2010) (S wells).

discussion focuses on the assessment of the source rock potential and depositional environments, the existence of organic facies variations, and on the maturity of both sources and migrant petroleum.

The organic carbon content of the Ksar Torchane (Atar I_4) and Tod (Atar I_5) shaly intervals along the northern margin of the Taoudeni Basin is highly variable, both vertically and laterally. Most samples from the northwestern area are essentially organic-lean or display poor to very poor total organic carbon (TOC) contents. Towards the northeastern edge of the Mauritanian sector of the basin, the richest outcrop and shallow core samples (W wells) reach TOC contents >50% (Fig. 6; see also Blumenberg *et al.* 2012). Rock-Eval bulk pyrolysis analyses show that the petroleum-generative potential is excellent in some of the richest samples, with hydrogen indices (HI) of up to 600 mgHC gTOC^{-1} (Fig. 6), indicating an oil-prone character. The source rock assessment data reported by Rooney *et al.* (2010) for other shallow stratigraphic wells along the northern margin of the basin (S1 and S2 wells; location on Fig. 1) also highlight the vertical variability and the excellent petroleum-generative

potential of certain intervals. Reported TOC values for these S wells reach values of up to 23% and high HI, in excess of 700 mgHC gTOC^{-1} (Fig. 6). Even higher TOC contents (up to 30%) and HI values (*c.* 800 mgHC gTOC^{-1}) are reported for several outcrop and water well samples by Lüning *et al.* (2009). Towards the centre of the basin, TOC and Rock-Eval pyrolysis data are only available from the Abolag-1 exploration well. The Atar Group section intersected by this well shows poor organic richness and remaining petroleum-generative potential, with TOC values <0.5% and very poor Rock-Eval pyrolysis yields. Because of the high maturity levels attained by this section, it is difficult to assess the original composition of the kerogen. Microscopic evidence suggests, however, that the Proterozoic section could have contained relatively high percentages of sapropelic kerogen and good hydrocarbon source rocks. These observations are consistent with previous reports of abundant amorphous organic matter in the Atar Group of Abolag-1 (Lottaroli *et al.* 2009).

These data emphasize the great uncertainties about the source rock potential in the Taoudeni

---

Fig. 5. Section of the Abolag-1 well in Mauritania. Electrical logs (gamma ray and sonic transit time) are provided. The intervals tested in the well are shown. Drill stem test 3 flowed gas and liquids at a rate of approximately 0.48×10^6 SCF per day. Intervals with significant gas peaks while drilling are indicated with arrows.

Basin, despite the encouraging evidence found along its northern margin. The data at our disposal are sparse and therefore uncertainties remain about the subsurface extension and thickness of the organic-rich intervals. The vertical and lateral variations in organic facies and richness are also largely unknown. Integration of the analysed and published data (Lüning *et al.* 2009; Rooney *et al.* 2010; Blumenberg *et al.* 2012) shows that the highest TOC contents and petroleum-generative potentials appear to be concentrated in the uppermost fine-grained interval of the Tod (Atar I_5) Formation, which displays a thickness of approximately 10 m in the W wells. Furthermore, regional trends in both maturity and remaining petroleum potential can be recognized. The lateral extension of the organic-rich intervals to the east into Algeria and Mali is currently unknown, although the Proterozoic carbonate–siliciclastic sequence (El Mreïti Group, age-equivalent of the Atar Group; Fig. 2) intersected in Yarba-1 contains highly radioactive black shale intervals. In addition, there are reports of organic-rich intervals in northern Mali (Dars 1961; Villemur 1967) and the occurrence of black shales in Proterozoic half-graben of the Algerian Taoudeni Basin (Lüning *et al.* 2009). In the northern margin of the Tindouf Basin in Morocco, the gas shows encountered within the Proterozoic sequence (well AZ-1) have allowed Ghori *et al.* (2009) and Lottaroli *et al.* (2009) to speculate about the existence of a similar petroleum play in this epicratonic basin, although the existence of potential source intervals is still unproved (Lüning *et al.* 2009).

The late Ordovician to earliest Silurian transgression that followed the Hirnantian glaciation in North Gondwana led to the deposition of world-class source rocks that are the origin of most of the North African petroleum. The basal Silurian shales ('hot' shales) are areally extensive and well developed in Algeria, Tunisia and western Libya; they also extend into Morocco and Niger and reach east to the Arabian Peninsula and Iraq, where they also play a major role in the regional petroleum systems (e.g. Lüning *et al.* 2000). Although the stratigraphy and the intra- and inter-basinal correlations of the Silurian Oued Chig and Toba el Hamar groups of the Taoudeni Basin are poorly constrained, mostly due to the patchy nature of the outcrops, it appears that the petroleum-generative potential of the fine-grained intervals within these units is not significant, at least in the Mauritanian sector of the Taoudeni Basin. This idea is supported by the results of the Atouila-1 well in the Malian side of the Taoudeni Basin, where the Palaeozoic sequence is best preserved (cf. Zhilong *et al.* 2008; Whenze 2009). The prolific Palaeozoic source rocks appear to be of low exploratory interest in the Taoudeni Basin, particularly

in its Mauritanian portion. This is probably the result of the relatively internal position occupied by the Taoudeni Basin within the WAC, which did not allow the establishment of environmental and chemical marine conditions comparable with those of other North African basins during the Silurian transgression.

Advances in instrumentation and analytical methods, especially in gas chromatography-mass spectrometry (GC-MS), have fostered research on the geochemistry of Precambrian organic matter. Hydrocarbon biomarkers in Precambrian rocks have the potential to provide important information about the prevailing organisms and metabolic pathways at that time and to shed light on ancient nutrient cycling in the biosphere. A recent concern with such studies arises from the fact that it is difficult, if not impossible, to establish with certainty the syngenicity of fossil molecules in many instances (e.g. Brocks *et al.* 2008; Brocks 2011).

Evidence exists, however, that indigenous soluble hydrocarbons have been preserved in some Archaean and Proterozoic sedimentary rocks (Summons & Walter 1990). Establishing correlations between the extracted hydrocarbons and kerogens has proved challenging, even in thermally well preserved rocks. These practical difficulties have been attributed to the lower organic richness of Mesoproterozoic and older source intervals, to the destructive effects of geological time on kerogen (Hoering & Navale 1987) and to differences in the composition of the Proterozoic and Phanerozoic precursor organic matter (Brocks *et al.* 2003). However, there are several published examples where such correlations have been possible, e.g. the MacArthur Basin of Australia and the Siberian Platform of Russia (Peters *et al.* 2005*b*), and these examples demonstrate that preservation of the indigenous biosignature is possible.

Blumenberg *et al.* (2012) critically assessed the syngenicity of biomarkers recovered from black shale samples from the Tod (Atar I_5) Formation in the W4 shallow coring well near El Mreïti, relatively close to other sampling locations considered in the present study. They concluded that it is unlikely that these hydrocarbons represent later stage contamination of the rocks. In contrast, our experience working in the area has yielded disparate results when it comes to establishing the syngenicity of isolated biomarkers. Although in some samples they seem to be indigenous to the host rock, some indications in other samples cast doubt over their genuine Proterozoic signature. This reiterates the great importance of a critical appraisal of the syngenetic origin of biomarkers in Precambrian rocks. The interested reader is referred to Brocks *et al.* (2008), Brocks (2011) and Blumenberg *et al.* (2012) for more detailed discussions on this topic.

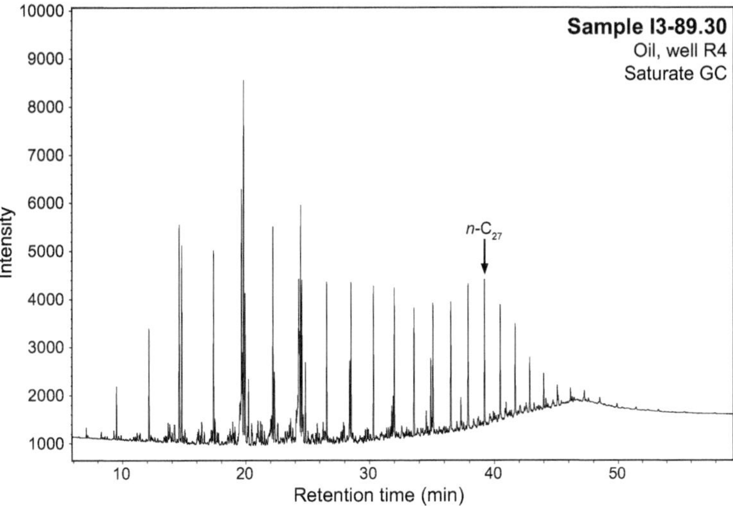

Fig. 7. Gas chromatograph of the saturate fraction of the oil sample I3-89.30 recovered from the Foum Chor (Char I₃) Formation clastics in well R4.

In all the samples analysed by us, aromatic hydrocarbons dominate in the organic rock extracts from black shale intervals in the Ksar Torchane (Atar I₄) and Tod (Atar I₅) formations, with relatively low saturates to aromatics ratios. The distributions of saturated hydrocarbons as seen in GC and GC-MS analyses of the source rock extracts and oils show the presence of n-alkanes ranging from n-C₁₁ up to n-C₃₆, with a maximum around n-C₁₇. The oil impregnations recovered from the Foum Chor (Char I₃) Formation in the R4 shallow core show expanded n-alkane distributions with respect to the source rock extracts, with high molecular weight n-alkanes up to n-C₂₉ (Fig. 7). These bimodal n-alkane distributions, showing high molecular weight n-alkanes at a maximum again near n-C₂₇ on GC analysis, could be related to an important microbial/algal lipid input to the organic matter and are consistent with Rock-Eval bulk pyrolysis data and organic petrography showing high HI and abundant amorphous organic matter in the Atar Group. Similar traits have also been reported in other analyses of organic extracts from organic-rich intervals in the the Atar Group (Blumenberg *et al.* 2012; Craig *et al.* 2013).

Terpanes and steranes are found at very low concentrations in all the analysed samples, with terpanes being much more abundant than steranes. The record of hopanes in Proterozoic sedimentary rocks is commonly associated with the dominance of prokaryotic organisms in Precambrian ecosystems (Summons & Walter 1990). In the case of steranes, both their syngenicity with the host rocks and their interpreted organic precursors have been subject to contention in many published

reports (e.g. George *et al.* 2007). However, fossil Precambrian steroids have been unequivocally identified in Neoproterozoic (Love *et al.* 2009) and Mesoproterozoic rocks (Summons *et al.* 2006). Furthermore, steranes preserved in Neoarchaean rocks have been interpreted as the first evidence of the early rise of the Eukarya (Brocks *et al.* 1999), predating by several hundred million years the first direct morphological evidence in the fossil record.

Special care was taken during sample handling and preparation to avoid contamination of the samples and the results of the biomarker analyses were evaluated in detail to establish with certainty their indigenous nature. As mentioned earlier, our results have been contrasting in this respect. In the case of the bitumen extracts from the Tod (Atar I₅) Formation in the R shallow wells, we consider that it is very unlikely that the recovered hopanes and steranes represent genuine Precambrian biomarkers. In these samples, the lowest extract yields correspond to the highest biomarker concentrations and the extracts all show the presence of the higher plant aromatic biomarkers cadalene, retene and 1,2,7-trimethylnaphthalene (Fig. 8). Although cadalene, retene and low levels of 1,2,7-trimethylnaphthalene can have microbial origins (Elias *et al.* 1997; Wen *et al.* 2000; Peters *et al.* 2005*b*), the co-occurrence of the three compounds together strongly suggests contamination. On the other hand, organic extracts from the W shallow wells and the oil impregnations sampled from those wells and R4 are considered to represent genuine Precambrian organic matter. These different interpretations could be explained by differences in maturity among the samples.

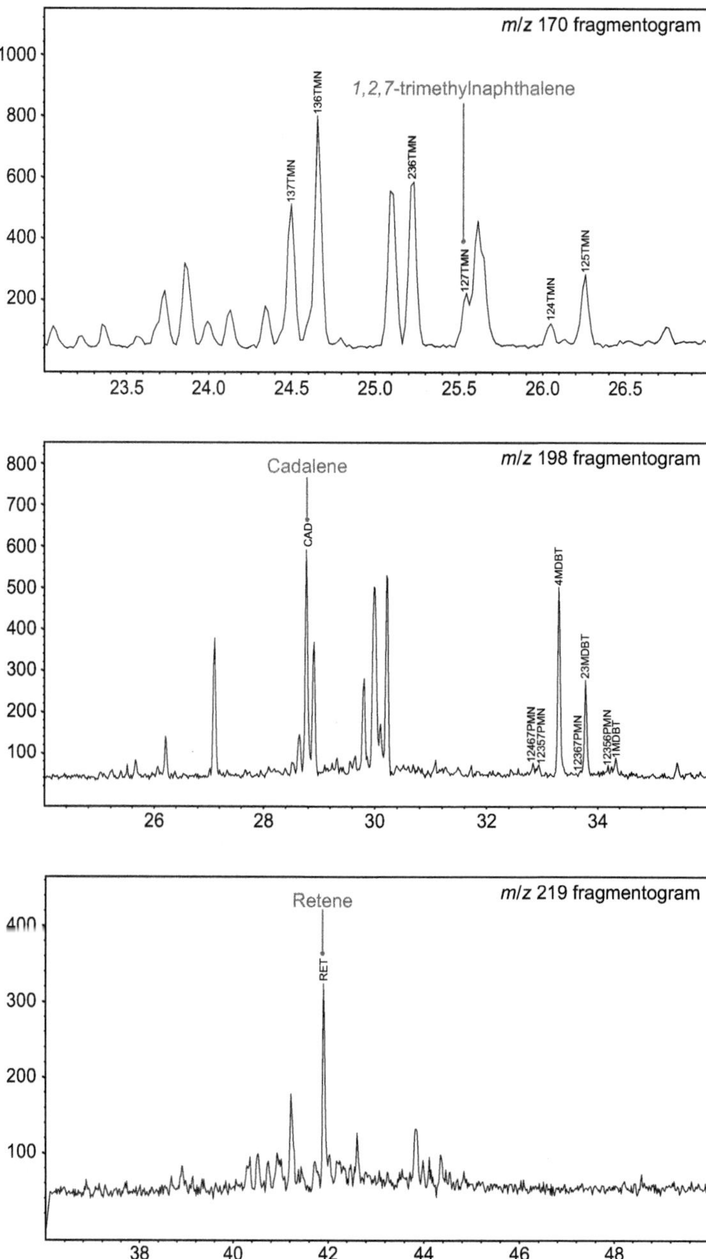

Fig. 8. GC-MS *m/z* 170, *m/z* 198 and *m/z* 219 fragmentograms of the aromatic fraction of an organic extract of sample R1-93 showing the occurrence of 1,2,7-trimethylnaphthalene, cadalene and retene. The co-occurrence of these compounds strongly suggests contamination of this sample, which yielded very low amounts of extract. The horizontal axis of each graph represents the retention time in minutes and the vertical axis is the relative intensity.

Hopanes are rarely preserved in high-maturity samples. The samples showing the lowest hopane concentrations, such as the organic extracts from the Tod (Atar I$_5$) Formation in the R wells, all tend to show a condensate-like pattern for the *n*-alkanes (as seen on the gas chromatograms and the *m/z* 57 traces from GC-MS analysis) and relatively abundant diamondoid contents.

The organic extracts from the Tod (Atar I_5) Formation in the W wells were derived from a marine source rock with important organic inputs from lipid-rich microbial precursors. The relative abundance of hopanes in these extracts indicates that the Taoudeni Basin ecosystems were dominated by prokaryotic organisms during the Mesoproterozoic. This is confirmed by the presence of 2α-methylhopanes reported by Blumenberg *et al.* (2012) and Craig *et al.* (2013), which are thought to be specific for oxygen-producing cyanobacteria such as those forming the microbial mats in Proterozoic stromatolites (Summons *et al.* 1999). Other lines of evidence also support the prokaryotic dominance in Archaean and Proterozoic seas (Knoll *et al.* 2006). Steranes are also present, albeit at much lower concentrations. Although hopanes are more resistant than steranes to thermal maturation and biodegradation processes (Peters *et al.* 2005*b*), this only holds true at relatively high levels of degradation. In consequence, the relative abundance of these compound classes in immature and moderately biodegraded samples appears to be indicative of the distribution of precursor organisms in ancient ecosystems. Blumenberg *et al.* (2012) suggested that the inferred environmental conditions for the Mesoproterozoic Taoudeni Basin, with limited nutrient availability and low oxygen conditions, may have only allowed a limited expansion of eukaryotic algae, in good agreement with other worldwide evidence from the same period (e.g. Anbar & Knoll 2002).

(a)

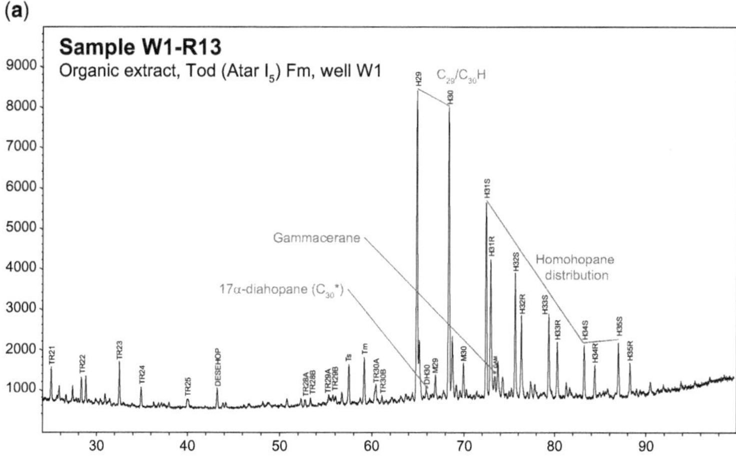

(b)

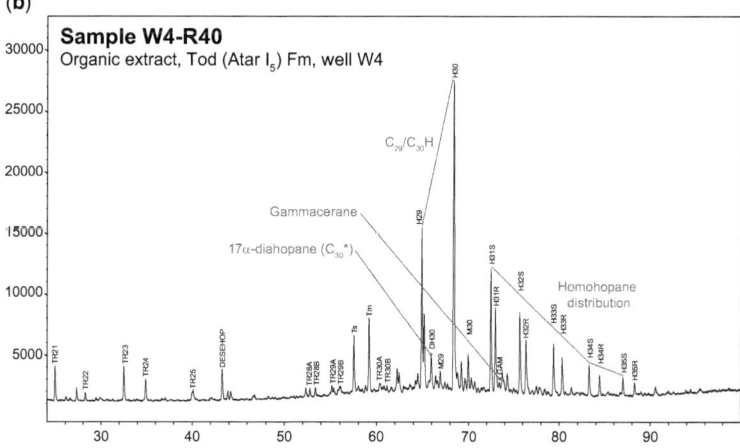

Fig. 9. Hopane distributions (m/z 191 fragmentograms) of the bitumen extracts (**a**) W1-R13 and (**b**) W4-R40 recovered from uppermost organic-rich interval of the Tod (Atar I_5) Formation in the W1 and W4 shallow wells, respectively. These m/z 191 fragmentograms illustrate lateral variations in the organofacies within this unit. The horizontal axis of each graph represents the retention time in minutes and the vertical axis is the relative intensity.

Comparing the hopane distributions (m/z 191) of two organic extracts from the uppermost fine-grained interval of the Tod (Atar I_5) Formation in wells W1 and W4 (Fig. 9), lateral variations in the organic facies within this unit become evident. High C_{29} nor-hopane/C_{30} hopane ratios, homohopane distributions in a V-shape and the presence of small quantities of gammacerane in the organic extract from well W1 (sample W1-R13) suggest a relatively clay-poor carbonate source rock deposited in anoxic marine sediments with restricted water circulation. On the other hand, extracts from the same unit in the W4 well (sample W4-R40) display relatively low C_{29} nor-hopane/C_{30} hopane ratios, linear homohopane distributions and the practical absence of gammacerane, indicating a clay-rich source rock deposited in a marine environment. In this well, the m/z 191 fragmentogram of the saturate fraction shows the significant presence of the rearranged hopane C_{30}* (17α-diahopane, DH30). Certain Mesoproterozoic crude oils have also been reported to contain C_{30}* (e.g. Summons *et al.* 1988*a, b*; Dutkiewicz *et al.* 2004). Blumenberg *et al.* (2012) argued that neither maturity nor biodegradation seems to explain the observed diahopane distributions, as had previously been suggested (Brocks & Summons 2003). The abundant presence of C_{30}* in the Taoudeni Basin probably originates from bacterial input to clay-rich sediments deposited under suboxic conditions in a shallow depositional environment (Peters *et al.* 2005*b*). The results presented here, compatible with dysoxic to anoxic shallow-marine depositional environments, are substantially consistent with conclusions drawn from other geochemical proxies (isotopic and elemental signatures), which also emphasize the highly heterogeneous character of the inferred water masses across the epicratonic Taoudeni Basin (Kah *et al.* 2012; Gilleaudeau & Kah 2013*a, b*).

The hopane (m/z 191) and sterane (m/z 217, 218) distributions of two organic extracts from the Tod (Atar I5) Formation of the W1 and W4 wells (samples W1-R13 and W4-R40) and one oil impregnation recovered from the W3 well (sample W3-R50) are presented in Figure 10. A good genetic correlation is observed between the organic extract from the W1 well and the oil impregnation from W3. Three oil impregnations have been sampled in W3. Two of these (samples W3-R46 and W3-R50) show good genetic correlation (Fig. 11). The Tod (Atar I_5) Formation drilled in the W1 area is therefore the source that originated these two W3 oils (or at least a lateral equivalent of same organic facies). A third oil impregnation sampled in W3 (sample W3-R51) is probably related to a different organic facies of the same source rock (Fig. 11). Good genetic correlation among different oil impregnations recovered from the Foum Chor

(Char I_3) Formation in the R4 shallow core is observed according to the hopane (m/z 191) and sterane (m/z 217, 218) distributions (samples I3–89.30, I3–90.70 and I3–92.30; Fig. 12). These oils show very low relative concentrations of tricyclic terpanes and this could imply a low maturity level for these samples. It is worth mentioning that relatively low concentrations of tetracyclic polyprenoids have been identified in the Foum Chor (Char I_3) Formation oils from R4. Although these tetracyclic polyprenoids compounds have been proposed as a diagnostic biomarker of lacustrine-sourced oils when present at relatively high levels, they have been identified in low concentrations in many marine samples, including marine-sourced oils of Precambrian age (Holba *et al.* 2003). The hopane (m/z 191) and sterane (m/z 217, 218) distributions of the Foum Chor (Char I_3) Formation oils from the R4 well, the W3 well oil impregnations and the organic extract from the Tod (Atar I_5) Formation of the W1 well are notably similar (Figs 10–12). Monoaromatic steroid distributions, indicative of eukaryotic species input to the organic matter and highly specific as a correlation tool, support the genetic relationship between the Foum Chor (Char I_3) Formation oils from R4 well and the organic extracts from the Tod (Atar I_5) Formation in W1, as they show only small differences (Fig. 13). The relative abundances of the C_{27}, C_{28} and C_{29} sterols are typical of a marine shale source rock (Moldowan *et al.* 1985) and do not show the relative enrichment in the C_{29} homologues that has been described for lower Palaeozoic and older samples, notwithstanding that this is not the first time that this exception has been noted (Peters *et al.* 2005*b*).

We interpret the close similarity in the distributions of the hopanes, steranes and monoaromatic steroids as a solid indication of genetic correlation between the migrated petroleum and the source rock, especially taking into account the fact that the levels of thermal maturity of the analysed organic extracts and oils are relatively similar. Oil–source correlation is one of the most critical pieces of knowledge that petroleum exploration can produce and is an essential prerequisite for the proper definition of a petroleum system. Establishing this genetic relationship between petroleum occurrence and the Tod (Atar I_5) Formation with certainty has thus contributed to significantly reducing the risk in petroleum exploration in the Taoudeni Basin.

Source rock maturity and burial history

The present day thermal maturity of the Taoudeni Basin source rocks can only be evaluated from the kerogen hydrogen and carbon elemental compositions, bulk pyrolysis (i.e. Rock-Eval) measurements and/or biomarker data. Conventional

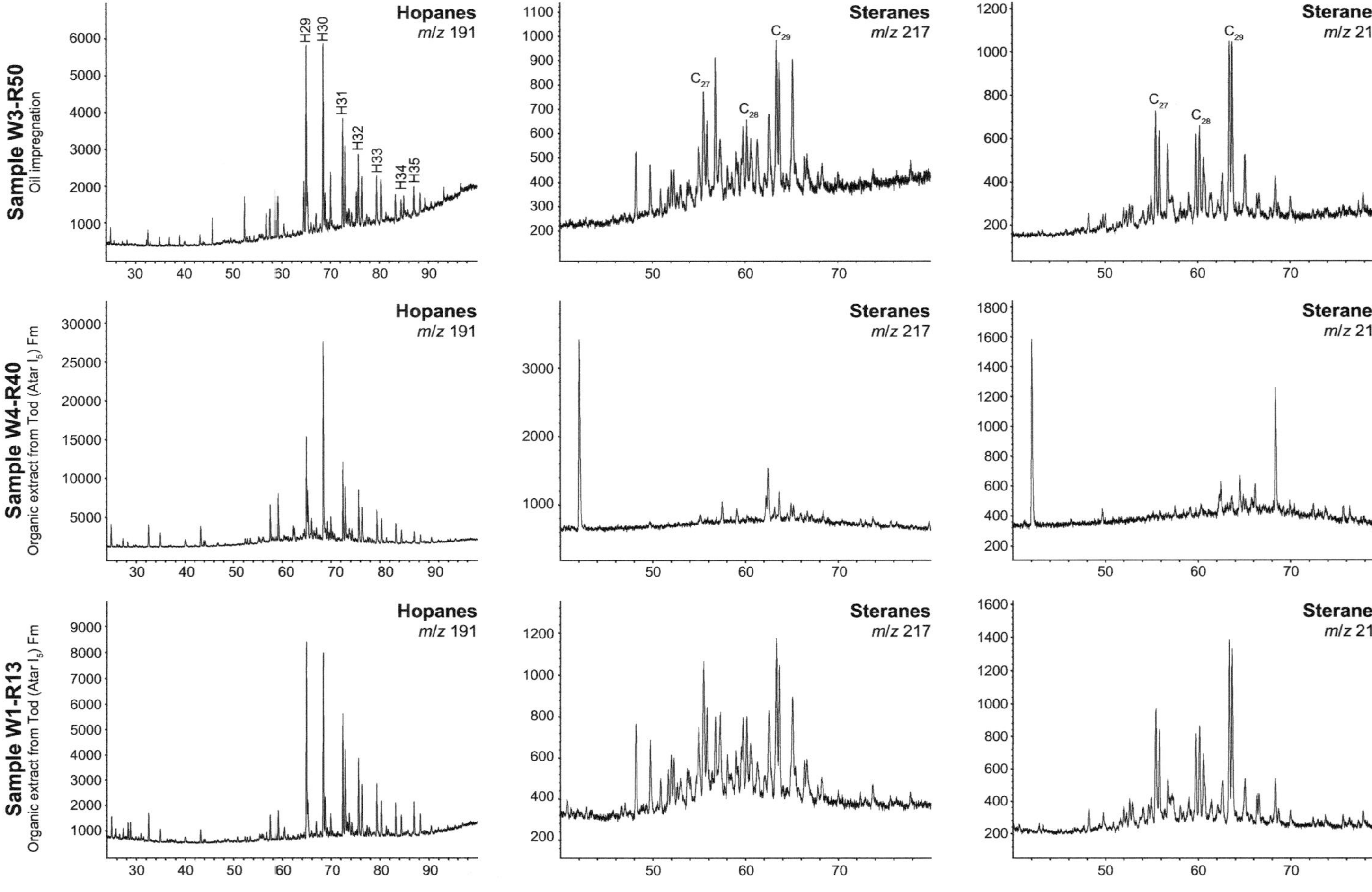

Fig. 10. Hopane (m/z 191 fragmentograms) and sterane (m/z 217 and 218 fragmentograms) distributions of the oil impregnation W3-R50 recovered from the W3 shallow well and the bitumen extracts W1-R13 and W4-R40. The horizontal axis of each graph represents the retention time in minutes and the vertical axis is the relative intensity.

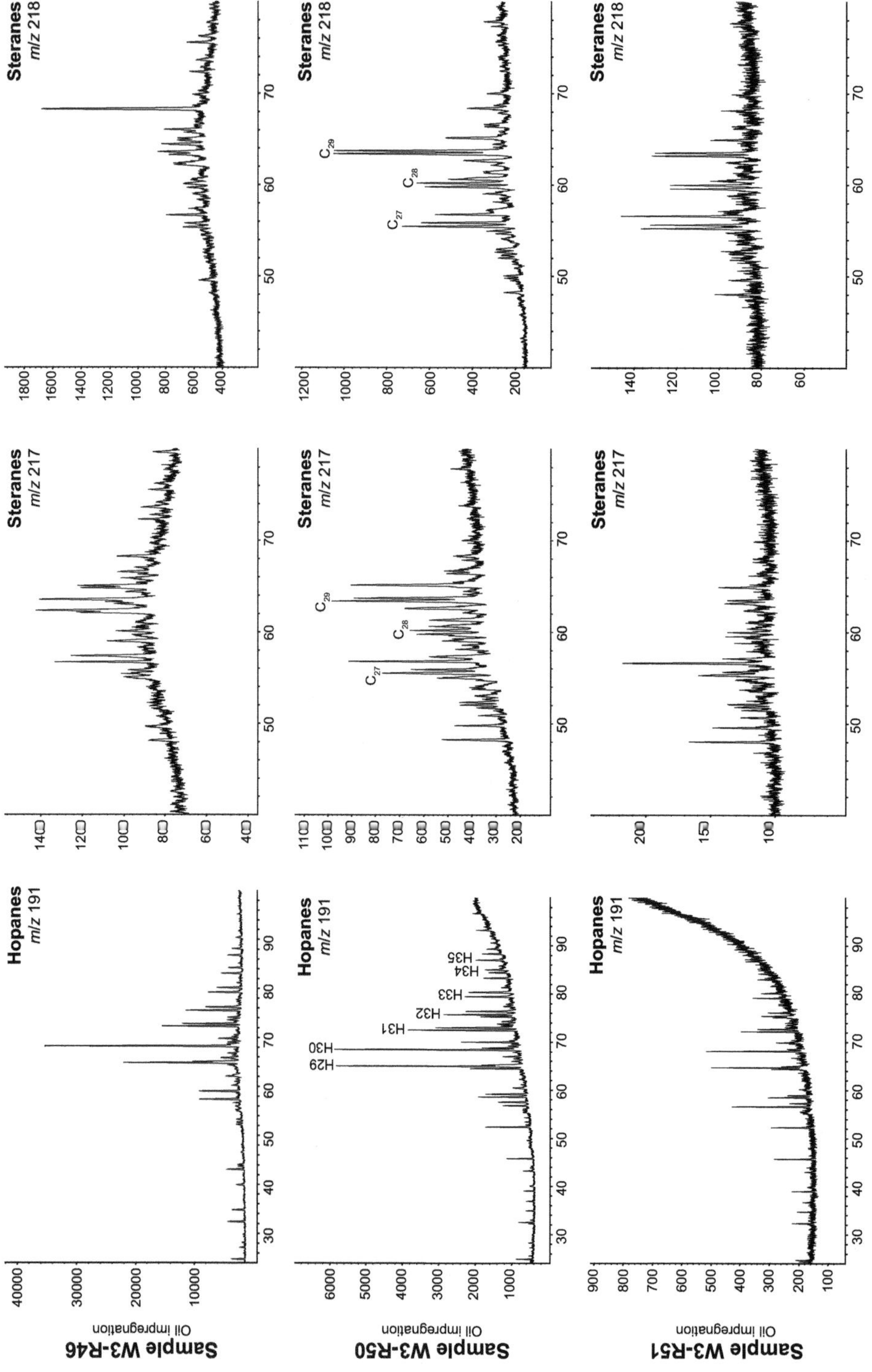

Fig. 11. Hopane (*m*/*z* 191 fragmentograms) and sterane (*m*/*z* 217 and 218 fragmentograms) distributions of the three oil impregnations recovered from the W3 shallow well. Samples W3-R46 and W3-R50 show a good genetic correlation, with very similar hopane and sterane distributions. The oil impregnation W3-R51 was probably sourced from a different organic facies of the same source rock. The horizontal axis of each graph represents the retention time in minutes and the vertical axis is the relative intensity.

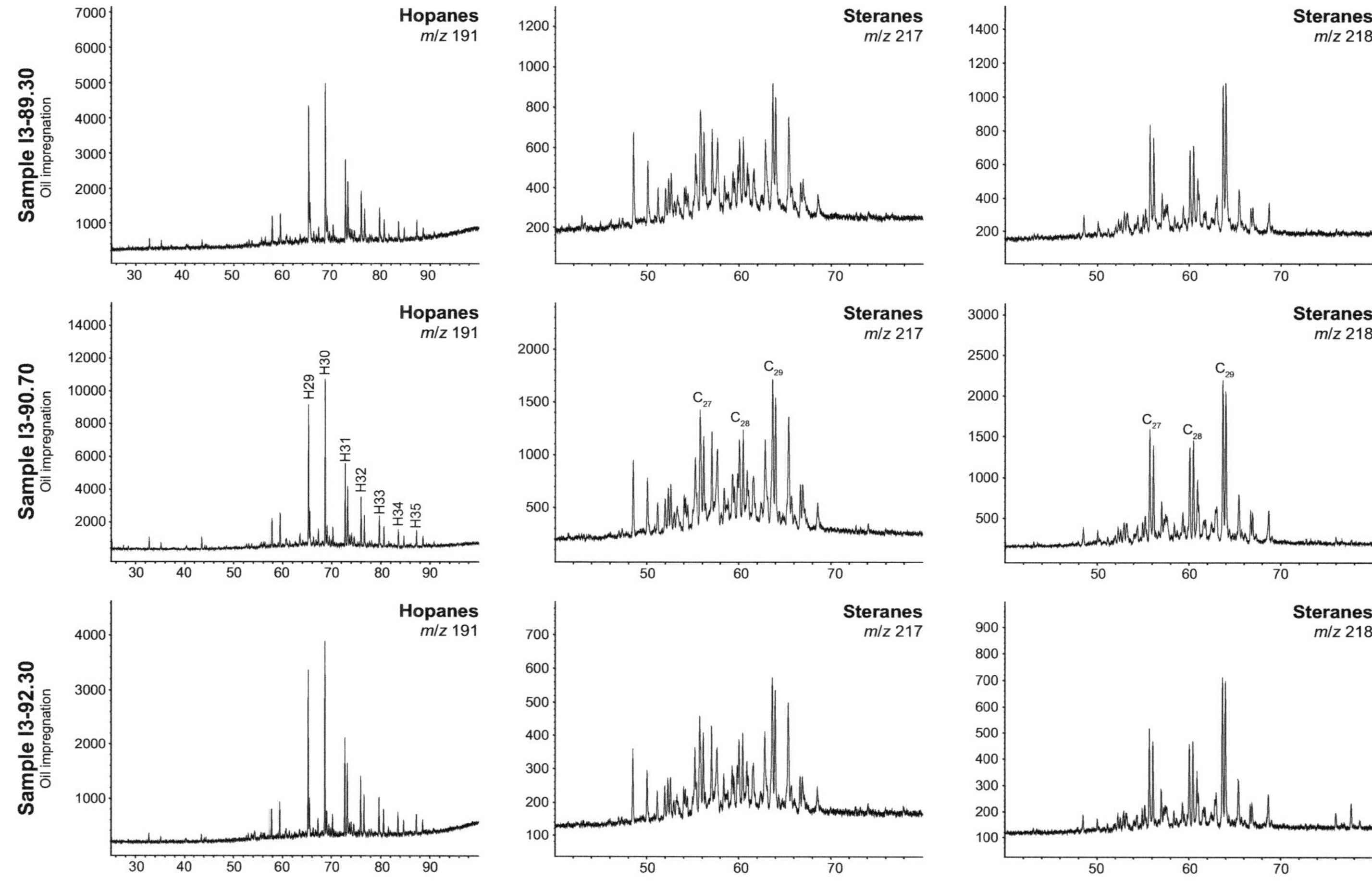

Fig. 12. Hopane (m/z 191 fragmentograms) and sterane (m/z 217 and 218 fragmentograms) distributions of the three oils sampled within the Char I₃ Formation clastics in the R4 shallow well. The three oils can be genetically correlated according to their hopane and sterane distributions. The horizontal axis of each graph represents the retention time in minutes and the vertical axis is the relative intensity.

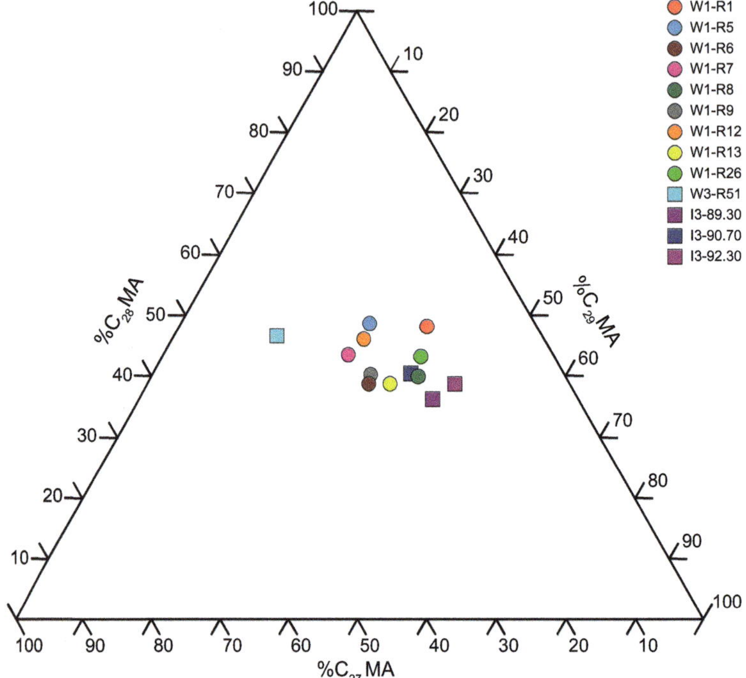

Fig. 13. Ternary diagram showing the relative abundance of the C$_{27}$, C$_{28}$ and C$_{29}$ C-ring monoaromatic steroids in the aromatic fractions of organic extracts from the W1 and W3 shallow wells (circles) and from oil impregnations from wells W3 and R4 (squares).

vitrinite reflectance measurements are not applicable for the main source rocks of the Taoudeni Basin as these predate the evolution of terrestrial plants. The maturity derived from Rock-Eval pyrolysis, reflectance measurements on bitumen and kerogen macerals other than vitrinite, and palynomorph coloration studies seems to decrease from the centre of the basin towards the northern margin, and from the NW towards the NE. Biomarker data, as we will show later, are consistent with these regional maturity trends.

These regional maturity variations can account, at least in part, for the observed lateral variations in the present day organic richness and remaining potential. However, if we try to reconstruct the original petroleum potential of the analysed samples (e.g. using the Claypool method described in Peters *et al.* 2005*a*), it also seems that the original richness was higher towards the NE and higher in the northern margin than in the centre of the basin. It is worth mentioning, however, that any conclusions about the organic content and facies variations towards the south are necessarily vague, as the only data from the centre of the basin are from the Abolag-1 well, where maturity has been severely affected by igneous activity (see Fig. 16) and yet petroleum generation is proved.

Organic maturity evaluations of Precambrian source rocks and oils based on biomarker data are best carried out using polyaromatic hydrocarbon ratios (e.g. George & Ahmed 2000). Polycyclic aromatic hydrocarbons – including dimethyl- and trimethylnaphthalene, phenanthrene, and dimethyl- and trimethylphenanthrene – were observed in relatively high concentrations in all the analysed samples. No higher molecular weight polycyclic aromatic hydrocarbon was detected. Detailed evaluation of the distributions of methylphenanthrene organic compounds indicates that extracts from the Tod (Atar I$_5$) Formation in the central and southwestern region of the Mauritanian sector of the Taoudeni Basin are more mature than the oils recovered from the Foum Chor (Char I$_3$) Formation in the W1 well (NE). However, these biomarkers indicate that Foum Chor (Char I$_3$) Formation oils from R4 well are more mature than the organic extracts and oils from the Tod (Atar I$_5$) Formation in the W shallow wells (Table 1; Fig. 14). The vitrinite reflectance equivalent, calculated from the methylphenanthrene index MPI-1, R_c^* (Radke *et al.* 1986; Radke 1988), indicates that the organic extracts from the Tod (Atar I$_5$) Formation located in the central and southwestern regions are in the main gas generation window (1.4–2% R_c^*) (Table 1;

Table 1. *Thermal maturity of bitumen extract samples and oil impregnations and stains from shallow wells along the northern margin of the Mauritanian sector of the Taoudeni Basin*

Sample	Formation	Well	MPI-1	R_c^*	Rock-Eval T_{max} (°C)
R1-87	Tod (Atar I_5)	R1	0.71	1.85	459
R1-93	Tod (Atar I_5)	R1	0.97	1.72	484
R2-184A	Tod (Atar I_5)	R2	1.20	1.58	420
R2-203	Tod (Atar I_5)	R2	0.76	1.84	422
R2-253	Tod (Atar I_5)	R2	1.52	1.39	448
R3-34	Aouleïgate (Atar I_{10})	R3	0.43	2.04	
R4-23	Ksar Torchane (Atar I_4)	R4	0.75	1.85	
R4-71	Ksar Torchane (Atar I_4)	R4	1.86	1.17	
W1-R1	Upper Tod (Atar I_5) Formation	W1	0.36	0.58	
W1-R4	Bitumen	W1	0.35	0.58	
W1-R5	Upper Ksar Torchane (Atar I_4) Formation	W1	0.29	0.54	
W1-R6	Upper Ksar Torchane (Atar I_4) Formation	W1	0.45	0.64	
W1-R7	Upper Ksar Torchane (Atar I_4) Formation	W1	0.15	0.46	
W1-R8	Upper Ksar Torchane (Atar I_4) Formation	W1	0.34	0.58	
W1-R9	Upper Ksar Torchane (Atar I_4) Formation	W1	0.27	0.53	
W1-R12	Lower Tod (Atar I_5) Formation	W1	0.50	0.67	
W1-R13	Lower Tod (Atar I_5) Formation	W1	0.33	0.57	
W1-R26	Upper Tod (Atar I_5) Formation	W1	0.43	0.63	
W4-R37	Lower Tod (Atar I_5) Formation	W4	0.24	0.51	435–440
W4-R38	Lower Tod (Atar I_5) Formation	W4	0.29	0.54	439
W4-R39	Lower Tod (Atar I_5) Formation	W4	0.32	0.56	437
W4-R40	Lower Tod (Atar I_5) Formation	W4	0.45	0.64	437
W4-R41	Lower Tod (Atar I_5) Formation	W4	0.16	0.47	438
W4-R42	Lower Tod (Atar I_5) Formation	W4	0.028	0.54	442
W3-R46	Oil	W3	0.13	0.45	
W3-R50	Oil	W3	0.34	0.58	
W3-R51	Oil	W3	0.09	0.42	

The MPI-1 ratio based on the distribution of methyl homologues of phenanthrene is given for all samples. A vitrinite reflectance equivalent (R_c^*) was calculated from MPI-1 using the empirical formula of Radke (1988). Rock-Eval T_{max} is shown for some samples for comparison.

Fig. 14). Other maturity indicators from the organic extracts from the R shallow wells are consistent with its inferred high maturity (condensate-like pattern of the *n*-alkanes, abundant diamonoids, no indigenous hopane or sterane, mature aromatics/naphthalenes, phenanthrenes and sulfur compounds, no aromatic steroids). The organic extracts from the Tod (Atar I_5) Formation in the W1 shallow well and oil stains sampled in W3 and the Foum Chor (Char I_3) Formation oils of R4 have all similar maturity levels associated with the early to middle oil window. The saturate biomarker parameters of these samples also indicate an early to middle oil window maturity. The C_{29} sterane isomerization ratios $\%\beta\beta$ and $\%20S$ did not reach equilibrium for these samples. The polyaromatic maturity-related parameters indicate that the Tod (Atar I_5) Formation bitumen extracts and oil stains from the W wells are early mature ($0.4-0.7\%R_c^*$) and the oil impregnations recovered from the Foum Chor (Char I_3) Formation in the R4 well (SW region) show a stage of maturity associated with the peak oil generation ($0.79-0.87\%R_c^*$). It is worth mentioning that a similar early stage of maturity was reported by Blumenberg

et al. (2012) for the organic matter of the W4 well based on both saturate and polyaromatic biomarker ratios. Other maturity indices based on biomarkers, such as the methyldibenzothiopene ratio and the methyladmantane and methyldiamantane indices (Chen *et al.* 1996), all show good correlations with methylphenanthrene indices and R_c^* on the highest maturity samples (Fig. 15).

Figure 16a shows the burial history of the Taoudeni Basin at the Abolag-1 well location. The evolution of the Tod (Atar I_5) Formation source rock can be summarized as follows. After deposition, the basin continued to gently subside until the Panafrican tectonic events. Although this long and complex orogeny involved several episodes of uplift, we can assume for modelling purposes that the effects in terms of maturation were comparable with those of a single uplift and erosion event finishing around 650 Ma and mainly affecting the southern and southwestern areas of the basin. We roughly estimate an eroded thickness ranging from 0 and 1400 m, although the assessment is very difficult from seismic data and thermal indicators. Subsidence resumed after the Panafrican events and

(a) Southwestern and central region

(b) Northeastern region

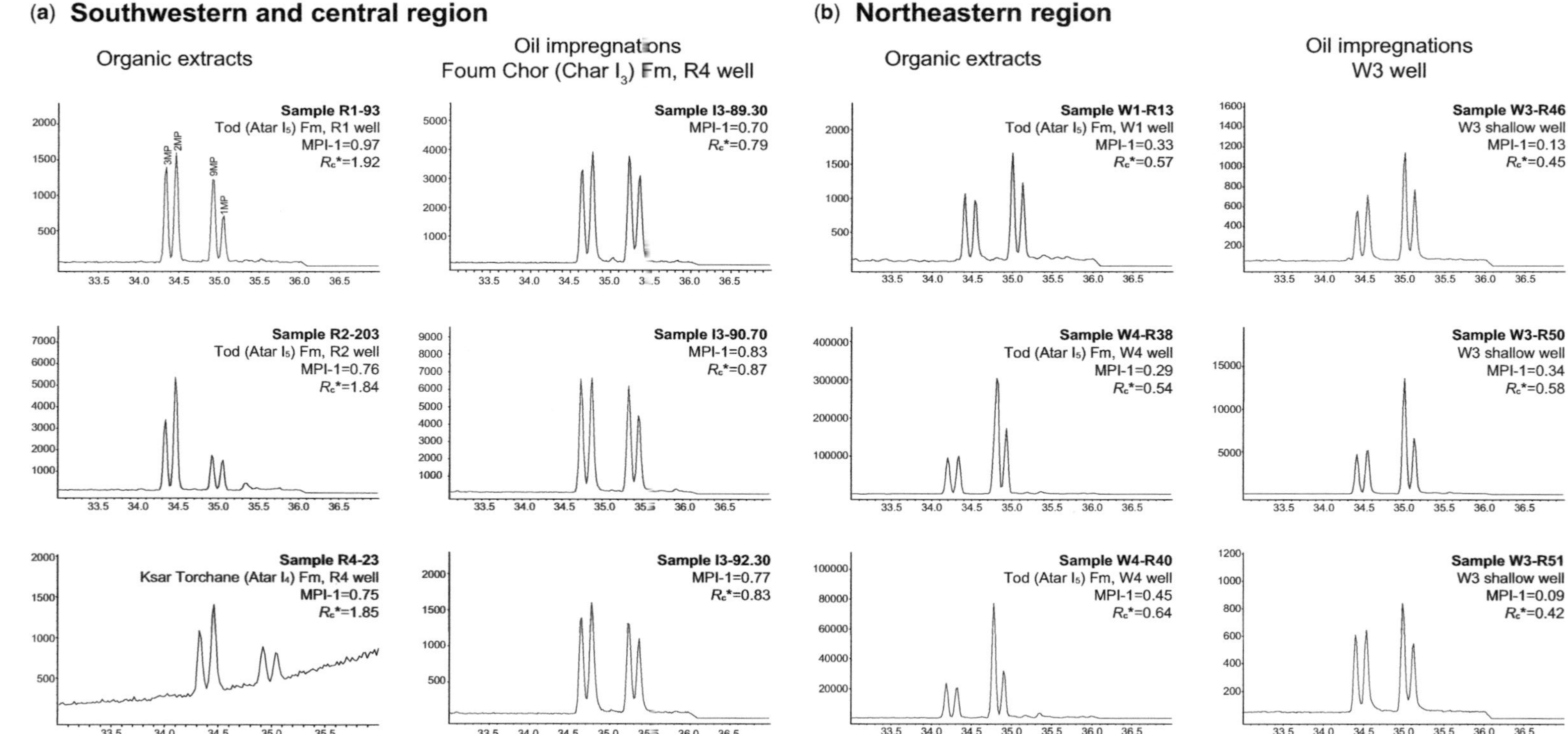

Fig. 14. Mass chromatograms m/z 192 from the aromatic fraction of bitumen extracts and oil impregnations from shallow wells across the northern edge of the Mauritanian sector of the Taoudeni Basin. The methylphenanthrene index MPI-1, measured using peak heights for phenanthrene and methylphenanthrenes, is shown for each sample. Also shown is the vitrinite reflectance equivalent, R_c^*, calculated from the MP-1 (Radke *et al.* 1986; Radke 1988). Lateral variations can be observed in both bitumen extracts and migrant petroleum from the SW towards the NE, in good agreement with other thermal maturity indicators. (**a**) Organic extracts and oils from the southwestern and central region of the northern margin of the basin. (**b**) Organic extracts and oil impregnations from the northeastern region. The horizontal axis of each graph represents the retention time in minutes and the vertical axis is the relative intensity.

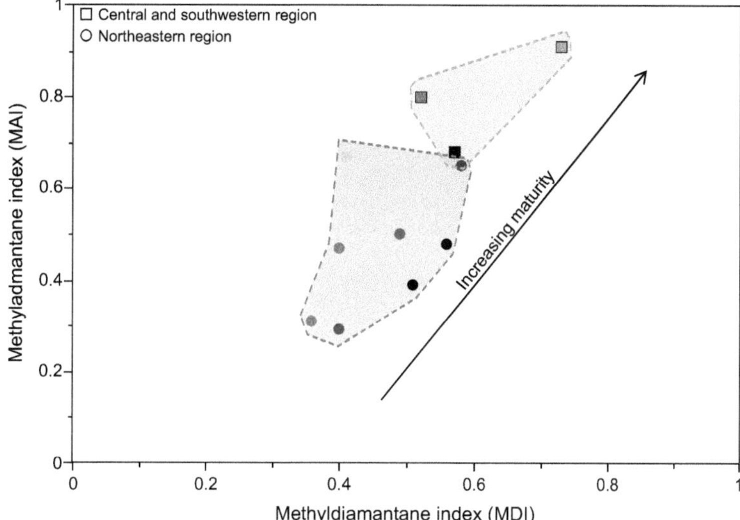

Fig. 15. Cross-plot of the diamondoid hydrocarbon ratios methyladmantane index versus methyldiamantane index of Chen *et al.* (1996) showing the thermal maturity trend along the northern margin of the Taoudeni Basin. The organic extracts from the southwestern and central areas (squares) are more mature than those from the northeastern region (circles).

continued until the Hercynian phase. Hercynian uplift and erosion were widespread across North and West Africa and affected the Taoudeni Basin mainly during the Permian. We estimate the maximum eroded thicknesses during this period to range between 200 and 1400 m in the centre of the basin. From our reconstruction, the Mesoproterozoic main source rock reached its deepest burial before the Hercynian uplift in most of the Taoudeni Basin. Slow subsidence resumed again and continued during the Mesozoic, possibly interrupted in the Cretaceous by the Austrian phase (although it is not clearly documented in the Taoudeni Basin), and until the Cenozoic when the Alpine orogeny took place. We estimate the corresponding amount of erosion to vary between 0 and 900 m. Regarding the burial history, we can assume that most of the maturity of the main source related to burial was certainly achieved before the Hercynian uplift.

Although the inferred importance of the thermal effect of the Mesozoic intrusions has already been highlighted, it is difficult to assess the consequences on maturity at a basin scale. Figure 16 shows the calibration of one-dimensional basin models of the Abolag-1 well obtained with maturity (vitrinite reflectance equivalent) and temperature (corrected bottom-hole temperature) data. Note that, in the absence of vitrinite, bitumen and zooclast (acritarchs) reflectance was measured. Measurements have been converted into an equivalent vitrinite reflectance ($\%R_{o,eq}$) using published techniques. The Cole (1994) formula was used for the

samples where the zooclast reflectance had been measured and the Riediger (1993) algorithm for the bitumen reflectance values. The spore coloration index, measured on acritarchs, was compared with the calculated equivalent vitrinite values. The spore coloration index values are between 8.5 and 9 at the base of the Abolag-1 well, corresponding to a range of $1.3-1.8\%R_{o,eq}$, in agreement with the equivalent vitrinite reflectance calculated from reflectance on bitumen and zooclasts.

The local effect of doleritic intrusions is clearly visible when comparing a model considering only burial (Fig. 16b) in a regional rifting thermal evolution (initial peak of heat flow during the Char Group deposition and then progressive decrease with an increase in temperature during the Jurassic magmatic phase) with the same regional history with local heat brought by high-temperature intrusions (*c.* 1100°C) (Fig. 16c). Far from the intrusions, in the deepest portion of the well, the same trend is observed with the Tod (Atar I_5) Formation source reaching an early gas window maturity. In the portion affected by doleritic sills and dykes, maturity can reach measured values as high as $9\%R_{o,eq}$, with a spiky trend of the calculated vitrinite reflectance curve clearly showing their local effect on maturation. It is difficult to evaluate the magnitude of this magmatic phase and its impact on regional maturity, or if the effect remained restricted to the intrusions themselves.

We consider the Mesozoic magmatic events to be of tremendous relevance for the petroleum

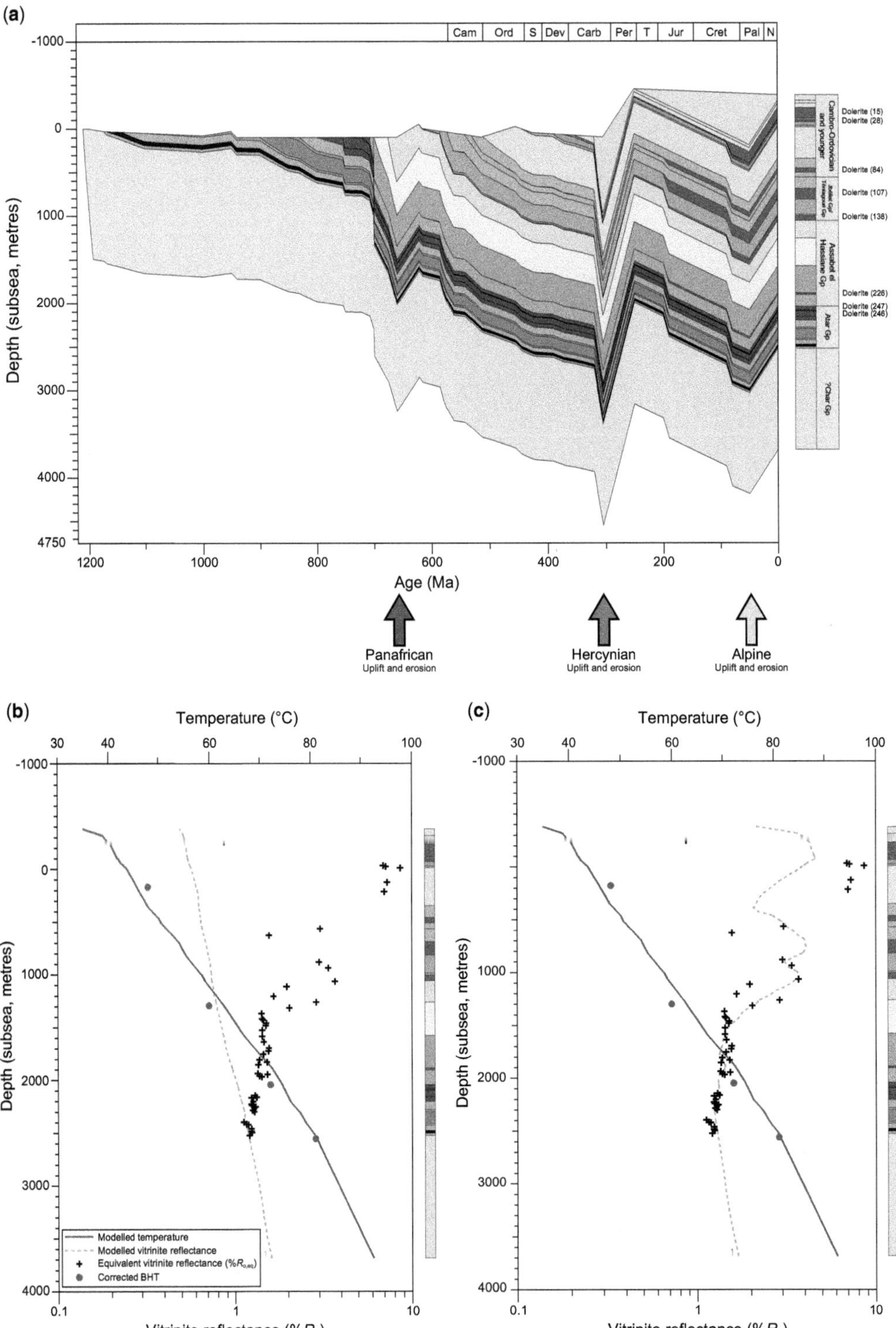

Fig. 16. (a) Burial history of the Taoudeni Basin at the Abolag-1 well location. These curves were calculated from tops identified in the well. The depth to basement was estimated from two-dimensional seismic data. Three main

prospectivity of the Taoudeni Basin (Baudino *et al.* 2014) for various reasons. First, the emplacement of the intrusive bodies could result in the creation of viable structural traps and the dolerite sills and dykes could also potentially provide seals. Second, the intrusions might have affected the preservation of potentially existing petroleum accumulations. Evidence of contact metamorphism in association with dolerite sills and dykes are common in outcrops. In the Abolag-1 well, doleritic intervals were emplaced within the basal part of the Assabet el Hassiane Group as well as within the main reservoir target of the Atar Group (Figs 5 & 16), where the presence of abundant bitumen indicates secondary processes of alteration of the precursor petroleum. Third, hydrocarbon generation and expulsion resulting from the local effect of intrusions on source rock maturation have been demonstrated in other basins (e.g. Baudino *et al.* 2004; Rodriguez Monreal *et al.* 2009). Finally, the Mesozoic magmatism might provide insights into the thermal evolution of the Taoudeni Basin that need to be incorporated into the assessment of its petroleum systems. The top-down control of shallow mantle temperatures invoked by the Coltice *et al.* (2007, 2009) model implies significant disturbances of the thermal state of the lithosphere. Although the precise magnitude and temporal evolution over the entire basin are difficult to predict, the Mesozoic magmatic phase might have triggered a basin-wide increase in temperature affecting maturation over several million years – possibly the so-called subregional- or regional-scale process identified in the AFTA study.

In conclusion, thermal stress indicators reveal a wide range of maturities, from the early oil window to overmature, for both source rocks and hydrocarbons, across the Taoudeni Basin, resulting from a long and complex thermal history where regional long-term controls (subsidence and uplift) have been complicated by local events (magmatic intrusions).

Reservoir potential

Reservoir potential of the Atar Group

The Atar Group records shallow-marine deposition over a regionally extensive epicratonic basin during Mesoproterozoic times. This unit is characterized by an alternation of stromatolite-bearing carbonates and fine-grained siliciclastic sediments with subordinate evaporites. This characteristic alternation may result primarily from regional climatic or tectonic controls leading to the episodic influx of siliciclastic material from source areas outside the Taoudeni Basin (Bertrand-Sarfati & Moussine-Pouchkine 1988; Moussine-Pouchkine & Bertrand-Sarfati 1997). Facies associations and sedimentary features observed in both the siliciclastic and carbonate strata indicate a subtidal to intertidal depositional setting for the Atar Group (Bertrand-Sarfati 1972; Bertrand-Sarfati & Moussine-Pouchkine 1985; Kah *et al.* 2009).

Carbonate reservoirs are proven within the Atar Group. The open hole drillstem test conducted in the lower part of the units stratigraphically equivalent to the Touiderguilt (Atar I_{11}/I_{12}) Formation in Abolag-1 (interval 2523–2536 mMD) flowed gas and possibly some condensate (Fig. 5). According to the final well report and composite log, this interval is composed of dolomitized limestones and dolostones with good porosity (average porosity from electrical logs *c.* 9%, reaching up to 11%). Previous workers have associated the producing intervals in Abolag-1 with heavily fractured and brecciated limestones (Lüning *et al.* 2009; Craig *et al.* 2010). However, the data at our disposal from that interval were limited to the original mud and electrical logs and thus we have not been able to confirm a fractured reservoir character in the tested interval in Abolag-1.

Fieldwork studies based on outcrops along the northern rim of the basin have allowed us to identify the intervals comprising the Tod (Atar I_5) to Tawaz (Atar I_7) formations and the Touiderguilt (Atar I_{11}/I_{12}) Formation as the most interesting in terms of reservoir potential within the Atar Group (Fig. 2). The lower interval is characterized by the fairly constant presence of microbial deposits, represented by stromatolite reefs. These microbial bioconstructions show up in impressive outcrops, interbedded with minor evaporitic layers, along the northern rim of the Taoudeni Basin and they are composed of columnar, conical and branching stromatolites (Kah *et al.* 2009). Microbial organisms dominated the Mesoproterozoic ecosystems and they could dominate most of the depositional settings. It is still a matter of discussion whether these bioconstructions

---

Fig. 16. (*Continued*) uplift and erosion phases were considered: Panafrican, with an estimated eroded thickness around 250 m; Hercynian, erosion of about 850 m; and Alpine, with eroded thickness about 500 m. Notice also the presence of several doleritic intrusions of Mesozoic age. (b and c) Thermal calibration of one-dimensional basin models of the Abolag-1 well (**b**) without considering the local effect of Mesozoic intrusions and (**c**) with additional heat from intrusions. Measured vitrinite reflectance values (crosses) are equivalent reflectance values calculated from bitumen and zooclast reflectance measurements. Bottom-hole temperatures (BHTs; solid circles) are corrected temperatures. The solid and dashed lines are the calculated temperature and maturity curves, respectively. The Tod (Atar I_5) Formation, considered as the main source rock of the Taoudeni Basin, is reaching the early gas window (*c.* 1.2%R_o) at this location.

formed alternations of low-relief biostromes or high-relief build-ups. In outcrops, vuggy porosity has been described in these biostromes between the stromatolite constructions (Kah *et al.* 2009). The upper potential reservoir interval (Touiderguilt Formation) is less known from outcrops, but it seems to be barren of palaeofauna according to regional depositional models describing stromatolite build-ups only in the Atar formations Ksar Torchane (Atar I_4) through Aouleïgate (Atar I_{10}). However, the Touiderguilt (Atar I_{11}/I_{12}) Formation has only been described from a few localities, where it is often largely removed by erosion of the overlying Assabet el Hassiane Group clastic sediments and is locally affected by intense silicification processes.

The shallow R wells drilled by Repsol along the northern margin of the basin can provide a useful insight into the reservoir quality of the Atar Group carbonates. The most interesting shallow wells are R2 and R4 as they intersect the lower carbonate reservoir interval consisting of the Tod (Atar I_5) to Tawaz (Atar I_7) formations. The R2 well drilled an almost complete section through this interval, whereas R4 drilled the fine-grained siliciclastic sediments of the Ksar Torchane (Atar I_4) Formation and reached the top of the lowermost biostrome of the Tod (Atar I_5) Formation. Well R3 intersected the Touiderguilt (Atar I_{11}/I_{12}) Formation, but it was mainly in the crushed zone and therefore very little information about the carbonate facies and reservoir properties could be obtained.

About 400 m of core were studied in detail from the R2 and R4 shallow cores. A diagenetic study of these cores has allowed the recognition of a complex paragenesis composed of early and burial processes, including, from older to younger, the following:

(1) Early post-depositional to early burial dolomitization: replacement of the initial limestone or of very early dolomite and the first creation of vuggy porosity. As a product of this process, we find micro- to mesocrystalline dolomite. This dolomitization does not completely obliterate the primary sedimentary structures (relict microbialite features; Fig. 17a).

(2) First petroleum migration, which was partially flushed away in some places (e.g. in larger vugs), whereas in others it occupied the free dolomite intercrystalline porosity and relict primary porosity.

(3) Transformation of this first petroleum liquid phase into bitumen, either due to thermal cracking or to biodegradation (Fig. 17b).

(4) Second important dolomitization, with dolomite cement (mesocrystals) precipitating

in large vugs, in some places over bitumen coating these vugs (Fig. 17b).

(5) Third burial dolomite cement resulting in macrocrystals (often saddle dolomite) further filling the space in larger vugs (Fig. 17b).

(6) Pressure solution originating subhorizontal stylolites (this also continues further in time; Fig. 17b).

(7) Intense late corrosion processes, possibly connected to acid and hydrothermal fluids, creating very large vugs (Fig. 17c).

(8) Partial or total filling of the large vugs by extremely large calcite, dolomite and quartz crystals, probably precipitating from the same hydrothermal fluids (homogeneization temperature from 150 to 200°C; Fig. 17c).

(9) Second petroleum migration phase, subsequent to late cements (no petroleum inclusions found), filling the porosity left in the large vugs and recently created porosity (e.g. stylolites, fractures; Fig. 17b) and, in some places, corroding the latest cements (Fig. 17d).

(10) Last degradation of liquid petroleum into bitumen, also in this case possibly related to a thermal event or to biodegradation in the telogenetic phase in the case of the shallow cores (Fig. 17d).

Bitumen seems to occur regionally in the Atar Group carbonates and has been observed at various stratigraphic levels in Abolag-1 and Ouguiya-1, towards the centre of the basin, as well as along its northern margin in shallow cores and outcrop samples. It is also present in siliclastic intervals of the Char Group. Bitumen has also been reported in the El Mreïti Group carbonates of the Yarba-1 well in Mali. The study of thin sections from the R shallow cores has revealed that bitumen appears to have migrated, or formed, in at least two different phases, one older and one younger than the late dolomite cements. Note that the difference between the bitumen filling earlier intercrystalline pores and the corroding bitumen (or bitumen filling late pore space created by corrosive fluids) is evident because the former does not cut through crystal faces (Fig. 17e), whereas the latter does (Fig. 17d). This observation is consistent with petroleum system models of the Taoudeni Basin that predict two phases of hydrocarbon charge (Craig *et al.* 2010; Baudino *et al.* 2014).

Organic solids are present in many major carbonate reservoirs around the world and particularly in many Proterozoic petroleum systems involving carbonate reservoirs, such as in the carbonate stringers of the Neoproterozoic Ara Group in the South Oman Salt Basin (Schoenherr *et al.* 2007), in some reservoirs of the East Siberia Basin (Kashirtsev

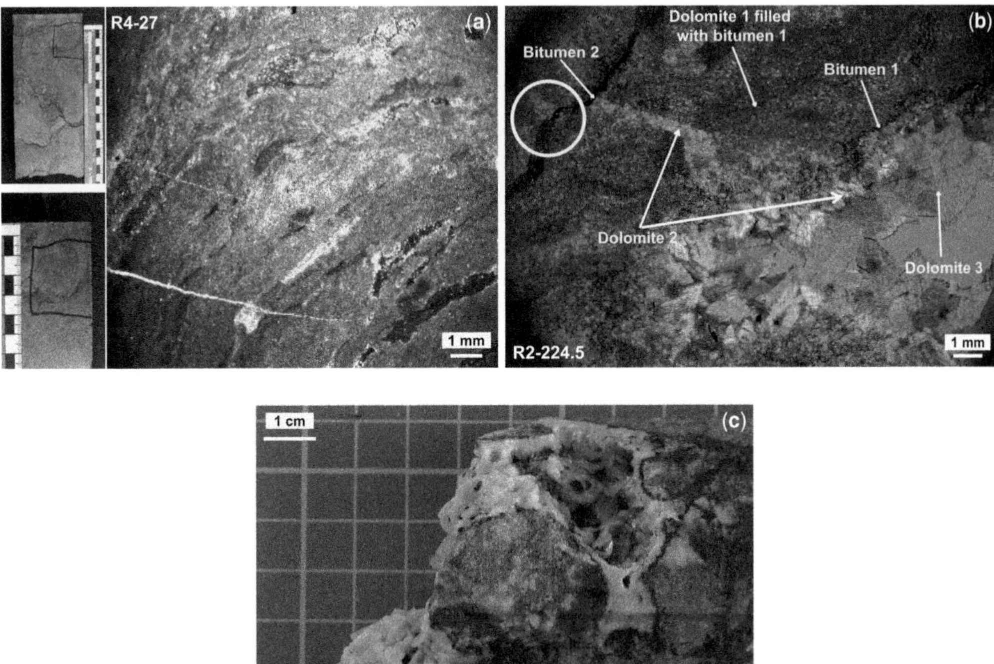

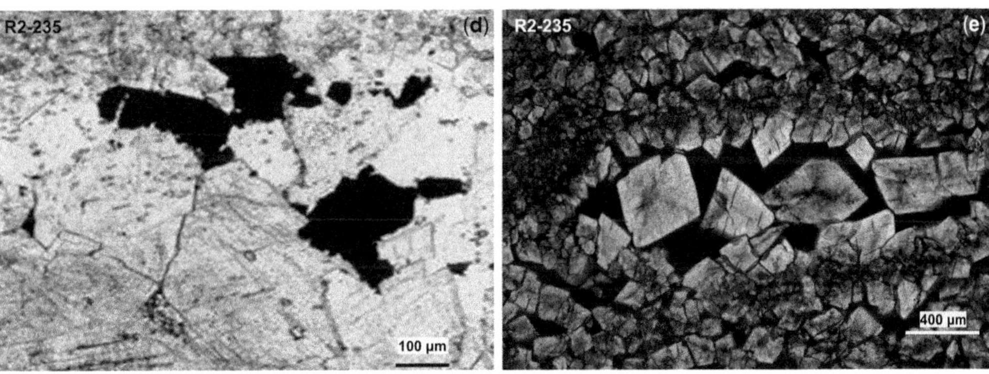

Fig. 17. (a) Core slab on the left and plane-polarized light, epoxy-impregnated and unstained thin section on the right from shallow core R4 at 27 m depth. Microcrystalline early dolomite replacement maintaining the original columnar stromatolite depositional structure can be seen. (b) Plane-polarized light, epoxy-impregnated and unstained thin section from shallow core R2 at 224.5 m depth. Three different dolomite and two different bitumen phases can be recognized. They occurred in this order: microcrystalline dolomite 1 replaced the original carbonate rock, then a first oil migration phase plugged the intercrystalline porosity and rimmed the large vug, degrading to bitumen 1. Mesocrystalline dolomite 2 cement precipitated over the vug edge, on top of bitumen 1 or dolomite 1, and filled an open fracture in the first quadrant of the photograph. Subsequently, large saddle dolomite crystals filled further space in the vug. Afterwards, pressure solution dislocated the vein filled by dolomite 2 (circle) through a stylolite that was later filled with a new hydrocarbon phase, which then degraded to bitumen 2. (c) Very large size vug partially filled with large dolomite and quartz crystals. In these intervals, cavernous porosity can reach values of 25%. (d) Plane-polarized light, not impregnated and unstained thin section from shallow core R2 at 235 m depth. Late dolomite macrocrystals show corrosion porosity filled by bitumen. (e) Plane-polarized light, epoxy-impregnated and unstained thin section from shallow core R2 at 230 m depth. The first bitumen phase fills early intercrystalline porosity without corroding the dolomite crystals.

et al. 2010) or the reservoirs of the Weyjuan gas field in the Sichuan Basin (Korsch *et al.* 1991). A variety of processes have been described that can lead to the formation of solid bitumens (Walters *et al.* 2006), essentially thermal cracking, thermochemical sulphate reduction, gas de-asphalting, reactive polar precipitation and biodegradation. The different temperatures at which each of these processes operates lead to differences in the maturity of the resulting bitumen. The origin of the solid hydrocarbons in the Atar Group carbonates is at present not well established, although we have observed some heterogeneity in bitumen maturity; organic solids in the Atar Group are probably the result of different local processes, mainly thermal cracking associated with igneous activity in Abolag-1 and other secondary hydrocarbon alteration processes (e.g. biodegradation) in other locations.

Several samples from the R shallow wells were collected and used for porosity estimation and permeability assessment. Of the 400 m studied, about 30 m of fair quality carbonate reservoir were observed, with porosity (mainly vuggy and secondly intercrystalline and residual from primary microbial structures) ranging from 4 to 12%, part of which was effective (Fig. 18). Another 40 m of core, mainly concentrated in the Tod (Atar I$_5$) Formation, appear to have been associated in the past with a high intercrystalline effective porosity (ranging from 7 to 21%) subsequently plugged by bitumen (Fig. 19).

Few data could be gathered from the Abolag-1 well, located several hundreds of kilometres away from the shallow wells and outcrops. Although originally recovered, no core sample now remains and we were only able to gather a few drilling cuttings from the Atar Group. The inspection of these

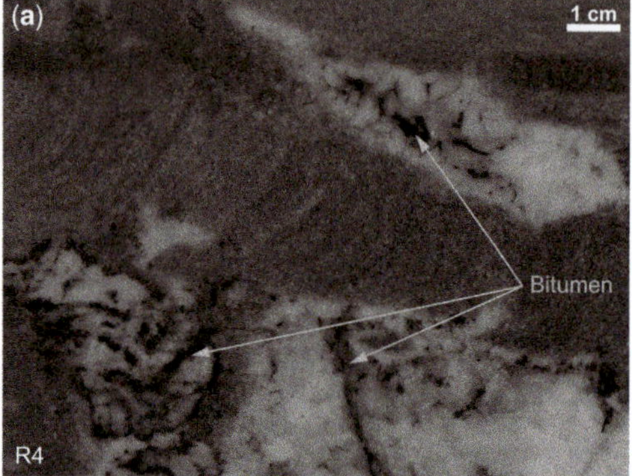

Fig. 19. Bitumen filling porosity in samples from the Tod (Atar I$_5$) Formation in the R4 shallow well as seen (**a**) at the core scale and (**b**) under the microscope in a plane-polararized micrograph. In (b), two phases of bitumen formation, or migration, can be observed, one older and one younger than the late dolomite cements. The thin section is resin-impregnated and not stained for carbonate species.

Fig. 18. Plane-polarized light micrograph of sample R4-20.1m from the R4 shallow well showing fair reservoir quality. Porosity is indicated by the blue resin-impregnated areas. The section was not stained for carbonate species. The macroporosity reached 5% in this sample, with an additional 6% intercrystalline microporosity.

cuttings was integrated with a detailed study of the original mud and composite logs of the well and with subsequent petrophysical and elastic properties analyses. Only the lower carbonate reservoir interval, consisting of the Tod (Atar I$_5$) to Tawaz (Atar I$_7$) formations, is represented among the available cuttings samples. The cuttings from this interval are all strongly dolomitized. They are composed of meso- and macrocrystalline dolomite, in which some microbial relict laminations are possibly visible; the pre-existing intercrystalline porosity (up to 20–25%) is completely plugged with bitumen. In contrast, free porosity, containing some brine and little gas, has been calculated for the upper carbonate reservoir interval (Touiderguilt Formation). In these upper units, a 3–10% range of free porosity was inferred from elastic impedance calculations, whereas petrophysical log analyses resulted in 43 m of net reservoir thickness (0.37 net to gross) with an average porosity of 11%. Lower values of porosity result from these analyses in the I_5 to I_7 interval, where bitumen fill occurs. According to elastic impedance analysis, most of the live fluid in the porous sequences is brine.

The nature of the potential carbonate reservoirs of the Atar Group was also investigated using seismic data. The characteristic carbonate–siliciclastic lithological alternation of this unit is well represented in seismic data, with thicker reflectors corresponding to the carbonate intervals and thinner reflectors corresponding to 'shaly' intervals. At the Abolag-1 location, the Touiderguilt (Atar I_{11}/I_{12}) Formation shows a well-defined alternation of high acoustic impedance reflectors, corresponding to tight carbonates, and low acoustic impedance reflectors, corresponding to more porous, fluid-bearing carbonates (Fig. 20). A thick high acoustic impedance reflector is observed in the interval consisting of the Tod (Atar I_5) Formation to the Tawaz (Atar I_7) Formation, representing dolomite filled with bitumen. This alternating seismic character is fairly uniform across the basin, although some significant lateral variations can be observed. Despite the seismic data being limited to two-dimensional lines, many of them vintage, and difficulties arising from the scarcity of calibration data, these variations in seismic character reflect the combined effect on the seismic response of variations in carbonate facies, fluid type and the presence of bitumen and may therefore be used to de-risk individual prospects. In terms of depositional geometry, lateral thickness changes can be easily appreciated in the Tarioufet (Atar I_6) to Tawaz (Atar I_7) interval across

the study area, whereas the thickness of the upper Atar carbonate reservoir (Touiderguilt Formation) is remarkably uniform. This suggests the occurrence of microbial build-ups with a certain relief in the lower carbonate reservoir interval.

In summary, the Atar Group seems to be fairly uniform along the Taoudeni Basin, at least for the carbonate intervals. The upper carbonate reservoir intervals (Touiderguilt Formation), composed of dolomitized shallow water carbonates barren of palaeofauna, tested hydrocarbons in the Abolag-1 well, but are poorly known from outcrops and shallow cores. The lower reservoir interval (Tod to Tawaz formations) is characterized by dolomitized stromatolite build-ups. The two intervals can show some sections with moderate porosity, which in the lower unit is mainly represented by vugs and residual primary porosity. Most of the intercrystalline pores of the lower reservoir are filled with bitumen, which seems to represent the main concern with respect to reservoir quality, at least for this lower interval.

Other potential reservoirs

The occurrence of significant gas peaks, oil impregnations and fluorescence on the clastic intervals suggests that reservoir targets other than the Atar Group carbonate intervals described here could exist in the Taoudeni Basin. In particular, significant gas peaks were recorded on the basal clastic intervals of the Assabet el Hassiane Group in Abolag-1 and Ouguiya-1. A direct yellow fluorescence was

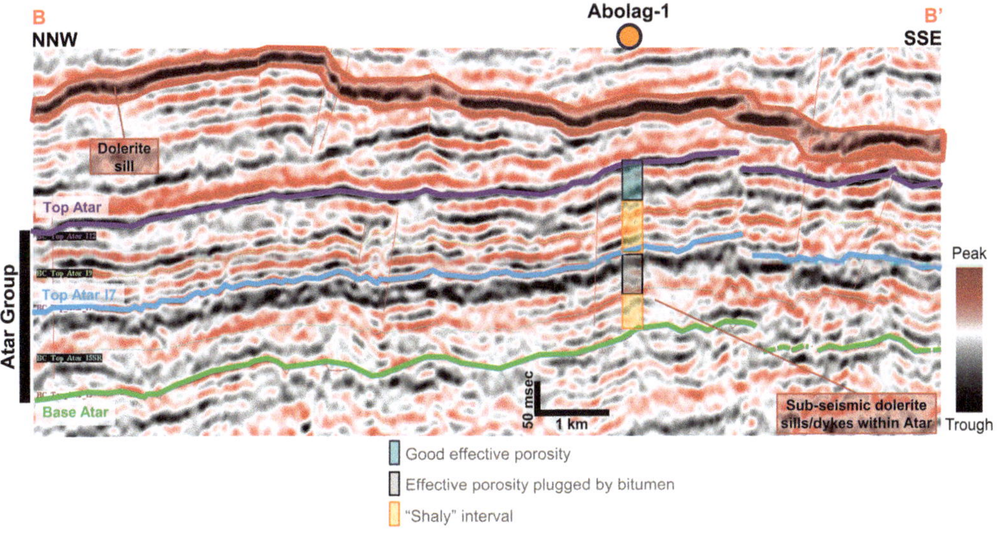

Fig. 20. Seismic line through the Abolag-1 location showing the alternation of high acoustic impedance (troughs) and low acoustic impedance (peaks) reflectors characteristic of the Atar Group sediments. See Figure 1 for location.

also observed in sidewall cores recovered from this unit in Ouguiya-1. Gas indications and migrant oil have been encountered in the Char Group clastics.

The Neoproterozoic Assabet el Hassiane Group, interpreted to have been deposited in fluvial–coastal, shelfal and coastal–deltaic environments, is considered to be another secondary reservoir objective in the Taoudeni Basin. The Assabet el Hassiane Group can be divided into three intervals, but it is only within the lowermost interval and, in particular, in the basal sandstone, that gas shows were recorded. This basal sandstone interval was penetrated in the Abolag-1 and Ouasa-1 exploration wells. A revision of cuttings from Abolag-1 indicates that the basal interval consists of fine-grained sandstones with minor primary and variable secondary porosity and the presence of bitumen post-dating the quartz overgrowths. In addition, traces of hydrocarbons were found both in stylolitic surfaces in the sandstones and in dolerite layers. The ubiquitous presence of solid organic residues reiterates a strong argument for the migration of large volumes of petroleum. The rest of the lower Assabet el Hassiane Group has been studied using cutting samples from Abolag-1. It is composed of well-sorted, fine- and medium-grained sandstones with variable porosity, either primary or secondary (oversized and mouldic). When textural parameters are favourable (e.g. well-sorted sandstones with medium to coarse grain size) and there is the presence of clay rims inhibiting quartz precipitation, the visual porosity values can reach 15%. However, when quartz overgrowths are well developed, the visual porosity is very low (up to 5%). Although visual estimations of the porosity can reach high values, this porosity is not effective as the pores are poorly connected and permeability is low.

The Agueni (Char I_1) and Foum Chor (Char I_3) formations of the Char Group contain a repetitive succession of fluvial and wave- and tide-dominated shallow-marine deposits (Benan & Deynoux 1998). These units constitute important siliciclastic reservoirs in terms of thickness (overall thickness of up to 400 m, including the fine-grained clastics and carbonates of the Azougui Char I_2 Formation) and lateral extent. It is unclear whether the quartzitic sandstone intersected at the base of the Abolag-1 well corresponds to the basal transgressive sandstone of the Ksar Torchane (Atar I_4) Formation or to the upper part of the Char Group, but gas indications were reported from it. The porosity calculated from petrophysical logs reaches values of up to 9%, but the interpreted net reservoir thickness is fairly small. In the R4 shallow well, the basal sandstone of the Ksar Torchane (Atar I_4) Formation shows a very poor reservoir quality despite being bitumen-bearing. It is composed of very fine-grained sandstones and conglomerates with an illitic matrix. Late

dolomite poikilotopic cements are present and no visual porosity can be identified.

In analogy with other Palaeozoic basins of North Africa, possible reservoirs could also be present within the Cambrian to Devonian sandstone intervals. The quartz-rich glaciogenic sandstones that were deposited during the Cambro-Ordovician over a vast region spanning from North Africa to Arabia is also represented in the Taoudeni Basin, although the early Palaeozoic sequence is fairly condensed compared with other coeval North African basins. In Abolag-1, Atouila-1 and Ouasa-1, the Cambro-Ordovician is represented by massive sandstones with few or no shale breaks within them. The petrophysical properties of this section show relatively good reservoir properties with up to several hundred metres net thickness and porosities in the range 8–15%, as derived from petrophysical interpretations. However, the lack of significant structure (apart from those already drilled and dry) and the apparent poor quality of the Silurian source make this play unattractive.

Traps and seals

The structural framework of the Taoudeni Basin is fairly simple, with several long-lived depocentres separated by basement highs and little deformation away from the Mauritanides mobile belt, located along the western margin of the basin in Mauritania and Senegal. In this fold–thrust belt and its foreland, kilometre-scale fault-related folds and large elongated anticlines parallel to the belt can be observed in the seismic data (Fig. 3).

Two main structural traps are identified in the central portion of the Taoudeni Basin in Mauritania: large anticline structures several tens of kilometres in diameter and smaller kilometre-scale fault-related structures. Large structural highs exist in the basin and apparently controlled the sedimentation since the Proterozoic, as suggested by the variations in thickness of the Atar Group (Fig. 3). They were probably formed at the early stages of basin evolution when a series of structural highs and rapidly subsiding sub-basins were created in a rifting tectonic environment. They might be locally complicated by later faulting, but apparently remained positive features. They were probably reactivated during the Panafrican, Hercynian and Alpine tectonic phases, which leads to uncertainty over preservation. This kind of structure was apparently pursued by the Abolag-1 well with limited seismic data before drilling.

Fault-related structures are mainly associated with reverse faults created during the Panafrican and mostly during the Hercynian and Alpine compressive events. It is difficult to separate fault families related to each phase in a very complex pattern of varying trends. However, several structures are associated with north–south and NE–SW faults that are sealed by late Palaeozoic deposits (Fig. 3) and can thus, in our opinion, be attributed to the Hercynian phase.

In addition to structural traps associated with the main tectonic phases that affected the basin, secondary traps might be considered. Domal structures developed as a consequence of the intrusion of dolerite sills and plugs. Although not yet identified, stratigraphic trapping might be associated with glacial erosional surfaces (both in Proterozoic and Palaeozoic deposits, in a play similar to that of the late Palaeozoic of North Africa) or to dolerite 'reservoirs' intruding shaly intervals.

The negative results of several exploration wells targeting relatively large structures might be considered of regional significance, but the validity of the drilled structures is questionable in most cases. The two-dimensional seismic lines acquired by Texaco in 1972 and 1973, prior to drilling the Abolag-1 well, are spaced by tens of kilometres. Newly acquired infill seismic data (Repsol's campaigns in 2007–2008 and 2010) have revealed that the well was not drilled on the actual culmination of the structure, a faulted three-way dip closure of the top Atar Group, located further to the south. Yarba-1 was drilled on a fairly large anticlinal structure, although it is also unclear whether it tested a valid closure or not. Finally, Ouasa-1 is located in a structurally low area and the validity of the drilled closure is also arguable.

Sealing lithologies exist within the Char, Atar and Assabet el Hassiane groups, the main potential reservoir targets of the Taoudeni Basin. The interbedded shale intervals of the Atar Group, in particular those within the Terioufet (Atar I_6), Terrarit (Atar I_8) and Aouleïgate (Atar I_{10}) formations, show relatively homogeneous thicknesses and are laterally extensive across the northern basin margin in Mauritania. In the producing interval of Abolag-1, seal is provided by the tight carbonate facies in the Touiderguilt (Atar I_{11}/I_{12}) Formation. Several fine-grained intervals within the Assabet el Hassiane Group show thicknesses and lateral extents favourable for a potential regional seal. The doleritic sills, preferentially intruding the fine-grained intervals within the Proterozoic and Palaeozoic successions of the Taoudeni Basin, could also provide potential seals locally.

Concluding remarks

We have focused on the identification and assessment of potential source rocks and reservoirs in the Proterozoic and Palaeozoic successions of the northwestern part of the Taoudeni Basin in Mauritania and in establishing a geochemical genetic link between known petroleum occurrences and their potential sources. Various disciplines and techniques were integrated to reach a better understanding.

This integrated study has resulted in significant advances towards the definition of a prospective petroleum system in the northern Taoudeni Basin. According to the convention of Magoon & Dow (1994), we can define the Atar-Atar (!) proven petroleum system considering that both the source rock

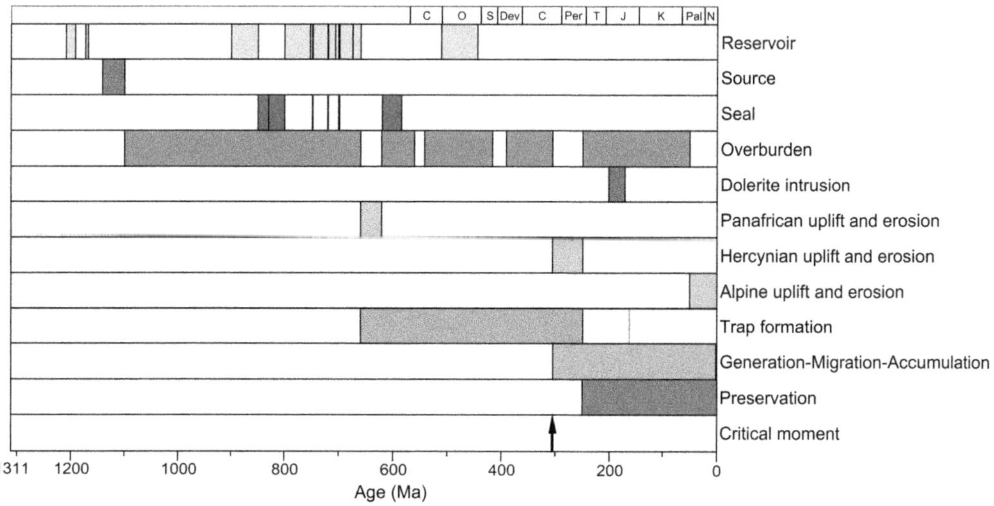

Fig. 21. Petroleum systems event chart summarizing the elements and processes at the basin scale of the Atar-Atar (!) petroleum system.

and main reservoirs lie within the Atar Group, both are effective and the petroleum recovered from the reservoir has been geochemically correlated with the source rock. Figure 21 is a petroleum systems event chart summarizing the elements and processes at a basin scale:

(1) *Source rock.* The Tod (Atar I_5) Formation and lateral equivalents of Mesoproterozoic age (*c.* 1100 Ma; Rooney *et al.* 2010; Rooney 2011) was identified as the main source rock in the northern Taoudeni Basin. Although significant original organic richness and facies variations have been observed, this unit could have a good to excellent oil-generative potential. Its maturity is highly variable at the surface and the early gas window was reached at the Abolag-1 well. Petroleum occurrences are widespread across the basin, including gas and liquids tested at Abolag-1, oil and gas shows in other exploration wells in Mauritania and Mali, and the pervasive presence of solid residual hydrocarbons throughout the basin and at various stratigraphic levels. Oil stains and impregnations identified in shallow stratigraphic wells have been genetically linked to organic extracts of the Tod (Atar I_5) Formation organic-rich sediments, indicating this as the most likely source rock for the observed shows.

(2) *Reservoirs.* Reservoirs are proven in Abolag-1, where gas and condensate were produced from the upper part of the Atar Group. Other potential reservoir targets include the carbonate build ups of the Tod (Atar I_5) Formation to Tawaz (Atar I_7) Formation interval and the clastic intervals within the Char, Atar and Assabet el Hassiane groups. Possible reservoir targets could also be present within the Cambrian to Devonian sandstone intervals.

(3) *Seal.* The thickness and lateral extent of several fine-grained intervals within the Assabet el Hassiane Group are favourable as a regional seal. Intraformational seals within the Atar Group are provided by shaly and tight carbonate intervals.

(4) *Overburden.* The overburden for source rock maturation is provided by a relatively thick sequence of late Proterozoic to Palaeozoic sediments and a more limited sequence of Mesozoic to Plio-Quaternary deposits. The burial history derived from our reconstructions suggests that most of the maturity related to burial was achieved before the Hercynian uplift.

(5) *Traps.* Two main structural traps are identified in the Taoudeni Basin: large anticline structures several tens of kilometres in diameter and smaller kilometre-scale fault-related structures. Large structural highs have existed in the basin since the Proterozoic and were probably formed at the early stages of basin evolution in a rifting tectonic setting. They might be complicated locally by later faulting, but apparently remained positive features, potentially draining hydrocarbons. Fault-related structures are mainly associated with reverse faults created during the Panafrican and mostly during the Hercynian and Alpine compressive events. Several structures associated with north–south and NE–SW faults could be attributed to the Hercynian phase. Other potential secondary traps could be related to domal structures resulting from magmatic intrusions or the stratigraphy (both in sediments and dolerites).

In terms of processes, namely trap formation and the cycle of generation–migration–accumulation of petroleum, uncertainties are much greater for the Taoudeni Basin. The timing of trap formation is difficult to elucidate in such a long and multi-phased structural evolution. However, our observations suggest that most of the prospective structures existed after the Hercynian tectonic phase; large positive structures have existed since the Proterozoic (probably formed at the early stage of basin evolution in a rifting tectonic setting) and other smaller structures are often controlled by reverse faults of Hercynian fabric (trending north and NE) and sealed by the Hercynian unconformity. We consider that the timing of trap formation at the basin scale can be bracketed between the Panafrican orogeny *lato sensu* (including all its different phases) and the Hercynian phase (Fig. 21). Alpine deformation apparently did not lead to the creation of major structures, at least in the centre of the basin. However, this late deformation has certainly reactivated part of the existing Hercynian and older prospective structures.

Petroleum generation–migration–accumulation is probably the most difficult process to assess for this basin. Thermal stress indicators and the variability in the occurrence of petroleum demonstrate the coexistence of products with very different maturities. Burial history reconstruction at the Abolag-1 well indicated that the Atar Group source rock reached its maximum depth and temperature just before the Hercynian uplift and that generation and expulsion occurred then, at least in the deepest portions of the basin. However, the Mesozoic magmatic episode probably resumed the generation of hydrocarbons long after the Hercynian and triggered a second pulse of petroleum generation, expulsion and migration (Baudino *et al.* 2014). We can therefore assume that the timing of this process would be bracketed between the Hercynian unconformity and the present.

This complexity demonstrates the competing regional versus local thermal effects in the basin. The relative importance of each in the generation and expulsion process is hard to assess and will, of course, greatly affect the timing of migration and trap availability.

In terms of preservation risk, the Alpine deformation may have triggered trap breaching of existing potential accumulations. Another main risk is related to the magmatic intrusions, which could provoke breaching, but also thermal alteration and secondary cracking (possibly a reason for the presence of residual solid hydrocarbons). Areas of intense magmatism should, therefore, be avoided in the search for hydrocarbon accumulations. However, many of the sills and dykes that have thicknesses greater than the seismic resolution thickness can be interpreted as they are oblique to the depositional pattern and display a high-amplitude response in the seismic data.

The study presented here is an example of exploring an unusual and high-risk play. Unusual because we are dealing with sedimentary rocks that are among the oldest on Earth, with specific depositional environments and faunas with no modern equivalent, a long and polyphase structural history and a complex thermal history in which burial might not be the only controlling factor. It is a high-risk play because it is a vast and under-explored basin with very little data, a carbonate main reservoir with a complex diagenetic evolution in which continuity and quality are extremely difficult to define areally and, finally, a play with large uncertainties on the timing of generation, trap formation and preservation.

We thank the organizers of the Petroleum Geoscience of the West African Margin conference for their invitation to contribute to this volume. We also thank Repsol Exploración's management and RWE Dea AG, our partners in the exploration of the Taoudeni Basin, for their permission to publish this work. Raquel Álvarez is acknowledged for her support in preparing the figures that illustrate the text. Paloma López-Guerrero (Naturhistorisches Museum Wien) provided useful comments and suggestions. We are grateful to two anonymous reviewers, whose constructive comments and suggestions greatly improved the quality of this paper.

References

ALLEN, P.A. & ALLEN, J.R. 2005. *Basin Analysis: Principles and Applications*. Blackwell, Malden.

ÁLVARO, J.J. & VIZCAÏNO, D. 2012. Proterozoic microbial reef complexes and associated hydrothermal mineralizations in the Banfora Cliffs, Burkina Faso. *Sedimentary Geology*, **263–264**, 144–156, http://doi.org/10.1016/j.sedgeo.2011.11.005

AMARD, B. 1986. Microfossiles (Acritarches) du Proterozoique superieur dans les shales de la formation d'Atar (Mauritanie). *Precambrian Research*, **31**, 69–95, http://doi.org/10.1016/0301-9268(86)90065-3

ANBAR, A.D. & KNOLL, A.H. 2002. Proterozoic ocean chemistry and evolution: a bioinorganic bridge? *Science*, **297**, 1137–1142, http://doi.org/10.1126/science.1069651

ANDERSON, D.L. 1982. Hotspots, polar wander, Mesozoic convection and the geoid. *Nature*, **297**, 391–393, http://doi.org/10.1038/297391a0

BAUDINO, R., RODRÍGUEZ MONREAL,F., ZENCICH, S. & CALEGARI, R. 2004. Generation of hydrocarbons by thermal effect of magmatic intrusions: a non-conventional petroleum system [abstract]. Paper presented at the IX Asociación Latinoamericana de Geoquímica Orgánica (ALAGO) Congress, 5–9 December 2004, Mérida, Mexico.

BAUDINO, R., MARTÍN-MONGE, A. *ET AL*. 2014. Assessing a petroleum system on the frontier of geological time: the Mesoproterozoic of the Taoudeni Basin (Mauritania) [abstract]. Paper presented at the AAPG International Conference and Exhibition 2014, 14–17 September 2014, Istanbul, Turkey.

BAYER, R. & LESQUER, A. 1978. Les anomalies gravimétriques de la bordure orientale du craton Ouest Africain: géométrie d'une suture panafricaine. *Bulletin de la Société Géologique de France*, **20**, 863–876, http://doi.org/10.2113/gssgfbull.S7-XX.6.863

BENAN, C.A.A. & DEYNOUX, M. 1998. Facies analysis and sequence stratigraphy of Neoproterozoic platform deposits in Adrar of Mauritania, Taoudeni basin, West Africa. *Geologische Rundschau*, **87**, 282–302, http://doi.org/10.1007/s005310050210

BERTRAND-SARFATI, J. 1972. Paléoécologie de certains stromatolites en récifs des formations du Précambrien supérieur du groupe d'Atar (Mauritanie, Sahara occidental): Création d'espèces nouvelles. *Palaeogeography, Palaeoclimatology, Palaeoecology*, **11**, 33–63, http://doi.org/10.1016/0031-0182(72)90036-3

BERTRAND-SARFATI, J. & MOUSSINE-POUCHKINE, A. 1985. Evolution and environmental conditions of *Conophyton–Jacutophyton* associations in the Atar dolomite (Upper Proterozoic, Mauritania). *Precambrian Research*, **29**, 207–234, http://doi.org/10.1016/0301-9268(85)90069-5

BERTRAND-SARFATI, J. & MOUSSINE-POUCHKINE, A. 1988. Is cratonic sedimentation consistent with available models – an example from the Upper Proterozoic of the West-African Craton. *Sedimentary Geology*, **58**, 255–276, http://doi.org/10.1016/0037-0738(88)90072-3

BERTRAND-SARFATI, J., MOUSSINE-POUCHKINE, A., AFFATON, P., TROMPETTE, R. & BELLION, Y. 1991. Cover sequences of the West African Craton. *In*: DALLMEYER, R.D. & LECORCHÉ, J.P. (eds) *The West African Orogens and Circum-Atlantic Correlatives*. Springer, Berlin, 65–82.

BLACK, R., CABY, R. *ET AL*. 1979. Evidence for late Precambrian plate tectonics in West Africa. *Nature*, **278**, 223–227, http://doi.org/10.1038/278223a0

BLUMENBERG, M., THIEL, V., RIEGEL, W., KAH, L.C. & REITNER, J. 2012. Biomarkers of black shales formed by microbial mats, Late Mesoproterozoic (1.1 Ga) Taoudeni Basin, Mauritania. *Precambrian Research*, **196–197**, 113–127, http://doi.org/10.1016/j.precamres.2011.11.010

BOOTE, D.R.D., DARDOUR, A., GREEN, P.F., SMEWING, J.D. & VAN HOEFLAKEN, F. 2008. Burial and unroofing history of the base Tanezzuft 'hot' shale source rock, Murzuq Basin, SW Libya: new AFTA constraints from basin margin outcrops [abstract]. Paper presented at the 4th Sedimentary Basins of Libya Symposium: the Geology of Southern Libya, 17–20 November 2008, Tripoli, Libya.

BRAY, R.J., GREEN, P.F. & DUDDY, I.R. 1992. Thermal history reconstruction using apatite fission track analysis and vitrinite reflectance: a case study from the UK East Midlands and Southern North Sea. *In*: HARTMAN, R.F.P. (ed.) *Exploration Britain: Geological Insights for the Next Decade*. Geological Society, London, Special Publications, **67**, 3–25, http://doi.org/10.1144/GSL.SP.1992.067.01.01

BROCKS, J.J. 2011. Millimeter-scale concentration gradients of hydrocarbons in Archean shales: live-oil escape or fingerprint of contamination? *Geochimica et Cosmochimica Acta*, **75**, 3196–3213, http://doi.org/10.1016/j.gca.2011.03.014

BROCKS, J.J. & SUMMONS, R.E. 2003. Sedimentary hydrocarbons, biomarkers for early life. *In*: HOLLAND, H.D. & TUREKIAN, K.K. (eds) *Treatise on Geochemistry*, **8**, 63–115, http://doi.org/10.1016/B0-08-043751-6/08127-5

BROCKS, J.J., LOGAN, G.A., BUICK, R. & SUMMONS, R.E. 1999. Archean molecular fossils and the early rise of Eukaryotes. *Science*, **285**, 1033–1036, http://doi.org/10.1126/science.285.5430.1033

BROCKS, J.J., LOVE, G.D., SNAPE, C.E., LOGAN, G.A., SUMMONS, R.E. & BUICK, R. 2003. Release of bound aromatic hydrocarbons from late Archean and Mesoproterozoic kerogens via hydropyrolysis. *Geochimica et Cosmochimica Acta*, **67**, 1521–1530, http://doi.org/10.1016/j.gca.2007.11.028

BROCKS, J.J., GROSJEAN, E. & LOGAN, G.A. 2008. Assessing biomarker syngeneity using branched alkanes with quaternary carbon (BAQCs) and other plastic contaminants. *Geochimica et Cosmochimica Acta*, **72**, 871–888, http://doi.org/10.1016/j.gca.2007.11.028

BRONNER, G., ROUSSEL, J., TROMPETTE, R. & CLAUER, N. 1980. Genesis and geodynamic evolution of the Taoudeni cratonic basin (Upper Precambrian and Paleozoic), western Africa. *In*: BALLY, A.W., BENDER, P.L., MCGETCHIN, T.R. & WALCOTT, R.I. (eds) *Dynamics of Plate Interiors*. American Geophysical Union & Geological Society of America, Geodynamics Series, **1**, 81–90, http://doi.org/10.1029/GD001p0081

CHEN, J., FU, J., SHENG, G., LIU, D. & ZHANG, J. 1996. Diamondoid hydrocarbon ratios: novel maturity indices for highly mature crude oils. *Organic Geochemistry*, **25**, 179–190, http://doi.org/10.1016/S0146-6380(96)00125-8

CLAUER, N. 1981. Rb–Sr and K–Ar dating of Precambrian clays and glauconites. *Precambrian Research*, **15**, 331–352, http://doi.org/10.1016/0301-9268(81)90056-5

CLAUER, N., CABY, R., JEANNETTE, D. & TROMPETTE, R. 1982. Geochronology of sedimentary and metasedimentary Precambrian rocks of the West African craton. *Precambrian Research*, **18**, 53–71, http://doi.org/10.1016/0301-9268(82)90036-5

COLE, G.A. 1994. Graptolite-chitinozoan reflectance and its relationship to other geochemical maturity indicators in the Silurian Qusaiba Shale, Saudi Arabia. *Energy & Fuels*, **8**, 1443–1459, http://doi.org/10.1021/ef00048a035

COLTICE, N., PHILLIPS, B.R., BERTRAND, H., RICARD, Y. & REY, P. 2007. Global warming of the mantle at the origin of flood basalts over supercontinents. *Geology*, **35**, 391–394, http://doi.org/10.1130/G23240A.1

COLTICE, N., BERTRAND, H., REY, P., JOURDAN, F., PHILLIPS, B.R. & RICARD, Y. 2009. Global warming of the mantle beneath continents back to the Archaean. *Gondwana Research*, **15**, 254–266, http://doi.org/10.1016/j.gr.2008.10.001

CRAIG, J., GRIGO, D., REBORA, A., SERAFINI, G. & TEBALDI, E. 2010. From Neoproterozoic to Early Cenozoic: exploring the potential of older and deeper hydrocarbon plays across North Africa and the Middle East. *In*: VINING, B.A. & PICKERING, S.C. (eds) *Petroleum Geology: From Mature Basins to New Frontiers – Proceedings of the 7th Petroleum Geology Conference*. Geological Society, London, Petroluem Geology Conference Series, **7**, 673–705, http://doi.org/10.1144/0070673

CRAIG, J., BIFFI, U. *ET AL*. 2013. The palaeobiology and geochemistry of Precambrian hydrocarbon source rocks. *Marine and Petroleum Geology*, **40**, 1–47, http://doi.org/10.1016/j.marpetgeo.2012.09.011

CULVER, S.J. & HUNT, D. 1991. Lithostratigraphy of the Precambrian-Cambrian boundary sequence in the southwestern Taoudeni Basin, West Africa. *Journal of African Earth Sciences (and the Middle East)*, **13**, 407–413, http://doi.org/10.1016/0899-5362(91)90105-8

DALLMEYER, R.D. & LÉCORCHÉ, J.P. 1989. ^{40}Ar/^{39}Ar polyorogenic mineral record within the central Mauritanide orogen, West Africa. *Geological Society of America Bulletin*, **101**, 55–70, http://doi.org/10.1130/0016-7606(1989)101<0055:AAPMAR>2.3.CO;2

DALLMEYER, R.D. & VILLENEUVE, M. 1987. ^{40}Ar/^{39}Ar polyorogenic mineral age record of a polyphased tectonothermal evolution in the southern Mauritanide orogen, southeastern Senegal. *Geological Society of America Bulletin*, **98**, 602–611, http://doi.org/10.1130/0016-7606(1987)98<602:AMAROP>2.0.CO;2

DARS, R. 1961. *Les formations sédimentaires et les dolérites du Soudan occidental (Afrique de l'Ouest)*. Mémoires du Bureau des Recherches Géologiques et Minières, **12**.

DECKART, K., FÉRAUD, G. & BERTRAND, H. 1997. Age of Jurassic continental tholeiites of French Guyana, Surinam and Guinea: implications for the initial opening of the Central Atlantic Ocean. *Earth and Planetary Science Letters*, **150**, 205–220, http://doi.org/10.1016/S0012-821X(97)00102-7

DEYNOUX, M. 1978. Les formations glaciaires du Précambrien terminal et de la fin de l'Ordovicien en Afrique de l'Ouest. Deux exemples de glaciation d'inlandsis sur plate-forme stable. *Travaux des Laboratoires des Sciences de la Terre*, **17**, 1–554.

DEYNOUX, M., AFFATON, P., TROMPETTE, R. & VILLENEUVE, M. 2006. Pan-African tectonic evolution and glacial events registered in Neoproterozoic to Cambrian cratonic and foreland basins of West Africa. *Journal of African Earth Sciences*, **46**, 397–426, http://doi.org/10.1016/j.jafrearsci.2006.08.005

DIA, O. 1984. *La chaîne panafricaine et hercynienne des Mauritanides face au bassin Protérozoïque supérieur à Dévonien de Taoudeni dans le secteur-clé de Mejeria (Taganet, Sud R.I.M.): lithostratigraphie et tectonique. Un exemple de tectoniques tangentielles superposées.* PhD thesis, Université d'Aix-Marseille.

DOSSO, L., VIDAL, P.H., SICHLER, B. & BONIFAY, A. 1979. Age précambrien de dolérites de la dorsale Réguibat (Mauritanie). *Comptes Rendus de l'Académie des Sciences de Paris, série D*, **288**, 739–742.

DUTKIEWICZ, A., VOLK, H., RIDLEY, J. & GEORGE, S.C. 2004. Geochemistry of oil in fluid inclusions in a middle Proterozoic igneous intrusion: implications for the source of hydrocarbons in crystalline rocks. *Organic Geochemistry*, **35**, 937–957, http://doi.org/10.1016/j.orggeochem.2004.03.007

ELIAS, V.O., DE BARROS, A.M.A., DE BARROS, A.B., SIMONEIT, B.R.T. & CARDOSO, J.N. 1997. Sesquiterpenoids in sediments of a hypersaline lagoon: a possible algal origin. *Organic Geochemistry*, **26**, 721–730, http://doi.org/10.1016/S0146-6380(97)00052-1

ERNST, R.E. & BUCHAN, K.L. 2002. Maximum size and distribution in time and space of mantle plumes: evidence from large igneous provinces. *Journal of Geodynamics*, **31**, 309–342, http://doi.org/10.1016/S0264-3707(02)00025-X

GEORGE, S.C. & AHMED, M. 2000. Organic maturity evaluation of the Proterozoic middle Velkerri Formation, Northern Territory, Australia: use of aromatic hydrocarbon distributions. *AAPG Bulletin*, **84**, 1429.

GEORGE, S.C., DUTKIEWICZ, A., VOLK, H., RIDLEY, J., MOSSMAN, D.J. & BUICK, R. 2007. Eukaryote-derived steranes in Precambrian oils and rocks: fact or fiction? [abstract O4]. Paper presented at the European Association of Organic Geochemists' 23rd International Meeting on Organic Geochemistry, 9–14 Septembre 2007, Torquay, UK.

GHIENNE, J.F. & DEYNOUX, M. 1998. Large-scale channel fill structures in Late Ordovician glacial deposits in Mauritania, Western Sahara. *Sedimentary Geology*, **119**, 141–159, http://doi.org/10.1016/S0037-0738(98)00045-1

GHORI, K.A.R., CRAIG, J., THUSU, B., LÜNING, S. & GEIGER, M. 2009. Global Infracambrian petroleum systems: a review. *In*: CRAIG, J., THUROW, J., THUSU, B., WHITHAM, A. & ABUTARRUMA, Y. (eds) *Global Neoproterozoic Petroleum Systems and the Emerging Potential in North Africa*. Geological Society, London, Special Publications, **326**, 109–136, http://doi.org/10.1144/SP326.6

GILLEAUDEAU, G.F. & KAH, L.C. 2013a. Carbon isotope records in a Mesoproterozoic epicratonic sea: carbon cycling in a low-oxygen world. *Precambrian Research*, **228**, 85–101, http://doi.org/10.1016/j.precamres.2013.01.006

GILLEAUDEAU, G.F. & KAH, L.C. 2013b. Oceanic molybdenum drawdown by epeiric sea expansion in the Mesoproterozoic. *Chemical Geology*, **356**, 21–37, http://doi.org/10.1016/j.chemgeo.2013.07.004

GROSJEAN, E., LOVE, G.D., STALVIES, C., FIKE, D.A. & SUMMONS, R.E. 2009. Origin of petroleum in the Neoproterozoic–Cambrian South Oman Salt Basin. *Organic Geochemistry*, **40**, 87–110, http://doi.org/10.1016/j.orggeochem.2008.09.011

HALVERSON, G.P., HOFFMAN, P.F., SCHRAG, D.P., MALOOF, A.C. & RICE, A.H.N. 2005. Toward a Neoproterozoic composite carbon-isotope record. *Geological Society of America Bulletin*, **117**, 1181–1207, http://doi.org/10.1130/B25630.1

HOERING, T.C. & NAVALE, V. 1987. A search for molecular fossils in the kerogen of Precambrian sedimentary rocks. *Precambrian Research*, **34**, 247–267, http://doi.org/10.1016/0301-9268(87)90003-9

HOLBA, A.G., DZOU, L.I. ET AL. 2003. Application of tetracyclic polyprenoids as indicators of input from fresh-brackish water environments. *Organic Geochemistry*, **34**, 441–469, http://doi.org/10.1016/S0146-6380(02)00193-6

JANNEY, P.E. & CASTILLO, P.R. 2001. Geochemistry of the oldest Atlantic oceanic crust suggests mantle plume involvement in the early history of the central Atlantic Ocean. *Earth and Planetary Science Letters*, **192**, 291–302, http://doi.org/10.1016/S0012-821X(01)00452-6

KAH, L.C., BARTLEY, J.K. & STAGNER, A.F. 2009. Reinterpreting a Proterozoic enigma: *Conophyton–Jacutophyton* stromatolites of the Mesoproterozoic Atar Group, Mauritania. *In*: SWART, P.K., EBERLI, G.P., MCKENZIE, J.A., JARVIS, I. & STEVENS, T. (eds) *Perspectives in Carbonate Geology: A Tribute to the Career of Robert Nathan Ginsburg*. International Association of Sedimentology, Special Publications, **41**, 277–295, http://doi.org/10.1002/9781444312065.ch17

KAH, L.C., BARTLEY, J.K. & TEAL, D.A. 2012. Chemostratigraphy of the Late Mesoproterozoic Atar Group, Taoudeni Basin, Mauritania: Muted isotopic variability, facies correlation, and global isotopic trends. *Precambrian Research*, **200–203**, 82–103, http://doi.org/10.1016/j.precamres.2012.01.011

KASHIRTSEV, V.A., KONTOROVICH, A.E., IVANOV, V.L. & SAFRONOV, A.F. 2010. Natural bitumen fields in the northeast of the Siberian Platform (Russian Arctic sector). *Russian Geology and Geophysics*, **51**, 72–82, http://doi.org/10.1016/j.rgg.2009.12.007

KNIGHT, K.B., NOMADE, S., RENNE, P.R., MARZOLI, A., BERTRAND, H. & YOUBI, N. 2004. The Central Atlantic Magmatic Province at the Triassic–Jurassic boundary: paleomagnetic and ^{40}Ar/^{39}Ar evidence from Morocco for brief, episodic volcanism. *Earth and Planetary Science Letters*, **228**, 143–160, http://doi.org/10.1016/j.epsl.2004.09.022

KNOLL, A.H., JAVAUX, E.J., HEWITT, D. & COHEN, P. 2006. Eukaryotic organisms in Proterozoic oceans. *Philosophical Transactions of the Royal Society B: Biological Sciences*, **361**, 1023–1038, http://doi.org/10.1098/rstb.2006.1843

KORSCH, R.J., HUAZHAO, M., ZHAOCAI, S. & GORTER, J.D. 1991. The Sichuan Basin, southwest China: a late

Proterozoic (Sinian) petroleum province. *Precambrian Research*, **54**, 45–63, http://doi.org/10.1016/0301-9268(91)90068-L

LAHONDÈRE, D., THIÉBLEMONT, D. *ET AL.* 2003. *Notice explicative des cartes géologiques et gîtologiques à 1/200 000 et 1/500 000 du nord de la Mauritanie.* DMG, Ministère des Mines et de l'Industrie, Nouakchott.

LAHONDÈRE, D., ROGER, J. *ET AL.* 2005. *Notice explicative des cartes géologiques à 1/200 000 et 1/500 000 de l'extrême sud de la Mauritanie.* DMG, Ministère des Mines et de l'Industrie, Nouakchott.

LOGAN, P. & DUDDY, I. 1998. An investigation of the thermal history of the Ahnet and Reggane Basins, Central Algeria, and the consequences for hydrocarbon generation and accumulation. *In*: MACGREGOR, D.S., MOODY, R.T.J. & CLARK-LOWES, D.D. (eds) *Petroleum Geology of North Africa*. Geological Society, London, Special Publications, **132**, 131–155, http://doi.org/10.1144/GSL.SP.1998.132.01.07

LOTTAROLI, F., CRAIG, J. & THUSU, B. 2009. Neoproterozoic–Early Cambrian (Infracambrian) hydrocarbon prospectivity of North Africa: a synthesis. *In*: CRAIG, J., THUROW, J., THUSU, B., WHITHAM, A. & ABUTARRUMA, Y. (eds) *Global Neoproterozoic Petroleum Systems and the Emerging Potential in North Africa*. Geological Society, London, Special Publications, **326**, 137–156, http://doi.org/10.1144/SP326.7

LOVE, G.D., GROSJEAN, E. *ET AL.* 2009. Fossil steroids record the appearance of Demospongiae during the Cryogenian period. *Nature*, **457**, 719–721, http://doi.org/10.1038/nature07673

LÜNING, S., CRAIG, J., LOYDELL, D.K., ŠTORCH, P. & FITCHES, B. 2000. Lower Silurian 'hot shales' in North Africa and Arabia: regional distribution and depositional model. *Earth-Science Reviews*, **49**, 121–200, http://doi.org/10.1016/S0012-8252(99)00060-4

LÜNING, S., KOLONIC, S., GEIGER, M., THUSU, B., BELL, J.S. & CRAIG, J. 2009. Infracambrian hydrocarbon source rock potential and petroleum prospectivity of NW Africa. *In*: CRAIG, J., THUROW, J., THUSU, B., WHITHAM, A. & ABUTARRUMA, Y. (eds) *Global Neoproterozoic Petroleum Systems and the Emerging Potential in North Africa*. Geological Society, London, Special Publications, **326**, 157–180, http://doi.org/10.1144/SP326.8

MAGOON, L.B. & DOW, W.G. 1994. The petroleum system. *In*: MAGOON, L.B. & DOW, W.G. (eds) *The Petroleum System: From Source to Trap*. AAPG Memoir, **60**, 3–24.

MARZOLI, A., RENNE, P.R., PICCIRILLO, E., ERNESTO, E., BELLIENI, G. & DE MIN, A. 1999. Extensive 200-million-year-old continental flood basalts of the Central Atlantic Magmatic Province. *Science*, **284**, 616–618, http://doi.org/10.1126/science.284.5414.616

MARZOLI, A., BERTRAND, H. *ET AL.* 2004. Synchrony of the Central Atlantic magmatic province and the Triassic–Jurassic boundary climatic and biotic crisis. *Geology*, **32**, 973–976, http://doi.org/10.1130/G20652.1

MCHONE, J.G. 2000. Non-plume magmatism and rifting during the opening of the Central Atlantic Ocean. *Tectonophysics*, **316**, 287–296, http://doi.org/10.1016/S0040-1951(99)00260-7

MOLDOWAN, J.M., SEIFERT, W.K. & GALLEGOS, E.J. 1985. Relationship between petroleum composition and depositional environment of petroleum source rocks. *AAPG Bulletin*, **69**, 1255–1268.

MOUSSINE-POUCHKINE, A. & BERTRAND-SARFATI, J. 1997. Tectonosedimentary subdivisions in the Neoproterozoic to Early Cambrian cover of the Taoudenni Basin (Algeria-Mauritania-Mali). *Journal of African Earth Sciences*, **24**, 425–443, http://doi.org/10.1016/S0899-5362(97)00073-0

MOUSSINE-POUCHKINE, A., BERTRAND-SARFATI, J., BALL, E. & CABY, R. 1988. Les séries sédimentaires et volcaniques anorogéniques protérozoïques impliquées dans la chaîne pan africaine: la région de l'Adrar Ahnet (NW Hoggar, Algérie). *Journal of African Earth Sciences (and the Middle East)*, **7**, 57–75, http://doi.org/10.1016/0899-5362(88)90053-X

NOMADE, S., KNIGHT, K.B. *ET AL.* 2007. Chronology of the Central Atlantic Magmatic Province: implications for the Central Atlantic rifting processes and the Triassic–Jurassic biotic crisis. *Palaeogeography, Palaeoclimatology, Palaeoecology*, **244**, 326–344, http://doi.org/10.1016/j.palaeo.2006.06.034

PETERS, K.E., WALTERS, C.C. & MOLDOWAN, J.M. 2005a. *The Biomarker Guide. Volume 1: Biomarkers and Isotopes in the Environment and Human History*. Cambridge University Press, Cambridge.

PETERS, K.E., WALTERS, C.C. & MOLDOWAN, J.M. 2005b. *The Biomarker Guide. Volume 2: Biomarkers and Isotopes in Petroleum Exploration and Earth History*. Cambridge University Press, Cambridge.

PITFIELD, P.E.J., KEY, R.M., WATERS, C.N., HAWKINS, M.P.H., SCHOFIELD, D.I., LOUGHLIN, S. & BARNES, R.P. 2004. *Notice explicative des cartes géologiques et gîtologiques au 1/200 000 et 1/500 000 du sud de la Mauritanie. Volume 1: Géologie*. DMG, Ministère des Mines et de l'Industrie, Nouakchott.

RADKE, M. 1988. Application of aromatic compounds as maturity indicators in source rocks and crude oils. *Marine and Petroleum Geology*, **5**, 224–236, http://doi.org/10.1016/0264-8172(88)90003-7

RADKE, M., WELTE, D.H. & WILLSCH, H. 1986. Maturity parameters based on aromatic hydrocarbons: influence of organic matter type. *Organic Geochemistry*, **10**, 51–63, http://doi.org/10.1016/0146-6380(86)90008-2

RIEDIGER, C.L. 1993. Solid bitumen reflectance and Rock-Eval T_{max} as maturation indices: an example from the Nordegg Member, Western Canada Sedimentary Basin. *International Journal of Coal Geology*, **22**, 295–315, http://doi.org/10.1016/0166-5162(93)90031-5

RODRIGUEZ MONREAL, F., VILLAR, H.J., BAUDINO, R., DELPINO, D. & ZENCICH, S. 2009. Modeling an atypical petroleum system: a case study of hydrocarbon generation, migration and accumulation related to igneous intrusions in the Neuquen Basin, Argentina. *Marine and Petroleum Geology*, **26**, 590–605, http://doi.org/10.1016/j.marpetgeo.2009.01.005

ROONEY, A.D. 2011. *Re–Os geochronology and geochemistry of Proterozoic sedimentary successions*. PhD thesis, Durham University.

ROONEY, A.D., SELBY, D., HOUZAY, J.-P. & RENNE, P.R. 2010. Re–Os geochronology of a Mesoproterozoic sedimentary succession, Taoudeni basin, Mauritania: implications for basin-wide correlations and Re–Os organic-rich sediment systematics. *Earth and Planetary Science Letters*, **289**, 486–496, http://doi.org/10.1016/j.epsl.2009.11.039

SCHOENHERR, J., LITTKE, R., URAI, J.L., KUKLA, P.A. & RAWAHI, Z. 2007. Polyphase thermal evolution in the infra-Cambrian Ara Group (South Oman Salt Basin) as deduced by maturity of solid reservoir bitumen. *Organic Geochemistry*, **38**, 1293–1318, http://doi.org/10.1016/j.orggeochem.2007.03.010

SCHOFIELD, D.I. & GILLESPIE, M.R. 2007. A tectonic interpretation of 'Eburnean terrane' outliers in the Reguibat Shield, Mauritania. *Journal of African Earth Sciences*, **49**, 179–186, http://doi.org/10.1016/j.jafrearsci.2007.08.006

SEBAI, A., FÉRAUD, G., BERTRAND, H. & HANES, J. 1991. ^{40}Ar/^{39}Ar dating and geochemistry of tholeiitic magmatism related to the early opening of the Central Atlantic rift. *Earth and Planetary Science Letters*, **104**, 455–472, http://doi.org/10.1016/0012-821X(91)90222-4

SHIELDS, G.A., DEYNOUX, M., STRAUSS, H., PAQUET, H. & NAHON, D. 2007. Barite-bearing cap dolostones of the Taoudéni Basin, northwest Africa: sedimentary and isotopic evidence for methane seepage after a Neoproterozoic glaciation. *Precambrian Research*, **153**, 209–235, http://doi.org/10.1016/j.precamres.2006.11.011

SUMMONS, R.E. & WALTER, M.R. 1990. Molecular fossils and microfossils of prokaryotes and protists from Proterozoic sediments. *American Journal of Science*, **290A**, 212–244.

SUMMONS, R.E., POWELL, T.G. & BOREHAM, C.J. 1988*a*. Petroleum geology and geochemistry of the Middle Proterozoic McArthur Basin, Northern Australia: III. Composition of extractable hydrocarbons. *Geochimica et Cosmochimica Acta*, **52**, 1747–1763, http://doi.org/10.1016/0016-7037(88)90001-4

SUMMONS, R.E., BRASSELL, S.C., EGLINTON, G., EVANS, E., HORODYSKI, R.J., ROBINSON, N. & WARD, D.M. 1988*b*. Distinctive hydrocarbon biomarkers from fossiliferous sediment of the Late Proterozoic Walcott Member, Chuar Group, Grand Canyon, Arizona. *Geochimica et Cosmochimica Acta*, **52**, 2625–2637, http://doi.org/10.1016/0016-7037(88)90031-2

SUMMONS, R.E., JAHNKE, L.L., HOPE, J.M. & LOGAN, G.A. 1999. 2-Methylhopanoids as biomarkers for cyanobacterial oxygenic photosynthesis. *Nature*, **400**, 554–557, http://doi.org/10.1038/23005

SUMMONS, R.E., BRADLEY, A.S., JAHNKE, L.L. & WALDBAUER, J.R. 2006. Steroids, triterpenoids and molecular oxygen. *Philosophical Transactions of the Royal Society B: Biological Sciences*, **361**, 951–968, http://doi.org/10.1098/rstb.2006.1837

TEAL, D.A.J. & KAH, L.C. 2005. Using C-isotopes to constrain intrabasinal stratigraphic correlations: Mesoproterozoic Atar Group, Mauritania [abstract]. Paper presented at the GSA Southeastern Section –54th Annual Meeting, 17–18 March 2005, Biloxi, MS, USA.

TOFT, P.B., TAYLOR, P.T., ARKANI-HAMED, J. & HAGGERTY, S.E. 1992. Interpretation of satellite magnetic anomalies over the West African Craton.

Tectonophysics, **212**, 21–32, http://doi.org/10.1016/0040-1951(92)90137-U

TROMPETTE, R. 1973. Le Précambrien supérieur et le Paléozoïque inférieur de l'Adrar de Mauritanie (bordure occidentale du bassin de Taoudenni, Afrique de l'Ouest), un exemple de sédimentation de craton. Étude stratigraphique et sédimentologique. *Travaux des Laboratoires des Sciences de la Terre, série B*, **7**, 1–702.

ULMISHEK, G.F. 2001. *Petroleum Geology and Resources of the Baykit High Province, East Siberia, Russia*. US Geological Survey Bulletin, **2201-F**, 1–18.

VERATI, C., BERTRAND, H. & FÉRAUD, G. 2005. The farthest record of the Central Atlantic Magmatic Province into West Africa craton: precise ^{40}Ar/^{39}Ar dating and geochemistry of Taoudenni basin intrusives (northern Mali). *Earth and Planetary Science Letters*, **235**, 391–407, http://doi.org/10.1016/j.epsl.2005.04.012

VERATI, C., RAPAILLE, C., FÉRAUD, G., MARZOLI, A., BERTRAND, H. & YOUBI, N. 2007. ^{40}Ar/^{39}Ar ages and duration of the Central Atlantic Magmatic Province volcanism in Morocco and Portugal and its relation to the Triassic–Jurassic boundary. *Palaeogeography, Palaeoclimatology, Palaeoecology*, **244**, 308–325, http://doi.org/10.1016/j.palaeo.2006.06.033

VILLEMUR, J.R. 1967. *Reconnaissance géologique et structurale du Nord du bassin de Taoudenni*. Mémoires du Bureau des Recherches Géologiques et Minières, **51**.

VILLENEUVE, M. 2005. Paleozoic basins in West Africa and the Mauritanide thrust belt. *Journal of African Earth Sciences*, **43**, 166–195, http://doi.org/10.1016/j.jafrearsci.2005.07.012

VILLENEUVE, M. & CORNÉE, J.J. 1994. Structure, evolution and palaeogeography of the West African Craton and bordering belts during the Neoproterozoic. *Precambrian Research*, **69**, 307–326, http://doi.org/10.1016/0301-9268(94)90094-9

WALTERS, C.C., KELEMEN, S.R., KWIATEK, P.J., POTTORF, R.J., MANKIEWICZ, P.J., CURRY, D.J. & PUTNEY, K. 2006. Reactive polar precipitation via ether crosslinkage: a new mechanism for solid bitumen formation. *Organic Geochemistry*, **37**, 408–427, http://doi.org/10.1016/j.orggeochem.2005.12.007

WATERS, C.N. & SCHOFIELD, D.I. 2004. Contrasting late Neoproterozoic to Ordovician successions of the Taoudeni Basin, Mauritania and Souss Basin, Morocco. *Journal of African Earth Sciences*, **39**, 301–309, http://doi.org/10.1016/j.jafrearsci.2004.07.038

WEN, Z., RUIYONG, W., RADKE, M., QINGYU, W., GUOYING, S. & ZHILI, L. 2000. Retene in pyrolysates of algal and bacterial organic matter. *Organic Geochemistry*, **31**, 757–762, http://doi.org/10.1016/S0146-6380(00)00064-4

WHENZE, G. 2009. Hydrocarbon generation conditions and exploration potential of the Taoudeni Basin, Mauritania. *Petroleum Science*, **6**, 29–37, http://doi.org/10.1007/s12182-009-0006-z

ZHILONG, H., BAOSHUN, Z., QINGCHUN, J., SONGPO, W. & BO, L. 2008. Petroleum systems of the Taoudeni Basin, West Africa. *Petroleum Science*, **5**, 24–30, http://doi.org/10.1007/s12182-008-0004-6

Structure of the ocean–continent transition, location of the continent–ocean boundary and magmatic type of the northern Angolan margin from integrated quantitative analysis of deep seismic reflection and gravity anomaly data

L. COWIE[1]*, R. M. ANGELO[1,2], N. KUSZNIR[1], G. MANATSCHAL[3] & B. HORN[4]

[1]*Earth and Ocean Sciences, University of Liverpool, Liverpool L69 3BX, UK*

[2]*Present address: ConocoPhillips, Houston, TX 77079, USA*

[3]*CNRS-EOST, Université de Strasbourg, 1 rue Blessing, F-67084 Strasbourg, France*

[4]*ION Geophysical/GX Technologies, Houston, TX, USA*

**Corresponding author (e-mail: leanne.cowie.87@gmail.com)*

Abstract: The crustal structure and distribution of crustal types on the northern Angolan rifted continental margin have been the subject of much debate. Hyper-extended continental crust, oceanic crust and exhumed serpentinized mantle have all been proposed to underlie the Aptian salt and the underlying sag sequence. Quantitative analysis of deep seismic reflection and gravity anomaly data, together with reverse post-break-up subsidence modelling, have been used to investigate the ocean–continent transition structure, the location of the continent–ocean boundary, the crustal type and the palaeobathymetry of Aptian salt deposition. Gravity inversion methods (used to give the depth to the Moho and the crustal thickness), residual depth anomaly analysis (used to identify departures from oceanic bathymetry) and subsidence analysis have all shown that the distal Aptian salt is underlain by hyper-extended continental crust rather than exhumed mantle or oceanic crust. We propose that the Aptian salt was deposited *c.* 0.2 and 0.6 km below global sea-level and that the inner proximal salt subsided by post-rift (post-tectonic) thermal subsidence alone, whereas outer distal salt formation was synrift, prior to break-up, resulting in additional tectonic subsidence. Our analysis argues against Aptian salt deposition on the Angolan margin in a 2–3 km deep isolated ocean basin and supports salt deposition on hyper-extended continental crust formed by diachronous rifting migrating from east to west and culminating in the late Aptian.

The northern Angolan rifted continental margin has been the subject of extensive seismic surveys (e.g. Contrucci *et al.* 2004; Moulin *et al.* 2005; Unternehr *et al.* 2010). Seismic imaging and interpretation of the sub-salt is difficult due to the presence of thick sedimentary packages that are impacted by a massive middle to upper Aptian salt sequence (up to 5 km thick in places). This presents major scientific and technical challenges to understanding the crustal structure and tectonic history. As a result, the structure of the ocean–continent transition (OCT) and the location of the continent–ocean boundary (COB) along the northern Angolan margin are still not fully understood. The presence and distribution of thinned continental crust, oceanic crust and exhumed mantle, the nature of the pre-salt sag basins, the tectonic context of the Aptian salt deposition and whether the salt is pre-break-up or post-break-up, the palaeo-water depths through the break-up period and the mechanisms responsible for the generation of accommodation space are all uncertain and much debated (Karner *et al.* 1997; Karner & Driscoll 1999; Jackson *et al.* 2000; Karner *et al.* 2003; Moulin 2003; Moulin *et al.* 2005). In particular, there is controversy about whether the salt was deposited in a deep isolated ocean basin or on thinned continental crust.

Our analysis along the offshore northern Angolan margin was focused along three profiles in the Kwanza region (Fig. 1a). The three profiles include the ION deep long-offset seismic reflection profile CS1-2400 (Fig. 1c), and the P3 (Fig. 1d) and P7 + 11 profiles (Fig. 1e) (Contrucci *et al.* 2004; Moulin *et al.* 2005). The aims of this work were to determine the structure of the OCT along the northern Angolan rifted continental margin, to provide an understanding of the palaeobathymetries of both proximal and distal base Aptian salt deposition and to examine the location of the salt within the broad framework of the OCT.

From: SABATO CERALDI, T., HODGKINSON, R. A. & BACKE, G. (eds) 2017. *Petroleum Geoscience of the West Africa Margin.* Geological Society, London, Special Publications, **438**, 159–176.
First published online January 25, 2016, updated January 27, 2016, http://doi.org/10.1144/SP438.6

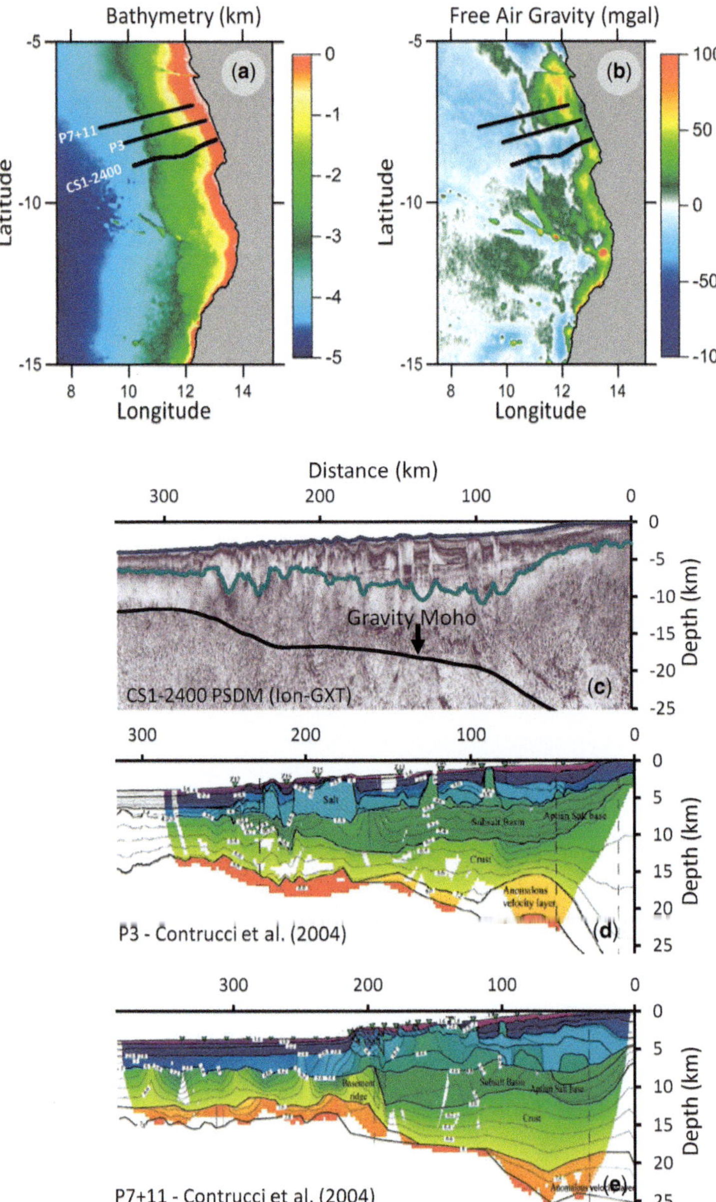

Fig. 1. Data used in the reverse post-break-up thermal subsidence modelling and gravity anomaly inversion for the northern Angolan rifted continental margin. (**a**) Bathymetry (km) (Amante & Eakins 2009) showing the location of profiles CS1-2400, P3 and P7 + 11. (**b**) Satellite-derived free air gravity anomalies (Sandwell & Smith 2009). (**c**) Deep long-offset seismic reflection depth section (PSDM) for the ION CS1-2400 profile. (**d**) Seismic velocity model along the P3 profile from seismic refraction data (modified from Contrucci *et al.* 2004, fig. 3b). (**e**) Seismic velocity model along the P7 + 11 profile from seismic refraction data (modified from Contrucci *et al.* 2004, fig. 6a).

Integrated quantitative analysis methodology

Integrated quantitative analysis of deep seismic reflection and gravity anomaly data has been applied to the Kwanza margin, offshore northern Angola, to determine the structure of the OCT, the location of the COB and the magmatic type using ION deep long-offset seismic reflection data. The integrated work flow and quantitative analytical techniques included gravity anomaly inversion, residual depth anomaly (RDA) analysis and subsidence analysis. The combined interpretation of these independent quantitative measurements was then used to determine the structure of the OCT, the location of the COB and the type of magmatic margin. This integrated approach has been validated previously on the Iberian margin (Cowie 2015), where Ocean Drilling Project boreholes provided ground-truth of the structure of the OCT, the location of the COB and the type of magmatic margin. In addition, we applied a joint inversion technique using deep seismic reflection and gravity anomaly data to determine the lateral variations in the density of the crustal basement and seismic velocity for the ION deep seismic reflection profile CS1-2400. The joint inversion of deep seismic and gravity data validated the thickness of the crustal basement interpreted from deep long-offset seismic reflection data and was used to further constrain the type of crustal basement.

Crustal basement thickness and thinning of the continental lithosphere determined from gravity anomaly inversion

Gravity anomaly inversion was used to determine the depth to the Moho, the thickness of the crustal basement and the thinning factors for the continental lithosphere $(1-1/\beta)$. The data used within our gravity anomaly inversion were bathymetry (Amante & Eakins 2009) (Fig. 1a), satellite-derived free air gravity (Sandwell & Smith 2009) (Fig. 1b), the 2D sediment thickness from pre-stacked depth-migrated (PSDM) seismic reflection data along the CS1-2400 profile (Fig. 1c) and ocean age isochrons from Müller et al. (1997). The gravity anomaly inversion methodology was described by Chappell & Kusznir (2008) and Greenhalgh & Kusznir (2007) and has been previously applied by Cowie & Kusznir (2012) and Alvey et al. (2008).

The gravity anomaly inversion method was carried out in the 3D spectral domain using the scheme of Parker (1972). A lithosphere thermal gravity anomaly correction was incorporated to account for the lithosphere mass deficiency due to the elevated geothermal gradient within oceanic and thinned continental margin lithosphere. Without the inclusion of the lithosphere thermal gravity anomaly correction at rifted continental margins, the predicted depth to the Moho and the crustal basement thickness are too large and the continental lithosphere thinning factors are too low. The thermal gravity anomaly correction is dependent on the thermal re-equilibration time since lithospheric stretching and thinning and therefore on the age of continental break-up. There is general agreement

(e.g. Karner & Gambôa 2007; Aslanian et al. 2009) that rifting on the Angolan margin started in the Neocomian and culminated with continental break-up in the late Aptian. However, there is no consensus on the rates of lithospheric stretching and thinning during this time interval, although there is some evidence that rates of deformation accelerated in the Barremian and Aptian (e.g. Crosby et al. 2011). Although a finite rifting model would be appropriate to determine the lithosphere thermal anomaly developed during lithospheric stretching and thinning leading to break-up, the history of rifting rates is not known. As a consequence, we used an instantaneous rift model to determine the thermal perturbation of the lithosphere and to explore the upper and lower bounds of rift age. We used 112 Ma, corresponding to the age of break-up (after Moulin et al. 2005), for the preferred thermal re-equilibration time to determine the lithosphere thermal gravity anomaly correction, but have also examined sensitivities to ages for thermal re-equilibration that span the period from the Berriasian (140 Ma) to the early Albian (110 Ma). This range corresponds to the start and end of the main rifting episode in the South Atlantic (Teisserenc & Villemin 1989).

Gravity anomaly inversion Moho depths have been calibrated against seismic Moho depths from the oceanic domain of the CS1-2400 profile using the clear Moho reflectors. Calibration suggests that a reference Moho depth of 35.5 km is required to predict crustal basement thicknesses consistent with those seen in the oceanic domain of the CS1-2400 seismic reflection profile. An uncertainty in the oceanic Moho depth on the CS1-2400 PSDM depth section, used for the calibration of the reference Moho depth, arises from uncertainty in the basement seismic velocity, but is estimated to be no more than ± 0.5 km.

A crustal cross-section along the CS1-2400 profile (Fig. 2a) was constructed using depths to the Moho predicted from gravity anomaly inversion, assuming the calibrated reference Moho depth of 35.5 km; the bathymetry and 2D sediment thickness are from the CS1-2400 seismic profile. The crustal cross-section highlights changes in crustal basement thicknesses along the CS1-2400 profile. At the western end of the profile, the gravity anomaly inversion predicts crustal basement thicknesses between 5 and 7 km; in the central region of the profile the crustal basement thicknesses increase to c. 11 km and at the eastern end of the profile the crustal basement is c. 25 km. The Moho depths determined from the gravity anomaly inversion are generally in good agreement with those seen on the CS1-2400 seismic profile.

The corresponding continental lithospheric thinning factor $(\gamma = 1 - 1/\beta)$ estimated for profile CS1-2400, derived from gravity anomaly inversion

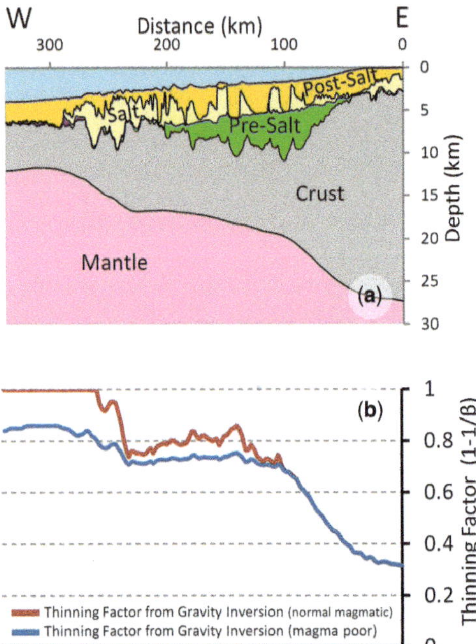

Fig. 2. (**a**) Crustal cross-section along the CS1-2400 profile showing the depth to the Moho from gravity anomaly inversion using the calibrated reference Moho depth of 35.5 km. (**b**) Continental lithospheric thinning profile, predicted from gravity anomaly inversion, along the CS1-2400 profile. Sensitivities to a normal magmatic solution and a magma-starved solution have been examined. A normal magmatic solution predicts thinning factors of 1.0 at the western end of the profile, whereas a magma-starved solution predicts thinning factors of *c.* 0.85.

and assuming depth-uniform stretching and thinning, are shown in Figure 2b. A continental lithospheric thinning factor of zero indicates that there has been no stretching or thinning of the continental lithosphere, whereas a continental lithospheric thinning factor of one indicates that there has been infinite stretching and thinning of the original continental lithosphere and that no continental crust or lithosphere remains. Stretching of the continental lithosphere leads to a decrease in the thickness of the crustal basement; however, decompression melting during rifting and seafloor spreading generates oceanic crust, seaward-dipping reflectors and magmatic underplating, which increases the thickness of the crustal basement. A correction for magmatic addition has been included within the gravity anomaly inversion method. This uses a parameterization of the decompression melting model of White & McKenzie (1989) to predict the thickness of the crustal magmatic addition (Chappell & Kusznir

2008). Decompression melting and the resulting volume and timing of magmatism during rifting and continental break-up are sensitive to the thermal structure of the continental lithosphere and asthenospheric mantle, the chemical composition (enriched or depleted), the rate of lithospheric stretching and thinning, and the amount of melt retention within the mantle. As a consequence, we do not believe that it is possible to apply a deterministic approach to the prediction of magmatic addition. Instead, we examined two end-members: normal magmatic addition and magma-starved. A 'normal' magmatic solution corresponds to 'normal' decompression melting that predicts a 7 km thick oceanic crust; this is initiated at a critical thinning factor of 0.7. In our magma-starved solution, there was no magmatic addition from decompression melting.

The distribution of continental lithospheric thinning factors can be used to help constrain the structure of the OCT and the location of the COB along the profile. At the western end of the profile the continental lithospheric thinning factors for a 'normal' magmatic solution are 1.0, whereas for a magma-starved solution the continental lithospheric thinning factors are *c.* 0.85. In the central section of the profile, the continental lithospheric thinning factors, for both solutions examined, reduce to between 0.7 and 0.85. We preferred the normal magmatic solution at the western end of the profile; however, we believe that the magma-starved solution is preferred in the central and eastern region of the profile.

Using an older age of 140 Ma for the age of lithospheric thermal perturbation (and thermal relaxation) reduces the magnitude of the lithosphere thermal gravity anomaly correction. As a consequence, this gives a slightly deeper Moho, thicker crust and lower thinning factors for the central and eastern part of the profile. This sensitivity to rift age is relatively minor and does not affect the interpretation of OCT structure or crustal type.

Residual depth anomaly analysis along the CS1-2400 profile

RDA analysis was applied to examine bathymetric anomalies in the OCT with respect to the expected oceanic bathymetries along the CS1-2400 profile (Fig. 3). RDAs are commonly used for oceanic regions to compare the observed bathymetry with that predicted from secular cooling models of oceanic lithosphere. An RDA, for the oceanic crust, is the difference between the observed bathymetry (b_{obs}) and the bathymetry predicted from ocean age ($b_{\mathrm{predicted}}$):

$$RDA = b_{\mathrm{obs}} - b_{\mathrm{predicted}} \qquad (1)$$

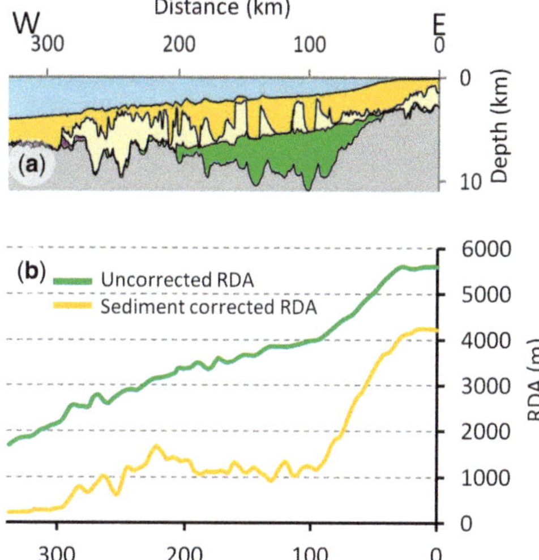

Fig. 3. (**a**) Bathymetry and depth to top-basement for the CS1-2400 profile. (**b**) Comparison of the uncorrected RDA results with the sediment-corrected RDA results along the CS1-2400 profile.

Zero oceanic RDAs correspond to oceanic crust of global average thickness (7 km) in the absence of mantle dynamic topography; positive RDAs correspond to thicker than average oceanic crust and negative RDAs correspond to thin oceanic crust or serpentinized exhumed mantle. We determined RDAs for the CS1-2400 profile to investigate where the RDA signal varied from that seen in the oceanic domain as a result of changes in crustal thickness and composition across the OCT. The age of the lithosphere thermal perturbation due to rifting and break-up in the OCT and its thermal re-equilibration time correspond to the break-up age and the age of the oldest oceanic lithosphere, respectively.

Age-predicted bathymetric anomalies were calculated from Crosby & McKenzie (2009). The age of the oceanic lithosphere was taken from the global ocean isochron model of Müller *et al.* (2008). The region inboard of the oldest ocean isochron (corresponding to the break-up age) may be given that age or the isochron gradient may be projected into the margin. The difference in the predicted RDA between these two approaches to defining the thermal age of the continental margin lithosphere is negligible and has no impact on the RDA interpretation. Sensitivities to the thermal plate model predictions from Parsons & Sclater (1977) and Stein & Stein (1992) were also examined; the RDA results computed using these different thermal plate model predictions did not differ significantly.

The RDAs were corrected for sediment loading. The present day bathymetry was corrected for sediment loading using flexural back-stripping and

decompaction (Kusznir *et al.* 1995), which consists of the removal of the sedimentary load, allowing for the flexural isostatic response and decompaction of the remaining sediments. Flexural back-stripping and decompaction assumes shaly sand compaction and density (Sclater & Christie 1980) during the removal of the sedimentary layer, whereas the salt layer is given a simple salt lithology (Hudec & Jackson 2007). Figure 3b shows a comparison of the uncorrected RDA and the sediment-corrected RDA along the CS1-2400 profile (Fig. 3a). At the western end of the profile there is *c.* 1500 m difference between the uncorrected RDA and the sediment-corrected RDA; the largest difference is seen in the central section of the profile. The sediment-corrected RDA along the CS1-2400 profile (Fig. 3b) is positive, with a magnitude between zero and +300 m at the western end of the profile.

The sediment-corrected RDA has a minor sensitivity to the effective elastic thickness used during flexural back-stripping to define the flexural isostatic response to the sediment unloading correction. We used an effective elastic thickness of 1.5 km, which is appropriate for shorter wavelength synrift sediment loads (see Roberts *et al.* 1998 for further discussion).

Continental lithospheric thinning from subsidence analysis along the CS1-2400 profile

Subsidence analysis was used to determine the distribution of continental lithospheric thinning and the distal extent of continental crust to constrain the structure of the OCT. Subsidence analysis involves the conversion of water-loaded subsidence into continental lithospheric thinning factors assuming the model of McKenzie (1978). Water-loaded subsidence, determined by flexural back-stripping, is interpreted as the sum of the initial (S_i) and thermal (S_t) subsidence in the context of the McKenzie (1978) intra-continental rift model. A correction for magmatic addition due to adiabatic decompression (White & McKenzie 1989) during continental rifting and seafloor spreading has been included (Roberts *et al.* 2013) and uses the same scheme as described earlier in the gravity anomaly inversion methodology. Magmatic addition from decompression melting increases the thickness of the crust thinned by lithospheric stretching due to intrusion and extrusion, which isostatically reduces the initial subsidence predicted by McKenzie (1978) and corresponds to the formation of oceanic crust.

Figure 4b shows the sensitivities to continental lithospheric thinning factors from subsidence analysis, including a 'normal' magmatic solution and a magma-starved solution, with reference to the

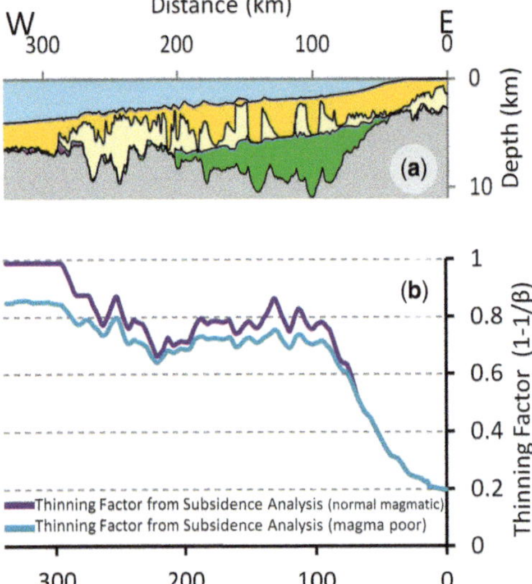

Fig. 4. (**a**) Bathymetry and depth to top-basement for the CS1-2400 profile. (**b**) Continental lithospheric thinning factors from subsidence analysis along the CS1-2400 profile. Sensitivities to a normal magmatic margin and a magma-starved solution are shown.

CS1-2400 profile (Fig. 4a). At the western end of the profile a 'normal' magmatic solution predicts thinning factors of 1.0, whereas a magma-starved solution predicts thinning factors of *c.* 0.9.

Joint inversion of deep seismic and gravity anomaly data: application to the CS1-2400 profile

Joint inversion of deep seismic reflection and gravity anomaly data was applied to the CS1-2400 profile to validate the seismic interpretation of the Moho and to determine the lateral variation in crustal basement density and seismic velocity. The ION deep seismic profile was interpreted in both the time (PSTM) and depth (PSDM) domains. The joint inversion process required that the seismic reflection data showed seismic reflectivity from the Moho; for the CS1-2400 profile this corresponds to most of the section apart from the area beneath the thick distal salt.

The depth to the Moho was first determined from gravity inversion using sediment thicknesses from the PSDM depth section. The crustal basement thickness determined from gravity inversion was converted to interval two-way travel time and then added to the seismic interpretation of top-basement on the PSTM time section to show the gravity Moho in the time domain. The basement seismic velocity V_p used for this conversion from interval depth to

interval two-way travel time was calculated from the crustal basement density used in the gravity inversion using the empirical linear relationship proposed by Birch (1964) and Ludwig *et al.* (1970). The initial value of basement density used in the gravity inversion was 2850 kg m^{-3} (Chappell & Kusznir 2008). Because the comparison of the gravity Moho depth with the seismic reflection image was carried out in the time domain, the joint inversion methodology was not affected by uncertainties in the basement seismic velocities used to produce the PSDM depth seismic image.

The gravity inversion Moho for the CS1-2400 profile, converted into the time domain, is shown in Figure 5a superimposed on the PSTM section. The seismic interpretation of the Moho two-way travel time and the gravity Moho depth taken into the time domain compare well. This suggests that the seismic interpretation of the Moho on the PSTM section is correct and validates the deep seismic interpretation.

The differences in two-way travel time between the seismic and gravity Mohos are assumed to arise from heterogeneities in the crustal basement density and the seismic velocity. The joint inversion solves for coincident seismic and gravity Mohos in the time domain and calculates the lateral variation in crustal basement density and seismic velocity along the profile (Fig. 5c, d). The joint inversion was carried out using the time domain (PSTM) seismic reflection data because these are the raw data and do not have the assumed basement seismic velocities of the depth domain (PSDM) seismic sections.

Joint inversion determines the combination of basement seismic velocities and densities required along the profile to match the Moho predicted from gravity anomaly inversion with the picked Moho in the time domain. Basement density and seismic velocity were assumed to be linked by Birch's empirical relationship (Birch 1964). During the joint inversion, changes in basement density changed the Moho predicted from gravity anomaly inversion in the depth domain; corresponding changes in seismic velocity changed the conversion of the new gravity Moho from the depth domain into the time domain.

Basement densities and seismic velocities from the joint inversion (Fig. 5c, d) showed lateral variations along the profile. The basement densities and seismic velocities for the western distal end of the CS1-2400 profile are significantly higher than those for the remaining profile. We suggest that the higher values in the west correspond to oceanic crustal basement, whereas the lower values in the centre and east correspond to continental crustal basement. The short-wavelength variations in basement density and seismic velocity arise from the inversion methodology and result from fault-controlled top-basement topography.

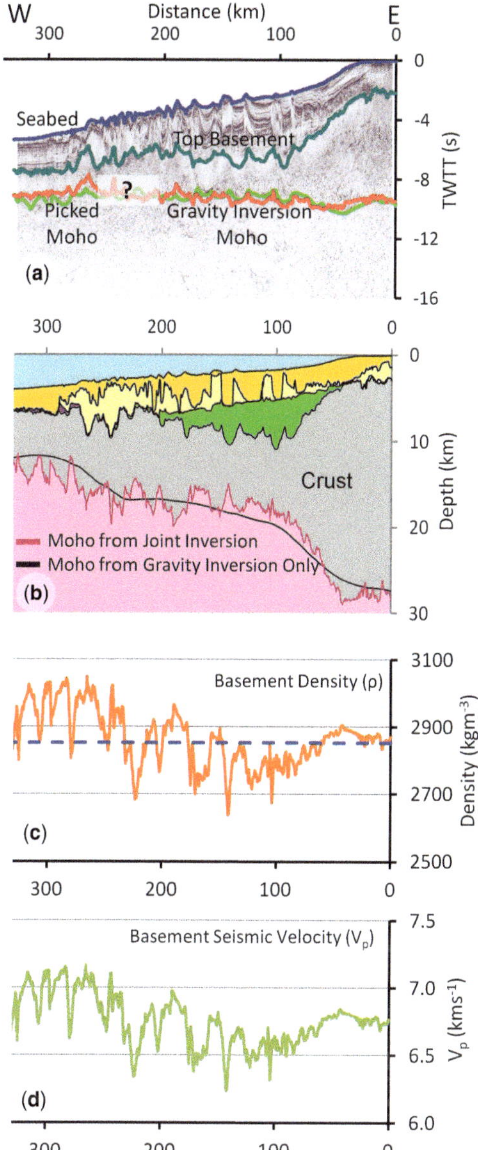

Fig. 5. (**a**) ION CS1-2400 PSTM deep long-offset seismic profile. The horizons for seabed, top-basement, the Moho predicted from gravity anomaly inversion and the picked seismic Moho are indicted. (**b**) Crustal cross-section along the CS1-2400 profile showing Moho depth from gravity anomaly inversion and the Moho depth determined from joint inversion; both are in good agreement, with some variation in magnitude. (**c**) Lateral variations in basement density along the CS1-2400 profile. The blue dashed line highlights a basement density of 2850 kg m^{-3}, which is the basement density used within the initial gravity anomaly inversion. Densities range between 2770 and 2970 kg m^{-3}. (**d**) The corresponding lateral variations in seismic velocity along the CS1-2400 profile.

Solving for the coincident seismic and gravity anomaly inversion predicted Moho in the time domain gave an improved estimate of the depth to the Moho. This improved estimate from joint inversion is shown in Figure 5b and compares with that derived from gravity inversion alone.

Structure of the OCT and location of the COB along the CS1-2400 profile

There is a range of different definitions of the OCT and COB (e.g. Whitmarsh & Miles 1995; Discovery 215 Working Group *et al.* 1998; Dean *et al.* 2000; Manatschal *et al.* 2001; Péron-Pinvidic *et al.* 2007; Manatschal *et al.* 2010). In this paper, we define the OCT as the region between unequivocal continental crust of 'normal' thickness and unequivocal oceanic crust. The lithosphere in this region is highly thinned with complex tectonics, variable magmatism and possible mantle exhumation. We define the COB as the distal limit of unequivocal continental crust; however, determining the location of the COB is made difficult by the presence of exhumed mantle and complex tectonics.

A composite analysis plot for profile CS1-2400 is shown in Figure 6, consisting of: (1) the crustal cross-section from gravity anomaly inversion (Fig. 6a); (2) a comparison of the sediment-corrected RDA and the RDA component from crustal thickness variations (RDA$_{CT}$) (Fig. 6b); (3) a comparison of the continental lithospheric thinning factors predicted from gravity anomaly inversion and subsidence analysis (Fig. 6c); (4) lateral variations in basement density (Fig. 6d); and (5) seismic velocity from joint inversion (Fig. 6e). The joint inversion results, including the depth to the Moho, crustal basement densities and seismic velocities have been 'smoothed' by computing a moving average using a spatial gate of 30 km.

As a result of the thick sedimentary cover and mobile salt (including salt diapirs and canopies), seismic imaging of the salt and pre-salt sedimentary units is difficult, which could lead to errors in our interpretation of the internal structure and thickness of the salt and the pre-salt sedimentary layers. We are, however, more confident in our pick of the base salt. To understand the implications of either over- or underestimating the thickness of the salt layer, we have examined the effect of treating the salt layer as a sedimentary layer with a shaly sand lithology within the gravity inversion; this results in a slightly deeper Moho and smaller continental lithospheric thinning factors from the gravity inversion. The inclusion or omission of the salt layer does not fundamentally change our interpretation of the crustal domains along the profile.

The composite analysis plot is interpreted as showing three distinct crustal zones along the CS1-2400 profile highlighted by the dashed lines: zone A, oceanic crust; zone B, hyper-extended continental crust; and zone C, continental crust. The dashed lines indicate the boundaries between each of these interpreted crustal domains. However, although these interfaces are shown as sharp lines, in reality they are likely to be transitional boundaries. The COB is identified as the oceanwards start of 'normal' oceanic crust and is identified by changes in crustal basement thickness, inflections in the RDA analysis signals and changes in continental lithospheric thinning from subsidence analysis and gravity anomaly inversion.

Zone A: oceanic crust

In zone A, the crustal basement thickness (Fig. 6a) predicted from gravity anomaly inversion ranges between 5 and 6 km, as expected for oceanic crust. Oceanic crust of normal thickness should have a sediment-corrected RDA of approximately zero, notwithstanding the contribution of mantle dynamic topography. The sediment-corrected RDA in this domain (Fig. 6b) is slightly positive, consistent with the presence of oceanic crust together with some mantle dynamic uplift; this is in agreement with the mantle dynamic uplift reported by Crosby & McKenzie (2009) for the Angolan margin. In addition to the RDA corrected for sediment loading, the RDA component from variations in crustal basement thickness (RDA_{CT}) has also been computed; this is the result of the presence of anomalously thick or thin crust. The RDA_{CT} is negative in this domain, which implies that the crustal basement is thinner than 7 km, in agreement with the crustal basement thickness predicted from gravity inversion. The continental lithospheric thinning factors predicted from gravity anomaly inversion and subsidence analysis are in good agreement (Fig. 6c) and predict continental lithospheric thinning factors of 1.0, a 'normal' magmatic solution, implying the presence of oceanic crust. Joint inversion of the deep seismic and gravity data calculates crustal basement densities for zone A between 2850 and 3035 kg m^{-3} (with an average basement density of *c.* 2940 kg m^{-3}) (Fig. 6d) and seismic velocities

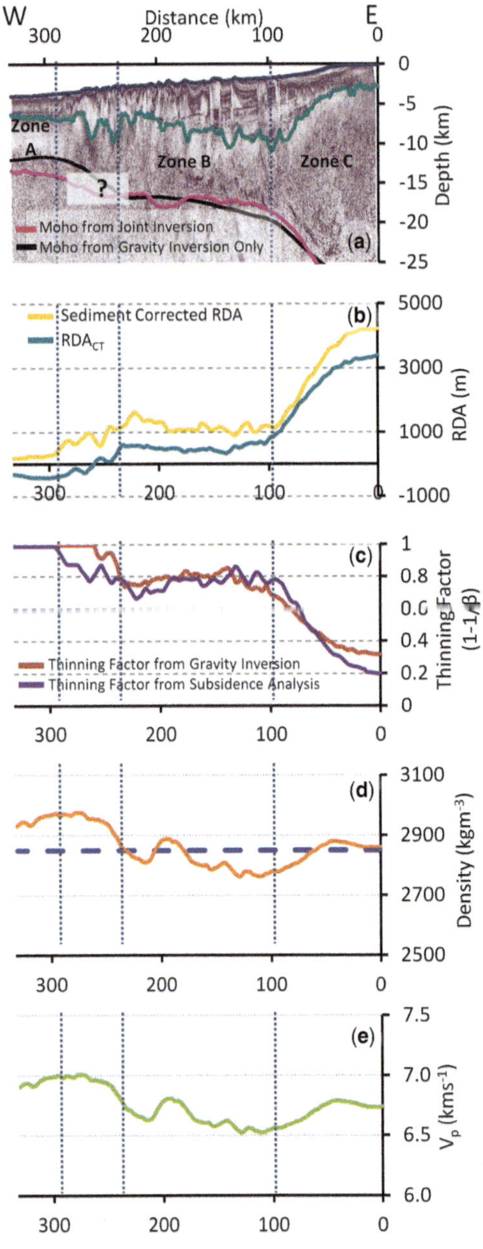

Fig. 6. Summary of the integrated quantitative analysis results for the CS1-2400 profile used to determine the structure of the OCT and the location of the COB. The vertical dashed lines indicate the interpreted boundaries between the predicted crustal domains. (**a**) Crustal cross-section along the CS1-2400 profile with the Moho depth from gravity anomaly inversion. (**b**) The sediment-corrected RDA and the RDA component from variations in crustal basement thickness; both have the same general trend along the profile, although the magnitudes differ. (**c**) Comparison of continental lithospheric thinning factors determined using subsidence analysis and gravity anomaly inversion assuming a normal magmatic solution; the same general trend is seen along the profile. (**d**) Smoothed crustal basement densities predicted from the joint inversion of deep seismic and gravity anomaly data. (**e**) Corresponding seismic velocities predicted from the joint inversion of deep seismic and gravity anomaly data.

between 6.7 and 7.1 km s^{-1} (Fig. 6e). These basement densities (and corresponding seismic velocities) are larger than the crustal basement density (2850 kg m^{-3}) used within the initial gravity anomaly inversion, which is expected as typical oceanic crustal densities range between 2860 and 2900 kg m^{-3} (Carlson & Raskin 1984; Carlson & Herrick 1990; Fowler 2006).

Between the oceanic domain and the hyper-extended continental crust domain, we interpret a domain of transitional crust. We believe that the crust is a mix of hyper-extended continental crust and magmatic addition. We interpret the edges of this transitional region as the inner and outer bounds of the COB. Within this region we see an increase in crustal basement thickness and both the sediment-corrected RDA and the RDA$_{CT}$, while the continental lithospheric thinning factors decrease. At the western end of the hyper-extended continental crust domain, thinning of the continental crust may increase, together with the start of magmatic addition, as oceanic crust is approached. Our interpretation of the presence of hyper-extended continental crust with the presence of magmatic material in this region is significantly different to that proposed by Unternehr *et al.* (2010), who proposed the presence of serpentinized exhumed mantle. Our quantitative analysis showed no evidence of exhumed mantle; exhumed mantle would show a thinner crust from gravity inversion, negative sediment-corrected RDAs and higher continental lithospheric thinning factors.

Zone B: hyper-extended continental crust

In our interpreted hyper-extended continental crust domain, gravity anomaly inversion predicted crustal basement thicknesses between 7 and 12 km (Fig. 6a). Both the sediment-corrected RDA and the RDA$_{CT}$ (Fig. 6b) plateau in this domain, at *c.* 1000 m for the sediment-corrected RDA and *c.* 500 m for the RDA$_{CT}$. The continental lithospheric thinning factors from gravity anomaly inversion and subsidence analysis are in good agreement in zone B (Fig. 6c) and range between 0.7 and 0.85, which is indicative of thinned continental crust. Basement densities and seismic velocities predicted from joint inversion are less than those calculated in zone A; the basement densities range between *c.* 2800 and 2900 kg m^{-3} (Fig. 6d) and the corresponding seismic velocities range between 6.5 and 6.75 km s^{-1} (Fig. 6e).

Zone C: continental crust

At the eastern end of the profile we interpret continental crust as the crustal basement thickness and both the sediment-corrected RDA and RDA$_{CT}$

increase, whereas the continental lithospheric thinning factors decrease to between 0.2 and 0.4. The predicted basement densities range between 2800 and 2855 kg m^{-3} and the seismic velocities range between 6.5 and 6.8 km s^{-1}. The average basement density for zone C is approximately 2830 kg m^{-3}, which is within the range proposed for the density of continental crust (Le Pichon & Sibuet 1981; Carlson & Herrick 1990; Christensen & Mooney 1995) and is similar to that used within the initial gravity anomaly inversion (2850 kg m^{-3}). Zone C includes the margin necking zone.

Palaeobathymetry of the base Aptian salt deposition from reverse post-break-up thermal subsidence modelling

The palaeobathymetry of the base Loeme salt (top Aptian) deposition on the Angolan rifted continental margin was determined using reverse post-break-up subsidence modelling (Kusznir *et al.* 1995; Roberts *et al.* 1998). We focused on the CS1-2400 profile, but also looked at two further profiles to the north: profiles P3 and P7 + 11 (Contrucci *et al.* 2004; Moulin *et al.* 2005). Reverse post-break-up subsidence modelling consists of the sequential flexural isostatic back-stripping of the post-break-up sedimentary sequences, decompaction of the remaining sedimentary units and reverse modelling of the post-break-up lithosphere thermal subsidence. The magnitude of the continental lithospheric stretching factor (β) (McKenzie 1978), which we predicted from gravity anomaly inversion, controls the amount of reverse post-break-up thermal subsidence and hence the restored model elevation relative to sea-level and the predicted palaeobathymetry (Roberts *et al.* 1998, 2009).

Flexural back-stripping and decompaction were applied to the CS1-2400 profile (Fig. 7a) to remove the salt and post-salt sedimentary layers to determine the bathymetry corrected for sedimentary loading to base salt (Fig. 7b). The complex salt movement in this region may appear to be problematic for flexural back-stripping. However, within the palaeobathymetric restoration we used the base salt as the target surface for back-stripping, which allowed us to ignore the salt movement, as we flexurally back-stripped through the salt to the time of deposition. We were able to disregard the salt movement because, as the salt moved, the lithosphere would have responded isostatically to compensate.

Flexural back-stripping and decompaction gives an incomplete palaeobathymetric restoration of the base salt; we also need to include reverse post-break-up thermal subsidence. We determined the magnitude of reverse post-break-up thermal subsidence by the continental lithospheric thinning factor

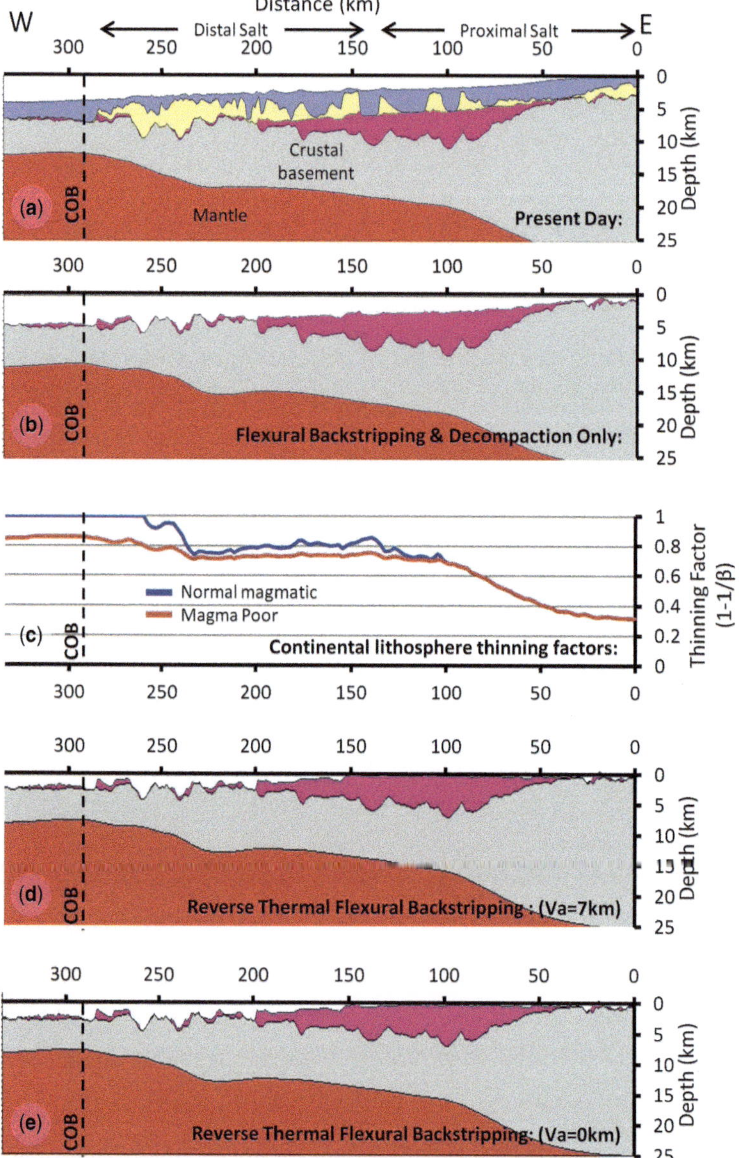

Fig. 7. Flexural back-stripping and reverse post-break-up thermal subsidence modelling along the ION CS1-2400 profile. (**a**) Digitized present day cross-section along the CS1-2400 profile. The post-salt sedimentary layer is coloured blue, the pre-salt sedimentary layer is pink, the salt layer is yellow, the crust is grey and the mantle is red. (**b**) Sediment-corrected bathymetry to base salt calculated from flexural back-stripping and decompaction using a T_e of 1.5 km. (**c**) Continental lithospheric thinning factor profile from gravity anomaly inversion for normal magmatic and magma-starved solutions. (**d**) Reverse post-break-up thermal subsidence modelling along the CS1-2400 profile assuming a normal magmatic solution. (**e**) Reverse post-break-up thermal subsidence modelling along the CS1-2400 profile assuming a magma-starved solution.

$(\gamma = 1 - 1/\beta)$ derived from gravity anomaly inversion. Lithospheric thinning factors from gravity inversion are shown in Figure 7c assuming a 112 Ma rift age for the thermal gravity anomaly correction in the gravity inversion; sensitivities for a normal magmatic solution and a magma-starved solution were examined. As discussed in previous sections, the continental lithospheric thinning factor for a normal magmatic solution is 1.0 at the western end of the CS1-2400 profile, whereas for a magma-starved solution the continental lithospheric thinning factor is *c.* 0.85. In the central section of the profile the continental lithospheric thinning factor is between 0.7 and 0.85 for both solutions examined.

Figure 7d shows the restored palaeobathymetry to base salt, including reverse thermal subsidence modelling, assuming a normal magmatic solution and a break-up age of 112 Ma. The proximal base salt restores to just below global sea-level with an average bathymetry of *c.* 0.2 km. In contrast, the distal base salt does not restore to near sea-level; the restored palaeobathymetries for the distal base salt (smoothing through fault-controlled topography) are between *c.* 0.9 and 2.5 km below global sea-level. In the deep fault-controlled troughs, palaeobathymetries for the distal base salt of *c.* 4.0 km below sea-level are predicted.

The restored palaeobathymetry to base salt, assuming a magma-starved solution (Fig. 7e), also shows that the proximal base salt restores to *c.* 0.2 km below global sea-level, while the distal base salt again does not. The predicted palaeobathymetries of the distal base salt range between *c.* 0.9 and 3 km below sea-level (smoothing through fault-controlled topography); the palaeobathymetries are greater (*c.* 4.5 km below sea-level) in the deep structural troughs.

We believe that the normal magmatic addition solution (Fig. 7d) is applicable to the oceanic part of the profile, whereas the magma-starved solution (Fig. 7e) is more applicable to the continental end of the profile, with a transitional region in between. In the oceanic domain, water depths at break-up of approximately 2.5 km (± 0.2 km depending on magmatic solution), consistent with a young oceanic ridge, are predicted for both a normal magmatic and a magma-starved solution.

Rifting on the Angola margin is believed to have commenced in the Neocomian (Teisserenc & Ville-min 1989). For the inner margin, beneath the proximal salt, the main rifting event may have been in the Barremian (Crosby *et al.* 2011). If a rift age of 130 Ma is used to give the lithosphere thermal correction in the gravity inversion and in the reverse post-break-up subsidence modelling, then the proximal base salt restores to *c.* 0.6 km below global sea-level. There still remains a substantial difference between the restored bathymetry of the base proximal and base distal salt.

It is possible that our interpretation has overestimated the thickness of the distal salt. We have examined the effect of treating the salt layer as a sedimentary layer within the gravity inversion and the reverse post-break-up subsidence modelling.

Reducing the thickness of the salt has a negligible effect on the predicted bathymetry of both the proximal and distal salt.

An additional sensitivity to the continental lithospheric thinning factors, used to drive reverse thermal subsidence, was examined along the CS1-2400 profile. A continental lithospheric thinning factor of 1.0 (corresponding to $\beta = \infty$), which gives an upper bound for the restored post-break-up thermal subsidence, was applied to the entire profile. The predicted bathymetry for the distal base salt remained almost unchanged at between 2 and 3 km below global sea-level. This implies that, if the distal base salt was deposited at or just below global sea-level, it has subsided not only due to post-break-up thermal subsidence and sediment loading alone but also syn-tectonic crustal thinning.

The location of the outer (or more distal) interpreted COB, determined from integrated quantitative analysis and identified by the dashed line on Figure 7, was used to examine where the salt is located within the OCT. We believe that the majority of the salt along the CS1-2400 profile is located to the east of the COB on hyper-extended continental crust.

In addition to the CS1-2400 profile, we also applied reverse post-break-up thermal subsidence modelling to the more northerly P3 and P7 + 11 profiles (Fig. 8). The results are comparable with those predicted for the CS1-2400 profile, with the proximal base salt restoring to approximately sea-level, whereas the distal base salt restores to between 2 and 3 km below global sea-level. Predicted thinning factors from gravity inversion, assuming normal magmatic addition, are 1.0 at the western end of both the P3 and P7 + P11 profiles, consistent with the presence of oceanic crust. Predicted palaeobathymetries for the western end of both profiles are, on average, 2.85 km, consistent with water depths on newly formed oceanic crust. For both profiles, the predicted palaeobathymetries for the base proximal salt are at or just below global sea-level, whereas the palaeobathymetry of the base distal salt is between 2 and 3 km.

Discussion

Crustal structure and location of the COB

Integrated quantitative analysis with gravity anomaly inversion, RDA analysis, subsidence analysis and joint inversion of deep seismic reflection and gravity data has been used to determine the structure of the OCT, the location of the COB and the magmatic type along the CS1-2400 profile in northern Angola. Our analysis showed changes in crustal structure along the profile, from which we have interpreted three well-defined crustal

P3 profile

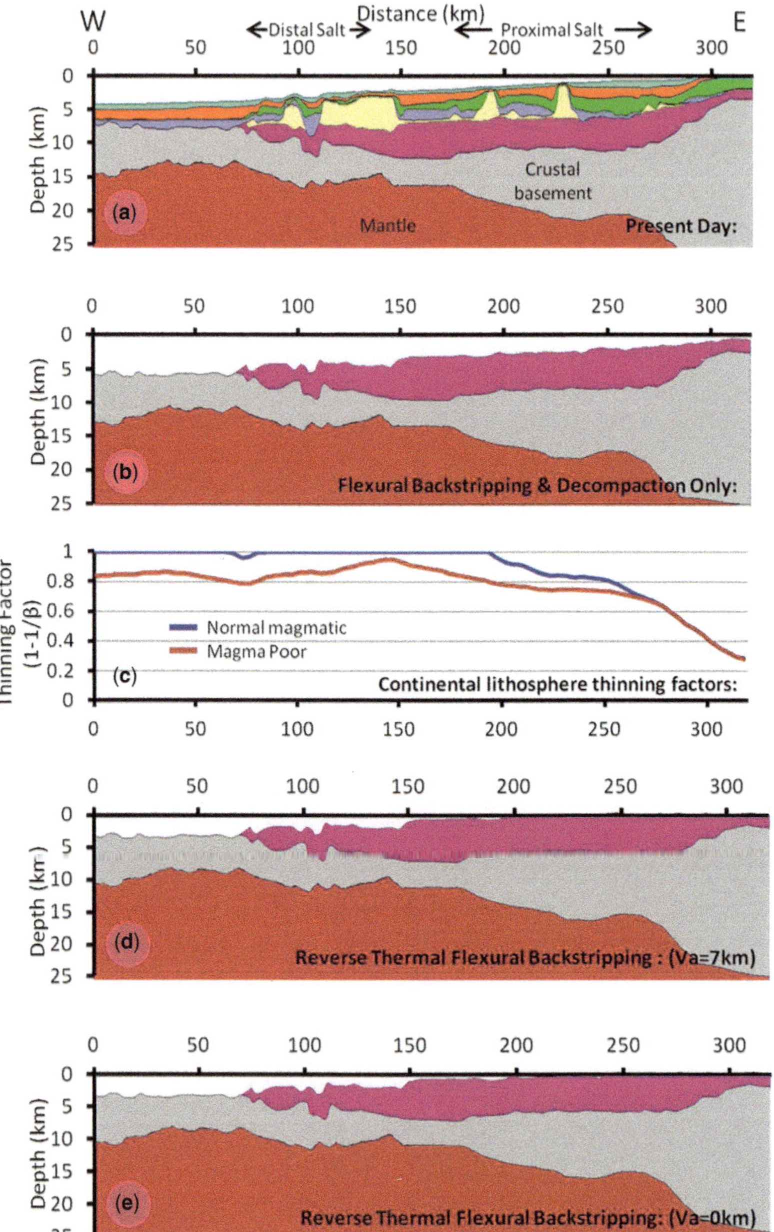

Fig. 8. Flexural back-stripping and reverse post-break-up thermal subsidence modelling along the P3 and P7 + 11 profiles (Contrucci *et al.* 2004; Moulin 2003). (**a**) Digitized present day cross-section along the P3 profile; (**f**) digitized present day cross-section along the P7 + 11 profile; the post-salt sedimentary layers are highlighted in turquoise, orange, green and blue; the pre-salt sedimentary layer is pink, the salt layer is yellow, the crust is grey and the mantle is red. (**b**, **g**) Sediment-corrected bathymetry to base salt calculated from flexural back-stripping and decompaction using a T_e of 1.5 km. (**c**, **h**) Continental lithospheric thinning factors from gravity anomaly inversion for a normal magmatic and a magma-starved solution. (**d**, **i**) Reverse post-break-up thermal subsidence modelling along the P3 profile assuming a normal magmatic solution. (**e**, **j**) Reverse post-break-up thermal subsidence modelling along the P3 profile assuming a magma-starved solution.

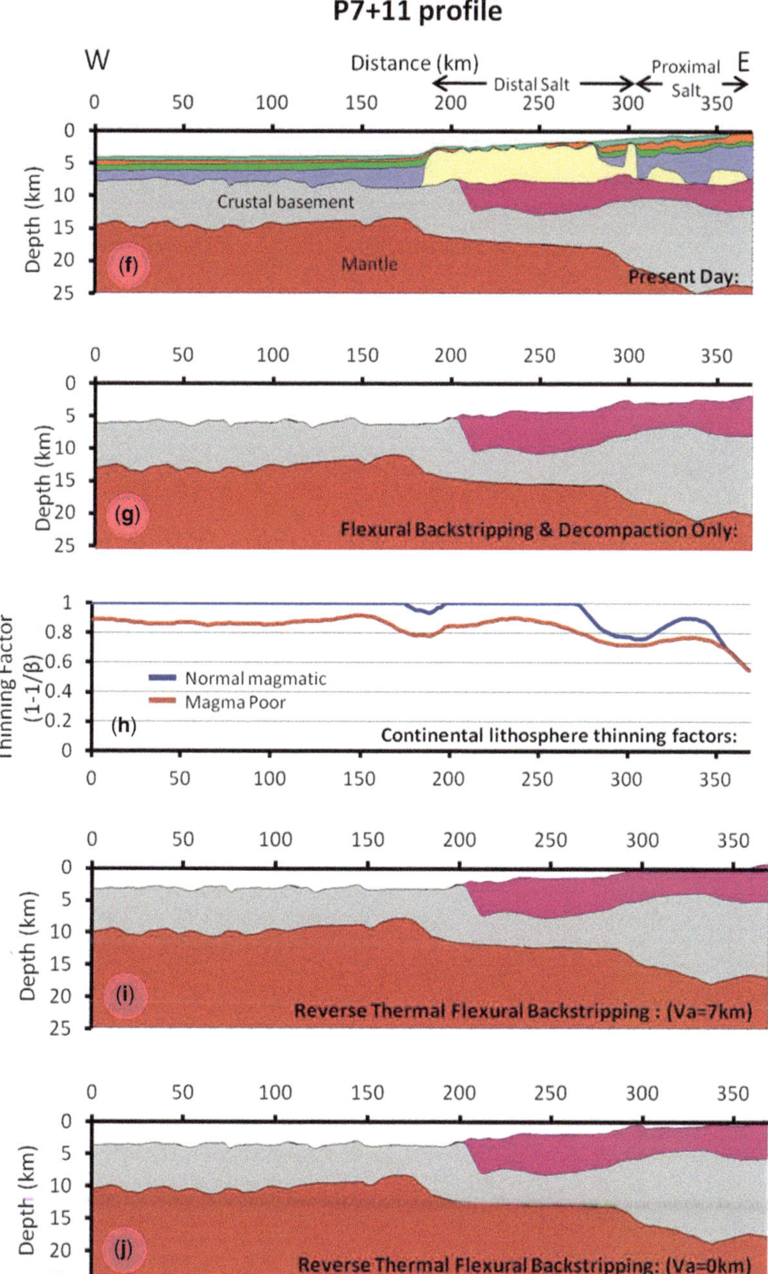

Fig. 8. *Continued.*

domains: oceanic crust, hyper-extended continental crust and continental crust (Fig. 9a). We have also interpreted a transitional region between the hyper-extended continental crust and the start of oceanic crust. The interpretation of our results suggests that a normal magmatic solution is applicable at the oceanic end of the profile, whereas a magma-starved solution is more applicable at the continental end of the profile. Considering the integrated quantitative analysis techniques together has enabled us

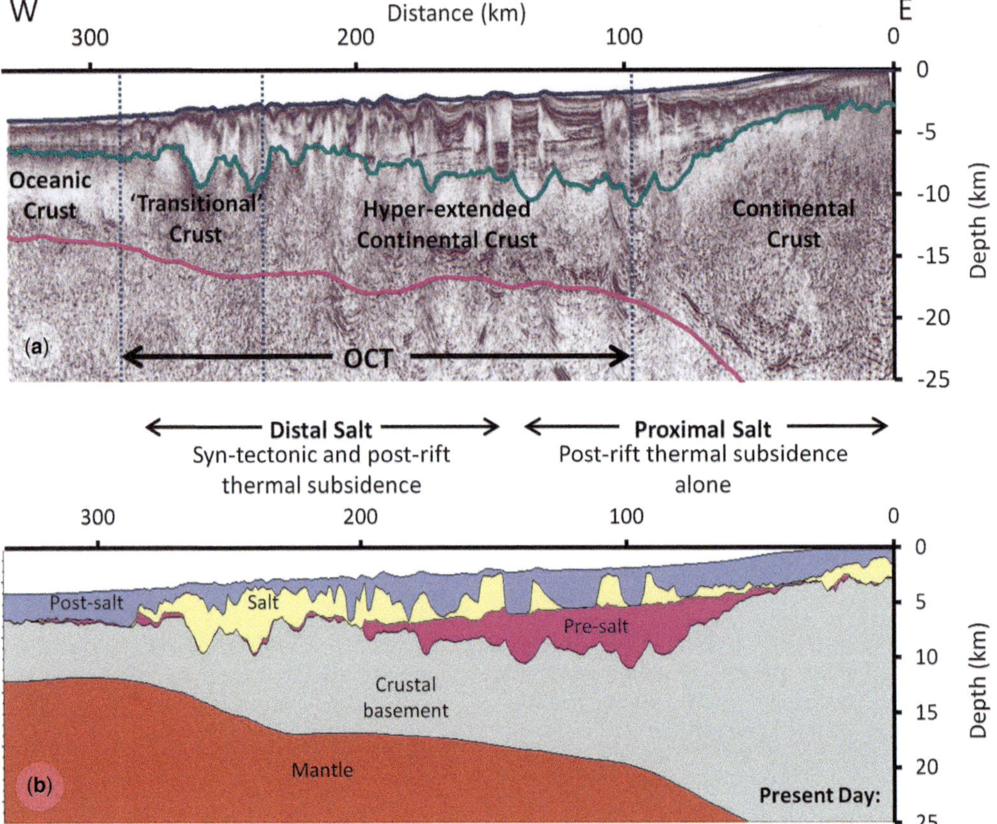

Fig. 9. (**a**) Interpretation of the integrated quantitative analysis results along the PSDM CS1-2400 profile. Seabed is shown in blue, top-basement in green, Moho from gravity anomaly inversion in black and Moho from the joint inversion in pink. Our interpreted boundaries between the predicted crustal domains are indicated by the dashed lines. (**b**) Digitized present day cross-section along the CS1-2400 profile. The locations of the distal and proximal salt and interpreted subsidence history are shown.

to make a robust geological interpretation of the OCT along the profile and a more accurate estimation of the location of the COB.

Our interpretation of the integrated quantitative analysis results along the CS1-2400 profile is shown in Figure 9a. Our analysis suggests that:

(1) Gravity and deep seismic reflection data predict that the earliest oceanic crust is *c.* 5–7 km thick.

(2) RDA analysis shows a slightly positive sediment-corrected RDA in the oceanic domain, consistent with the presence of mantle dynamic uplift; this is in agreement with that reported by Crosby & McKenzie (2009).

(3) Gravity inversion, RDA analysis and subsidence analysis all suggest that both the proximal and distal salt are underlain by

hyper-extended continental crust, not by oceanic crust.

(4) We interpret transitional crust between the oceanic crust and the hyper-extended continental crust domains. We believe this to be a mix of hyper-extended continental crust and the addition of magmatic material. Our interpretation of this transitional crust is significantly different from the interpretation of serpentinized exhumed mantle of Unternehr *et al.* (2010).

(5) The gravity anomaly inversion, RDA analysis and subsidence analysis results show that the OCT along profile CS1-2400 is fairly wide, with the distance between the COB and the margin necking zone measuring *c.* 180 km.

(6) Joint inversion of deep seismic reflection and gravity data shows a contrast in basement

density and seismic velocity between oceanic and continental crustal basement that is consistent with our domain of transitional crust between the oceanic crust and the hyper-extended crust.

Palaeobathymetry and depositional environment of base Aptian salt

Predicted palaeobathymetries have been determined for the base Loeme salt using 2D-flexural back-stripping and decompaction, together with reverse modelling of the post-break-up thermal subsidence. Continental lithospheric thinning factors derived from gravity anomaly inversion have been used to determine the reverse post-break-up thermal subsidence. For profile CS1-2400, thinning factors, derived from both the normal magmatic and magma-starved gravity inversion solutions assuming a rift age of 112 Ma and used to drive the reverse post-rift thermal subsidence modelling, restore the proximal autochthonous base salt to c. 0.2 km below global sea-level at the time of break-up. In contrast, reverse post-break-up subsidence modelling restores the distal base salt to between 2 and 3 km below global sea-level. Similar palaeobathymetries for the base salt are also calculated for the more northerly P3 and P7 + 11 profiles. If we apply a continental lithospheric thinning factor of 1.0 to drive the reverse post-rift thermal subsidence along the full length of the three profiles (which is unreasonable), the palaeobathymetries of the base distal salt still do not restore to sea-level, demonstrating that it is not possible to generate the subsidence of the base salt by post-rift subsidence alone. The predicted bathymetries at break-up of the first unequivocal oceanic crust are c. 2.5 km, as expected for newly formed oceanic crust of normal thickness. Using a rift age of 130 Ma for the proximal margin increases the predicted palaeobathymetry of the base proximal salt to c. 0.6 km below global sea-level, consistent with the analysis of Crosby et al. (2011).

Our preferred interpretation of the palaeobathymetric restoration of the distal and proximal base salt is that all the Aptian salt was deposited between 0.2 and 0.6 km below global sea-level, but that the distal salt was emplaced during late synrift while the continental crust under it was being actively thinned, resulting in additional tectonic subsidence. This is consistent with the seismic evidence, which shows that the distal base salt is extensionally faulted. This is also in agreement with the observation by Karner & Gambôa (2007) that the rate of subsidence required to generate the accommodation space for the distal salt is too large to be generated by thermal post-break-up subsidence. Crustal basement thicknesses from gravity inversion, RDA and

subsidence analysis, summarized in Figure 6, suggest that the distal salt is underlain by hyper-extended continental crust (Fig. 9b) rather than oceanic crust or exhumed mantle, as previously suggested (Reston 2010; Unternehr et al. 2010). In contrast with the distal salt, the proximal salt formed in a region where crustal thinning had taken place from the Neocomian or Barremian, but had ceased by the late Aptian; this is consistent with the pre-salt sag sequence under the proximal salt being post-rift (Unternehr et al. 2010; Crosby et al. 2011). Our interpretation requires that the distal salt subsided by synrift crustal thinning and post-rift thermal subsidence, whereas the proximal salt subsided by post-rift thermal subsidence alone. Diachronous thinning of the continental crust from inboard to outboard is to be expected from both observation and modelling and is consistent with the break-up tectonic models proposed by Péron-Pinvidic & Manatschal (2008), Pindell & Kennan (2007), Ranero & Perez-Gussinye (2010) and Brune et al. (2014).

An alternative explanation for the different subsidence styles of the proximal and distal salt has been proposed by Karner & Gambôa (2007), who suggested that a sag-style subsidence of the proximal salt but a syn-tectonic style for distal salt are the result of depth-dependent lithospheric stretching and thinning. Although depth-dependent lithospheric stretching and thinning have been reported at rifted margins (Driscoll & Karner 1998; Davis & Kusznir 2004; Kusznir & Karner 2007), diachronous rifting and thinning of the Angolan margin lithosphere from inboard to outboard may provide a simpler explanation for the differing subsidence styles of the proximal and distal salt.

An alternative interpretation is that the distal salt is parautochthonous and moved downslope to its present position during break-up. If the salt is parautochthonous (or allochthonous) in the distal regions, then this suggests that it was not deposited in deep water and that the salt should not restore to sea-level. This interpretation is similar to that advocated in the Gulf of Mexico by Hudec et al. (2013) and Rowan & Vendeville (2006).

An interpretation that is often invoked (e.g. Burke & Sengör 1988; Burke et al. 2003) to explain the palaeobathymetry of the base Aptian salt along the northern Angolan margin is that distal Aptian salt deposition occurred in confined environmental conditions (e.g. in a Messinian-type basin, isolated from global sea-level). Although a structural barrier in the south and north is not dismissed (and is indeed likely), we believe that there is no definite requirement to invoke an isolated ocean basin with local sea-level between 2 and 3 km below global sea-level for the deposition of the Aptian salt on the Angolan rifted margin. Furthermore, our analysis suggests that both the proximal and distal salt on the Angolan margin are underlain by hyper-extended continental crust, not by oceanic crust. The restored bathymetries of the base proximal salt from this study (and also Crosby *et al.* 2011) are no more than 0.6 km below global sea-level. A deep isolated ocean basin between 2 and 3 km deep for the deposition of distal salt would require a substantial difference in the depth of salt deposition, which we consider to be unlikely. A similar observation has been made by Moulin *et al.* (2005) and Aslanian *et al.* (2009). Additional strong arguments against the isolated basin interpretation are also presented by Pindell *et al.* (2014) with reference to the Gulf of Mexico. They argue that an isolated basin hypothesis is unlikely as it requires a complicated scenario of inter-related events to occur.

Our integrated quantitative analysis predicts the presence of oceanic crust at the western end of the profile, whereas in the centre of the profile, beneath the majority of the Aptian salt, we interpret hyper-extended continental crust. We believe that both the proximal and distal Aptian salt on the Kwanza margin were deposited at a datum 0.2–0.6 km below global sea-level, but that the distal salt was deposited during late synrift while the crust under it was being actively thinned, which resulted in additional tectonic subsidence. It is possible that some of the distal salt is parautochthonous and moved downslope to its present day position. It is also possible that syn-tectonic (pre-break-up) extension continued after salt deposition in the distal region.

References

ALVEY, A., GAINA, C., KUSZNIR, N.J. & TORSVIK, T.H. 2008. Integrated crustal thickness mapping and plate reconstructions for the high Arctic. *Earth and Planetary Science Letters*, **274**, 310–321.

AMANTE, C. & EAKINS, B.W. 2009. *ETOPO1 1 Arc-Minute Global Relief Model: Procedures, Data Sources and Analysis*. NOAA Technical Memo. NES-DIS NGDC-24, NOAA, Silver Spring.

ASLANIAN, D., MOULIN, M. *ET AL.* 2009. Brazilian and African passive margins of the Central Segment of the South Atlantic Ocean: kinematic constraints. *Tectonophysics*, **468**, 98–112.

BIRCH, F. 1964. Density and composition of mantle and core. *Journal of Geophysical Research*, **69**, 4377–4388.

BRUNE, S., HEINE, C., PÉREZ-GUSSINYÉ, M. & SOBOLEV, S.V. 2014. Rift migration explains continental margin asymmetry and crustal hyper-extension. *Nature Communications*, **5**, article no. 4014, http://doi.org/10.1038/ncomms5014

BURKE, K. & SENGÖR, A.M.C. 1988. Ten metre global sea-level change associated with South Atlantic Aptian salt deposition. *Marine Geology*, **83**, 309–312.

BURKE, K., MACGREGOR, D.S. & CAMERON, N.R. 2003. Africa's petroleum systems: four tectonic 'Aces' in the past 600 million years. *In*: ARTHUR, T., MACGREGOR, D.S. & CAMERON, N.R. (eds) *Petroleum Geology of Africa: New Themes and Developing Technologies*. Geological Society, London, Special Publications, **207**, 21–60, http://doi.org/10.1144/GSL.SP.2003.207.3

CARLSON, R.L. & HERRICK, C.N. 1990. Densities and porosities in the oceanic crust and their variations with depth and age. *Journal of Geophysical Research: Solid Earth*, **95**, 9153–9170.

CARLSON, R.L. & RASKIN, G.S. 1984. Density of the ocean crust. *Nature*, **311**, 555–558.

CHAPPELL, A.R. & KUSZNIR, N.J. 2008. Three-dimensional gravity inversion for Moho depth at rifted continental margins incorporating a lithosphere thermal gravity anomaly correction. *Geophysical Journal International*, **174**, 1–13.

CHRISTENSEN, N.I. & MOONEY, W.D. 1995. Seismic velocity structure and composition of the continental crust: a global view. *Journal of Geophysical Research: Solid Earth*, **100**, 9761–9788.

CONTRUCCI, I., MATIAS, L. *ET AL.* 2004. Deep structure of the West African continental margin (Congo, Zaïre, Angola), between 5°S and 8°S, from reflection/refraction seismics and gravity data. *Geophysical Journal International*, **158**, 529–553.

COWIE, L. 2015. *Determination of ocean continent transition structure, continent ocean boundary location and magmatic type at rifted continental margins*. PhD thesis, University of Liverpool.

COWIE, L. & KUSZNIR, N. 2012. Mapping crustal thickness and oceanic lithosphere distribution in the Eastern Mediterranean using gravity inversion. *Petroleum Geoscience*, **18**, 373–380, http://doi.org/10.1144/petgeo2011-071

CROSBY, A.G. & MCKENZIE, D. 2009. An analysis of young ocean depth, gravity and global residual topography. *Geophysical Journal International*, **178**, 1198–1219.

CROSBY, A.G., WHITE, N.J., EDWARDS, G.R.H., THOMPSON, M., CORFIELD, R. & MACKAY, L. 2011. Evolution of deep-water rifted margins: testing depth-dependent extensional models. *Tectonics*, **30**, TC1004, http://doi.org/10.1029/2010TC002687

DAVIS, M. & KUSZNIR, N. 2004. Depth dependent lithospheric stretching at rifted continental margins. *In*: KARNER, G.D., TAYLOR, B., DRISCOLL, N.W. & KOHLSTEDT, D.L. (eds) *Rheology and Deformation of the Lithosphere at Continental Margins*. Columbia University Press, New York.

DEAN, S.M., MINSHULL, T.A., WHITMARSH, R.B. & LOUDEN, K.E. 2000. Deep structure of the ocean-continent transition in the southern Iberia Abyssal Plain from seismic refraction profiles: the IAM-9 transect at 40°20′N. *Journal of Geophysical Research: Solid Earth*, **105**, 5859–5885.

DISCOVERY 215 WORKING GROUP, MINSHULL, T.A., DEAN, S.M., WHITMARSH, R.B., RUSSELL, S.M., LOUDEN, K.E. & CHIAN, D. 1998. Deep structure in the vicinity of the ocean-continent transition zone under the southern Iberia Abyssal Plain. *Geology*, **26**, 743–746.

DRISCOLL, N.W. & KARNER, G.D. 1998. Lower crustal extension across the Northern Carnarvon basin, Australia: evidence for an eastward dipping detachment. *Journal of Geophysical Research: Solid Earth*, **103**, 4975–4991.

FOWLER, C.M.R. 2006. *The Solid Earth: An Introduction to Global Geophysics*. Cambridge University Press, Cambridge.

GREENHALGH, E.E. & KUSZNIR, N.J. 2007. Evidence for thin oceanic crust on the extinct Aegir Ridge, Norwegian Basin, NE Atlantic derived from satellite gravity inversion. *Geophysical Research Letters*, **34**, L06305.

HUDEC, M.R. & JACKSON, M.P.A. 2007. Terra infirma: understanding salt tectonics. *Earth-Science Reviews*, **82**, 1–28.

HUDEC, M.R., NORTON, I.O., JACKSON, M.P.A. & PEEL, F.J. 2013. Jurassic evolution of the Gulf of Mexico Salt Basin. *AAPG Bulletin*, **97**, 1683–1710.

JACKSON, M.P.A., CRAMEZ, C. & FONCK, J.-M. 2000. Role of subaerial volcanic rocks and mantle plumes in creation of South Atlantic margins: implications for salt tectonics and source rocks. *Marine and Petroleum Geology*, **17**, 477–498.

KARNER, G.D. & DRISCOLL, N.W. 1999. Tectonic and stratigraphic development of the West African and eastern Brazilian Margins: insights from quantitative basin modelling. *In*: CAMERON, N.R., BATE, R.H. & CLURE, V.S. (eds) *The Oil and Gas Habitats of the South Atlantic*. Geological Society, London, Special Publications, **153**, 11–40, http://doi.org/10.1144/GSL.SP.1999.153.01.02

KARNER, G.D. & GAMBÔA, L.A.P. 2007. Timing and origin of the South Atlantic pre-salt sag basins and their capping evaporites. *In*: SCHREIBER, B.C., LUGLI, S. & BAÇBEL, M. (eds) *Evaporites Through Space and Time*. Geological Society, London, Special Publications, **285**, 15–35, http://doi.org/10.1144/SP285.2

KARNER, G.D., DRISCOLL, N.W., McGINNIS, J.P., BRUMBAUGH, W.D. & CAMERON, N.R. 1997. Tectonic significance of syn-rift sediment packages across the Gabon-Cabinda continental margin. *Marine and Petroleum Geology*, **14**, 973–1000.

KARNER, G.D., DRISCOLL, N.W. & BARKER, D.H.N. 2003. Syn-rift regional subsidence across the West African continental margin: the role of lower plate ductile extension. *In*: ARTHUR, T., MACGREGOR, D.S. & CAMERON, N.R. (eds) *Petroleum Geology of Africa: New Themes and Developing Technologies*. Geological Society, London, Special Publications, **207**, 105–129, http://doi.org/10.1144/GSL.SP.2003.207.6

KUSZNIR, N.J. & KARNER, G.D. 2007. Continental lithospheric thinning and breakup in response to upwelling divergent mantle flow: application to the Woodlark, Newfoundland and Iberia margins. *In*: KARNER, G.D., MANATSCHAL, G. & PINHEIRO, L.M. (eds) *Imaging, Mapping and Modelling Continental Lithosphere Extension and Breakup*. Geological Society, London, Special Publications, **282**, 389–419, http://doi.org/10.1144/SP282.16

KUSZNIR, N.J., ROBERTS, A.M. & MORLEY, C.K. 1995. Forward and reverse modelling of rift basin formation. *In*: LAMBIASE, J.J. (ed.) *Hydrocarbon Habitat in Rift Basins*. Geological Society, London, Special Publications, **80**, 33–56, http://doi.org/10.1144/GSL.SP.1995.080.01.02

LE PICHON, X. & SIBUET, J.-C. 1981. Passive margins: a model of formation. *Journal of Geophysical Research: Solid Earth*, **86**, 3708–3720.

LUDWIG, W.J., NAFE, J.E. & DRAKE, C.L. 1970. Seismic refraction. *In*: MAXWELL, A.E. (ed.) *The Sea*, **4**. Wiley-Interscience, New York, 53–84.

MANATSCHAL, G., FROITZHEIM, N., RUBENACH, M. & TURRIN, B.D. 2001. The role of detachment faulting in the formation of an ocean–continent transition: insights from the Iberia Abyssal Plain. *In*: WILSON, R.C.L., WHITMARSH, R.B., TAYLOR, B. & FROITZHEIM, N. (eds) *Non-Volcanic Rifting of Continental Margins: A Comparison of Evidence from Land and Sea*. Geological Society, London, Special Publications, **187**, 405–428, http://doi.org/10.1144/GSL.SP.2001.187.01.20

MANATSCHAL, G., SUTRA, E. & PÉRON-PINVIDIC, G. 2010. The lesson from the Iberia-Newfoundland rifted margins: how applicable is it to other rifted margins? *In*: PENA DOS REIS, R. & PIMENTEL, N. (eds) *Proceedings, 2nd Central & North Atlantic Conjugate Margins: Rediscovering the Atlantic, New Insights, New Winds for an Old Sea*, **2**. Outubro, Lisbon, 27–37.

McKENZIE, D. 1978. Some remarks on the development of sedimentary basins. *Earth and Planetary Science Letters*, **40**, 25–32.

MOULIN, M. 2003. *Etude géologique et géophysique des marges continentales passives: exemple du Zaire et de l'Angola*. PhD, University de Bretagne Occidentale.

MOULIN, M., ASLANIAN, D. ET AL. 2005. Geological constraints on the evolution of the Angolan margin based on reflection and refraction seismic data (ZaïAngo project). *Geophysical Journal International*, **162**, 793–810.

MÜLLER, R.D., ROEST, W.R., ROYER, J.-Y., GAHAGAN, L.M. & SCLATER, J.G. 1997. Digital isochrons of the world's ocean floor. *Journal of Geophysical Research*, **102**, 3211–3214.

MÜLLER, R.D., SDROLIAS, M., GAINA, C., STEINBERGER, B. & HEINE, C. 2008. Long-term sea-level fluctuations driven by ocean basin dynamics. *Science*, **319**, 1357–1362.

PARKER, R.L. 1972. The rapid calculation of potential anomalies. *Geophysical Journal of the Royal Astronomical Society*, **31**, 447–455.

PARSONS, B. & SCLATER, J.G. 1977. An analysis of the variation of ocean floor bathymetry and heat flow with age. *Journal of Geophysical Research*, **82**, 803–827.

PÉRON-PINVIDIC, G. & MANATSCHAL, G. 2008. The final rifting evolution at deep magma-poor passive margins from Iberia-Newfoundland: a new point of view. *International Journal of Earth Sciences*, **98**, 1581–1597.

PÉRON-PINVIDIC, G., MANATSCHAL, G., MINSHULL, T.A. & SAWYER, D.S. 2007. Tectonosedimentary evolution of the deep Iberia-Newfoundland margins: evidence for a complex breakup history. *Tectonics*, **26**, TC2011.

PINDELL, J.L. & KENNAN, L. 2007. Rift models and the salt-cored marginal wedge in the northern Gulf of Mexico: implications for dep water Paleogene Wilcox deposition and basinwide maturation. *In*: KENNAN, L., PINDELL, J. & ROSEN, N.C. (eds) *The*

Paleogene of the Gulf of Mexico and Caribbean Basins. GCSSEPM 27th Annual Bob F. Perkins Research Conference, 2–5 December, Houston, Texas, 146–186.

PINDELL, J., GRAHAM, R. & HORN, B. 2014. Rapid outer marginal collapse at the rift to drift transition of passive margin evolution, with a Gulf of Mexico case study. *Basin Research*, **26**, 701–725, http://doi.org/10.1111/bre.12059

RANERO, C.R. & PEREZ-GUSSINYE, M. 2010. Sequential faulting explains the asymmetry and extension discrepancy of conjugate margins. *Nature*, **468**, 294–299.

RESTON, T.J. 2010. The opening of the central segment of the South Atlantic: symmetry and the extension discrepancy. *Petroleum Geoscience*, **16**, 199–206, http://doi.org/10.1144/1354-079309-907

ROBERTS, A.M., KUSZNIR, N.J., YIELDING, G. & STYLES, P. 1998. 2D flexural backstripping of extensional basins; the need for a sideways glance. *Petroleum Geoscience*, **4**, 327–338, http://doi.org/10.1144/petgeo.4.4.327

ROBERTS, A.M., CORFIELD, R.I., KUSZNIR, N.J., MATTHEWS, S.J., HANSEN, E.-K. & HOOPER, R.J. 2009. Mapping palaeostructure and palaeobathymetry along the Norwegian Atlantic continental margin. Møre and Vøring basins. *Petroleum Geoscience*, **15**, 27–43, http://doi.org/10.1144/1354-079309-804

ROBERTS, A.M., KUSZNIR, N.J., CORFIELD, R.I., THOMPSON, M. & WOODFINE, R. 2013. Integrated tectonic basin modelling as an aid to understanding deep-water rifted continental margin structure and location. *Petroleum Geoscience*, **19**, 65–88, http://doi.org/10.1144/petgeo2011-046

ROWAN, M.G. & VENDEVILLE, B.C. 2006. Foldbelts with early salt withdrawal and diapirism: physical model and examples from the northern Gulf of Mexico and the Flinders Ranges, Australia. *Marine and Petroleum Geology*, **23**, 871–891.

SANDWELL, D.T. & SMITH, W.H.F. 2009. Global marine gravity from retracked Geosat and ERS-1 altimetry: ridge segmentation v. spreading rate. *Journal of Geophysical Research*, **114**, B01411.

SCLATER, J.G. & CHRISTIE, P.A.F. 1980. Continental stretching: an explanation of the Post-Mid-Cretaceous subsidence of the central North Sea Basin. *Journal of Geophysical Research: Solid Earth*, **85**, 3711–3739.

STEIN, C.A. & STEIN, S. 1992. A model for the global variation in oceanic depth and heat flow with lithospheric age. *Nature*, **359**, 123–129.

TEISSERENC, P. & VILLEMIN, J. 1989. Sedimentary basin of Gabon – geology and oil systems, *In*: EDWARDS, J.D., & SANTOGROSSI, P.A. (eds) *Divergent/Passive Margin Basins*. AAPG Memoir, **48**, 117–199.

UNTERNEHR, P., PÉRON-PINVIDIC, G., MANATSCHAL, G. & SUTRA, E. 2010. Hyper-extended crust in the South Atlantic: in search of a model. *Petroleum Geoscience*, **16**, 207–215, http://doi.org/10.1144/1354-079309-904

WHITE, R. & MCKENZIE, D. 1989. Magmatism at rift zones: the generation of volcanic continental margins and flood basalts. *Journal of Geophysical Research*, **94**, 7685–7729.

WHITMARSH, R.B. & MILES, P.R. 1995. Models of the development of the West Iberia rifted continental margin at 40°30′N deduced from surface and deep-tow magnetic anomalies. *Journal of Geophysical Research: Solid Earth*, **100**, 3789–3806.

Interaction of crustal heterogeneity and lithospheric processes in determining passive margin architecture on the southern Namibian margin

M. MOHAMMED[1,2], D. PATON[1]*, R. E. L. COLLIER[1], N. HODGSON[3] &
M. NEGONGA[4]

[1]*Basin Structure Group, School of Earth and Environment, University of Leeds,
Leeds LS2 9JT, UK*

[2]*Department of Petroleum Reservoir Engineering, Mosul University, Mosul, Iraq*

[3]*Spectrum, Dukes Court, Woking GU21 5BH, UK*

[4]*Namcor, Windhoek, Namibia*

**Corresponding author (e-mail: d.a.paton@leeds.ac.uk)*

Abstract: The influence of pre-rift crustal heterogeneity and structure on the evolution of a continental rift and its subsequent passive margin is explored. The absence of thick Aptian salts in the Namibian South Atlantic allows imaging of sufficient resolution to distinguish different pre-rift basement seismic facies. Aspects of the pre-rift basement geometry were characterized and compared with the geometries of the Cretaceous rift basin structure and with subsequent post-rift margin architectural elements. Half-graben depocentres migrated westwards within the continental synrift phase at the same time as basin-bounding faults became established as hard-linked arrays with lengths of *c.* 100 km. The rift–drift transition phase, marked by seaward-dipping reflectors, gave way to the early post-rift progradation of clastic sediments off the Namibian coast. In the Late Cretaceous, these shelf clastic sediments were much thicker in the south, reflecting the dominance of the newly formed Orange River catchment as the main entry point for sediments on the South African–Namibian margin. Tertiary clastic sediments largely bypassed the pre-existing shelf area, revealing a marked basinwards shift in sedimentation. The thickness of post-rift megasequences does not vary simply according to the location of synrift half-graben and thinned continental crust. Instead, the Namibian margin exemplifies a margin influenced by a complex interplay of crustal thinning, pre-rift basement heterogeneity, volcanic bodies and transient dynamic uplift events on the evolution of lithospheric strain and depositional architecture.

The South Atlantic passive margin provides an exceptional region in which to study the mechanisms involved in lithospheric stretching, continental break-up and margin development (Hirsch *et al.* 2007; Heine *et al.* 2013; Peron-Pinvidic *et al.* 2013; Sibuet & Tucholke 2014). However, despite a recent focus on the evolution of passive margins and the South Atlantic in particular (Blaich *et al.* 2011; Franke *et al.* 2010; Moulin *et al.* 2013), fundamental questions remain unanswered with regard to the degree that crustal heterogeneity controls passive margin formation and the influence that this structure may have on early margin accommodation space and on the development of oceanic transform segmentation. Addressing these questions is important not only for improving our understanding of lithospheric processes, but also in constraining our knowledge of the petroleum system within such settings. For example, spatial and temporal variations in lithospheric thickness across a margin and along its strike may influence the heat flux both in the synrift phase and during post-rift thermal relaxation. Source rock maturation histories are therefore sensitive to variations in the underlying structural and thermal evolution of the lithosphere (e.g. Paton *et al.* 2007, 2008; Beglinger *et al.* 2012).

In this study, we focused on the Namibian margin for two reasons. First, this region provides an opportunity to investigate the temporal evolution of a margin from the initiation of rifting through break-up and seafloor spreading into the full development of a passive margin. Second, as it is located to the south of the Walvis Ridge, it is not complicated by the presence of salt. Therefore we can establish the architecture of the post-rift margin without either having to account for salt tectonics or being constrained by poor sub-salt imaging of the synrift.

This study set out to test a base case hypothesis that, although the configuration of passive margins

From: Sabato Ceraldi, T., Hodgkinson, R. A. & Backe, G. (eds) 2017. *Petroleum Geoscience of the West Africa Margin*. Geological Society, London, Special Publications, **438**, 177–193.
First published online March 16, 2016, http://doi.org/10.1144/SP438.9

is dominated by synrift structures and often by post-rift deformation, e.g. gravitational collapse, the pre-rift configuration plays an often poorly understood role in defining the structure and evolution of such basins.

Regional setting

The West African margin is closely associated with the grain of the Proterozoic to early Palaeozoic Pan-African fold belt along much of its length (Fig. 1; Clemson 1997) and the onshore part is dominated by the broadly north–south-trending Neoproterozoic Gariep Belt (Coward 1981; Miller 1983). As a consequence of limited well data in the offshore region and the absence of significant Palaeozoic outcrops onshore, the stratigraphy of the post-Pan-African basement is uncertain on the Namibian margin. It is generally considered that the basement is overlain by a Carboniferous–Permian stratigraphy associated with the Karoo foreland basin (Maslanyj *et al.* 1992; Light *et al.* 1993). This was subsequently overlain by deposits associated with the early stages of continental rifting, although again the timing of this is poorly constrained. Clemson (1997) proposed that the early rift portion consists of Triassic to mid-Jurassic Karoo sediments that were deposited in a major rift basin along much of the present day passive margin. Regardless of the timing of this phase of rift initiation, there is a consensus that the Pan-African fabrics played an important role in the subsequent South Atlantic rift phase along the margin (e.g. Uchupi 1989; Maslanyj *et al.* 1992). A significant rift onset unconformity has been identified in a number of previous seismic reflection studies (e.g. Gerrard & Smith 1982; Light *et al.* 1993) and has been attributed to the initiation of the main South Atlantic rift phase in the Late Jurassic.

The basin-fill of the main rift phase (Fig. 2), which has been penetrated by boreholes in the South African part of the Orange Basin, consists of continental clastic sediments (fluvial claystones, sandstones and pebble beds) and claystones interbedded with volcaniclastics and volcanics (McMillan 2003), passing into significant thicknesses of volcanics towards the top. Light *et al.* (1993) recognized a number of intra-synrift sequences with distinct seismic facies and also noted a marked variation in dip across the margin, with westwards dips to the west of a structural 'hinge line' and eastwards dips to its east. Of particular note is the development of a seaward-dipping reflection (SDR) complex that developed in the west just prior to continental break-up (Gladczenko *et al.* 1997). The timing of break-up is constrained by the M4 magnetic anomaly (126.5 Ma), indicating an Early Cretaceous age (Nürnberg & Müller 1991) or, more specifically, a Barremian age according to the Walker *et al.* (2012) timescale.

The break-up unconformity, which sits above the seaward-dipping reflector package, is overlain across most of the basin by a Barremian–early Aptian package, which denotes marine inundation of much of the continental margin and has been considered to be part of the transition phase (Light *et al.* 1993). This package is important for the presence of organic-rich source rocks. An unconformity at the top of the transition phase marks the development of the main post-rift phase (Fig. 2), which is represented by a thick succession of Cretaceous and Tertiary age sediments with a number of internal unconformities. These unconformities in the northern Orange Basin in southern Namibia, and the correlatable unconformities in the South African portion of the Orange Basin, have a variety of horizon names, but are consistent in identifying the key basin-wide events (Emery *et al.* 1975; Gerrard & Smith 1982; Light *et al.* 1993; Brown *et al.* 1995; Paton *et al.* 2008). The early part of the post-rift stratigraphy consists of an Aptian–Turonian progradational succession that contains deltaic and fluvio-marginal deposits >3500 m in thickness in the inner part of the margin. The deposition of this package was punctuated in the Cenomanian and Turonian with organic-rich shales, which form significant, basin-wide and correlatable reflections (Light *et al.* 1993; Brown *et al.* 1995). The top of the Cretaceous is identifiable with an additional basin-wide unconformity and marks a major shift from deposition on the inner margin towards sediment accumulation predominantly on the outer margin during the Tertiary.

Data and methods

This study combines 2D reflection data from four surveys (ECL-89, SN-1996, GNA-1997 and Vernob-2003) that were acquired between 1998 and 2003. The data cover the northern parts of the Orange Basin in southern Namibia and in this study we utilized pre-stack time migration data that had a maximum recording time between 5 and 7 s two-way travel time. The lines in the southern area of the dataset intersect the Kudu gas field (wells Kudu 9A-1, 9A-2 and 9A-3) and these have been used to provide age and lithological constraints for the seismic interpretation. A megasequence approach has been taken for the interpretation of the data based on the identification of regional and local unconformities using reflection termination, cut-offs and onlap relationships with the sequence stratigraphic system based on Muntingh & Brown (1993) (see Fig. 1 for locations of lines A, B and C shown as Fig. 3a–c, respectively).

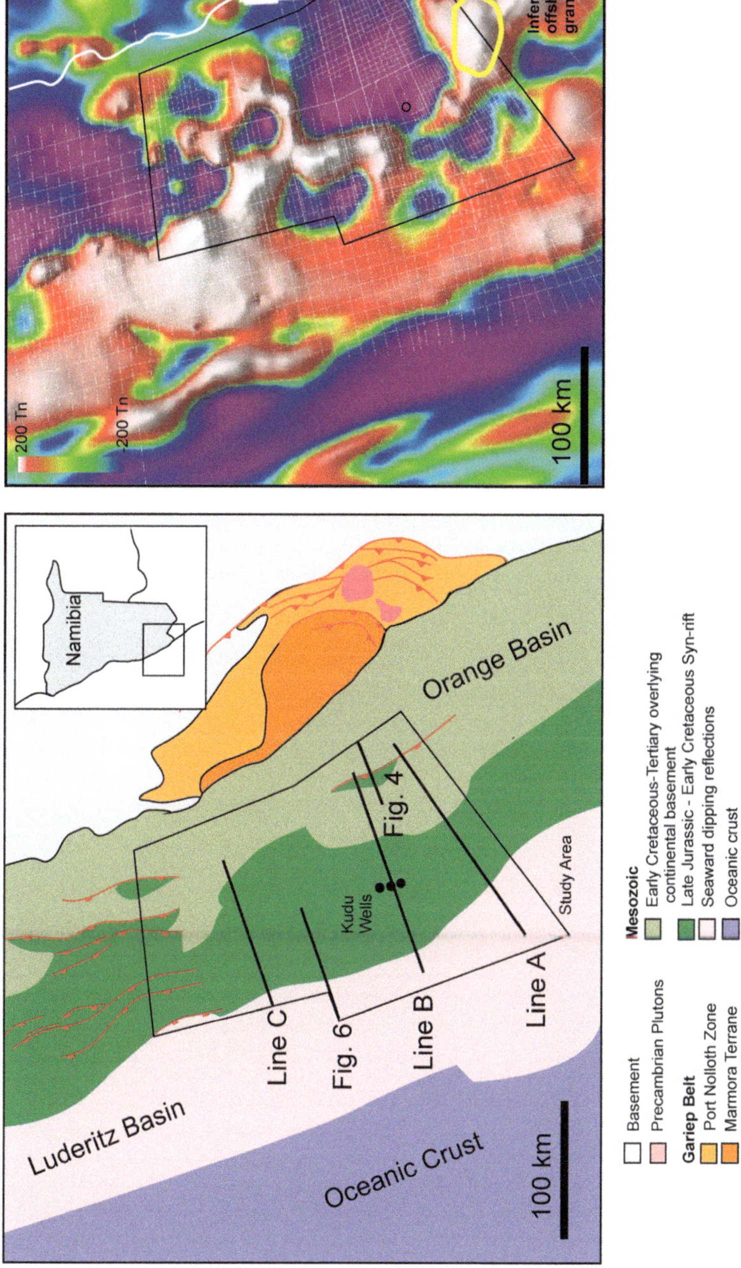

Fig. 1. Regional map of the Namibian South Atlantic margin (redrawn from namcor.com.na). This study focused on the northern part of the Orange Basin, which includes the Kudu Field. The positions of the seismic sections in Figure 3 are shown.

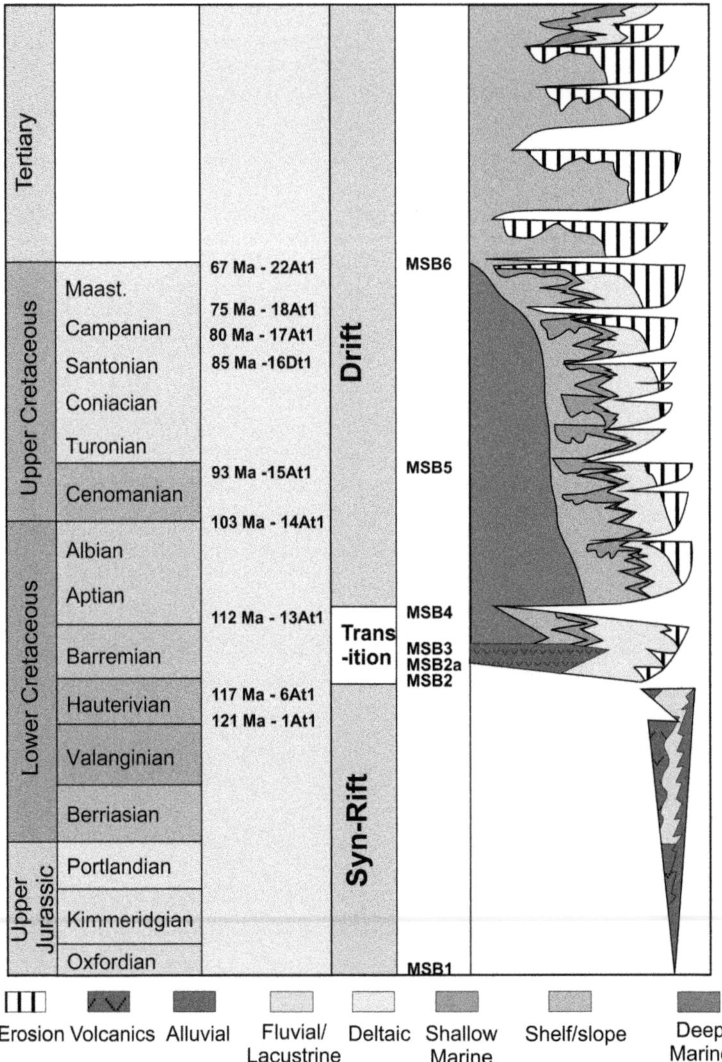

Fig. 2. Tectonostratigraphic scheme for the Namibian continental margin with megasequence boundaries (MSB1–MSB6) identified in this study. Previous nomenclature of Brown *et al.* (1995) from 1At1 to 22At1 for key seismic reflections is included for comparison. Geological timescale is from Walker *et al.* (2012).

Megasequence architecture

Pre-rift

The reflection character of the intra-basement pre-rift in the offshore data is highly variable and is consistent with the complex Gariep Belt directly onshore of the study area. In places, continuous high-amplitude reflections are evident and are demonstrably beneath the divergent reflections that are indicative of the main rift phase (Fig. 4). Given the presence of abundant mid-crustal reflectivity and of west-dipping and folded reflections in the eastern Namibian margin, these are likely to be imaging portions of the nearshore Gariep Belt. This is in contrast with the well-imaged, concordant reflections identified on other parts of the Namibian margin that have been inferred to be Karoo sediments (e.g. Clemson 1997). Elsewhere the basement is much more akin to a transparent acoustic basement with no internal reflectivity or character and this seismic facies is present along much of the margin (Fig. 3). This transparent acoustic basement

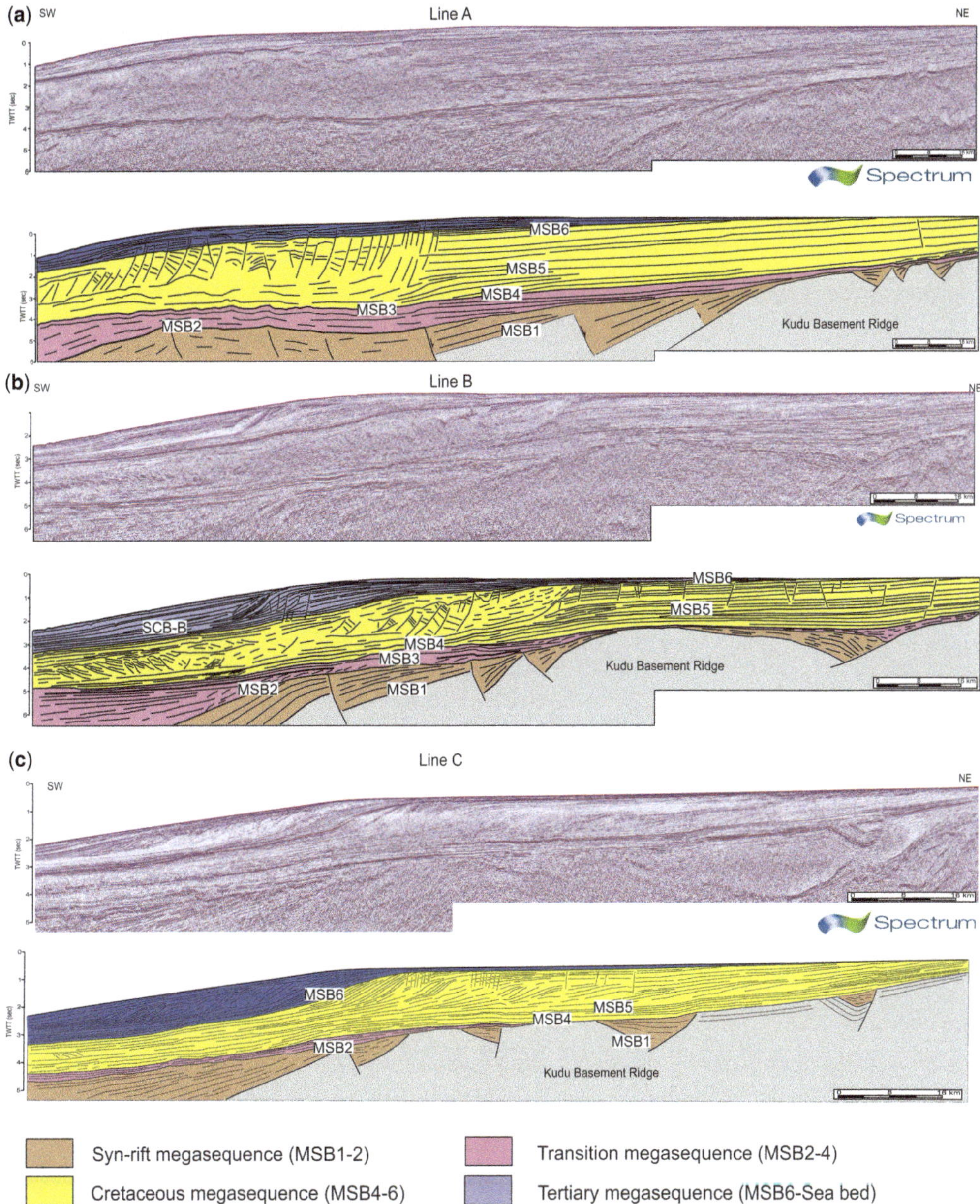

Fig. 3. Regional east–west sections across the Namibian Orange Basin. (**a**) This section, which is in the south, is taken as the type section and shows the Kudu Basement Ridge, the synrift and transition phase, and the Cretaceous and Tertiary post-rift sequences. (**b**, **c**) North of section (**a**) showing how the basin changes configuration towards the north (see Fig. 1 for location of lines A, B and C).

forms a 30 km wide continuous zone that trends approximately north–south. The feature is often referred to as the hinge line (e.g. Light *et al.* 1993), which implies a discrete fulcrum and by association conveys a kinematic process. We will discuss the nature of this zone, but prefer to call it the Kudu Basement Ridge (Fig. 3) to differentiate between the feature and its origin.

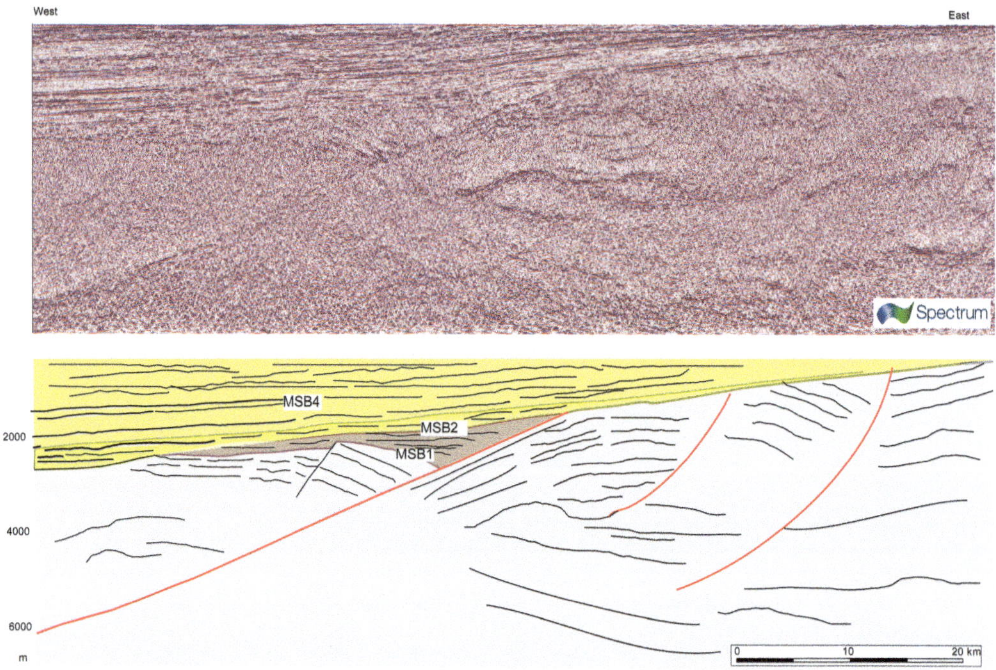

Fig. 4. This section (and its interpretation on the lower panel) is located to the east of the Kudu Basement Ridge and shows reflectivity within the crust. This corresponds to the offshore equivalent of the Gariep Belt and shows how Mesozoic faulting reactivated the pre-existing heterogeneity (see Fig. 1 for location).

Synrift megasequence (MSB1 to MSB2)

Although not penetrated by boreholes in this area, the synrift intervals are identifiable from the divergence of reflections in a half-graben geometry; the fault locations and timings were mapped and constrained from the termination of these divergent reflections. Of particular note for this interval is the spatial distribution of faulting across the Kudu Basement Ridge, which separates the basin into two distinct structural domains (Fig. 3).

To the east of the basement ridge small wedge-shaped half-graben with typical widths of between 5 and 12 km are present and are consistently controlled by west-dipping normal faults (Fig. 4). The geometry of the rift basin-fill, which is predominantly contained within the lower synrift interval, is dominated by low to medium amplitudes and semi-continuous reflectivity; the controlling faults reveal a highly asymmetrical system with faults that possibly decolle onto a more regionally correlatable west-dipping crustal reflection within the pre-rift/basement sequence. The internal seismic character of the rift basins is consistent with the style of deposition typical of early continental rift basins, with isolated depocentres that are likely to be fluvial-dominated with occasional lacustrine

basin-fill (e.g. Gawthorpe & Leeder 2000). Comparable synrift intervals have been identified and penetrated to the south in well A-J1 in the South African Orange Basin (Jungslager 1999).

The basins to the west of the basement ridge are also characterized by a divergent reflection geometry, but these reflectors onlap onto the basement ridge to the east, thicken towards the west and are controlled by east-dipping normal faults. In the type sections (Fig. 3a, b), three diverging packages are imaged and each of these is bounded by an east-dipping fault. These western faults are more widely spaced than those to the east of the Kudu Basement Ridge, with half-graben widths of up to 50 km. It is likely that smaller faults that were active early in the synrift phase are also present, but these are not resolved seismically. The seismic character of the basin-fill to the west of the basement ridge is divided into a lower and an upper package. The lower package, which is similar to that to the east of the ridge, is also noticeably different as it consists of higher amplitude, more continuous reflections. The early synrift interval, which is present to the east and west of the basement high, has a variable amplitude and semi-continuous reflectivity, whereas the later synrift interval shows divergence into the western normal faults; reflections are not cross-cut by the eastern faults, indicating that the late synrift faulting progressively migrated towards the west.

The isochron maps of the lower synrift interval (Fig. 5a) reveal that sedimentation is controlled by NNW–SSE-trending fault systems separated by the Kudu Basement Ridge. The continuity of areas of sediment accumulation allows us to correlate the faults between seismic lines. Discrete faults and associated depocentres are evident, typically 20–30 km in length, typical of fault segments in early rift systems (Gupta *et al.* 1998; Ebinger *et al.* 1999; Cowie *et al.* 2000). This is the deepest resolvable synrift interval and faults, or linked fault arrays, show lateral continuities of between

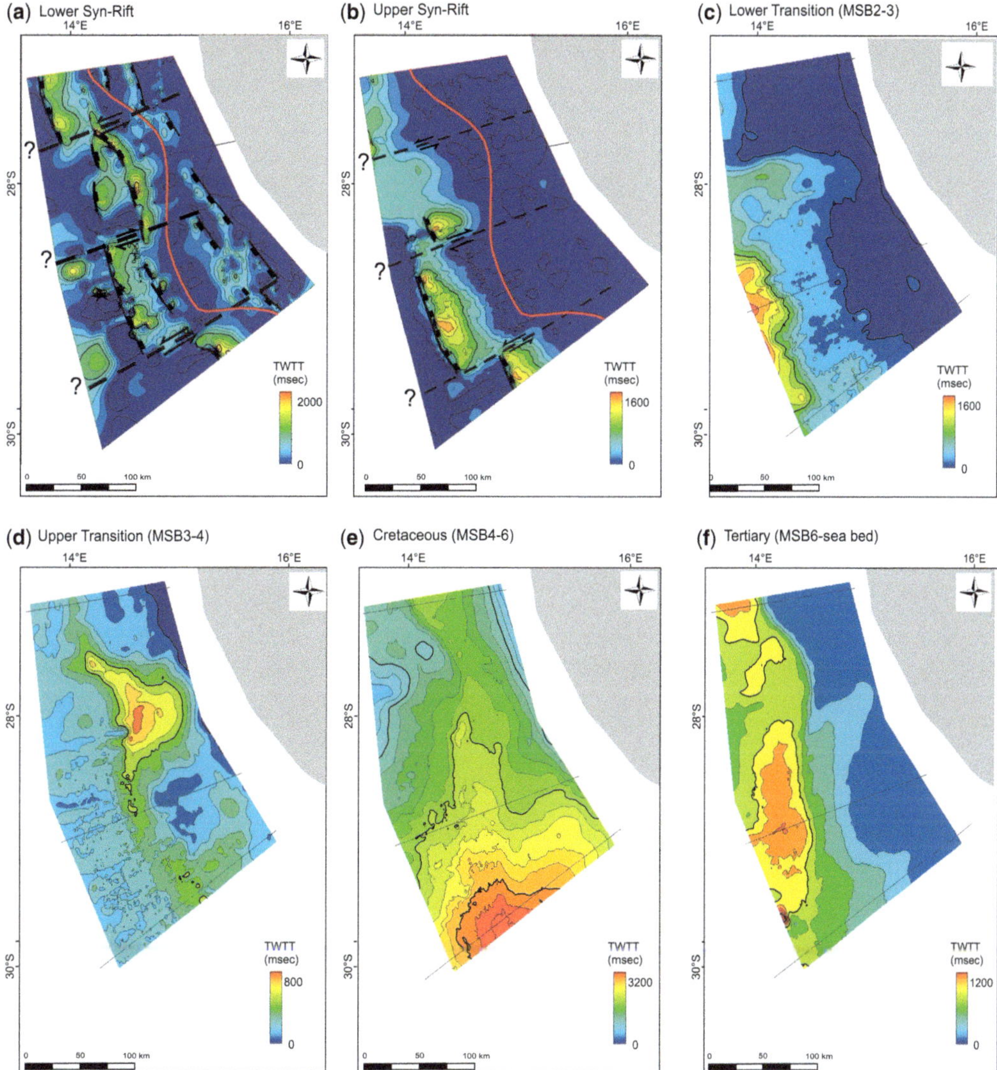

Fig. 5. Isopach maps of the main tectonostratigraphic intervals in the study area. (**a**) Deposition of the lower synrift is controlled by a number of north–south-trending normal faults on both sides of the Kudu Basement Ridge (location shown by red line) that are interpreted here as being dissected by SW–NE-trending transform faults. (**b**) In contrast, the upper synrift interval is controlled by fewer faults, located in the west of the basin, indicating the localization of strain towards the incipient spreading centre. The continued localization of accommodation space towards the west is reflected in the lower transition phase (**c**), whereas accommodation space is dramatically reorganized in the upper transition phase (**d**). The change in post-rift thickness variations between a single point source in the south in the Cretaceous (**e**), compared with bypass deposition in the outer margin (**f**) in the Tertiary, is evident.

20 and 100 km; again this is comparable with other rift systems. Three areas are also evident where the fault traces abruptly terminate and are offset from neighbouring arrays (Fig. 5a). These correspond either to transfer/accommodation zones (Morley *et al.* 1990) or conceivably represent later strike-slip faults that dissect the rift basin. In contrast, the isochron map for the upper synrift interval (Fig. 5b) reveals a significant reorganization of the rift system. The inboard faults are not present as they have switched off and accommodation space is focused progressively onto the outboard faults. Fault arrays and the associated isochron 'thicks' are more continuous, indicating that the western fault sets had by this stage evolved into hard-linked rift segments of approximately 100 km in length. The transfer zones are also more clearly apparent at this time.

Transitional megasequences (MSB2 to MSB4)

The lower part of the megasequence above MSB2 is enigmatic (MSB2 to MSB2a; Fig. 6). It sits beneath a seaward-dipping reflector package and shows an increase in thickness to the west across the margin, but there is no evidence, at least in the current data, of significant faulting. This observation therefore disagrees with our previous understanding of the system, although our current dataset is not extensive enough to define its geometry. Further analysis should be undertaken on subsequent datasets to determine whether the package is, or is not, controlled by faults.

The transitional phase (MSB2a to MSB4), as a whole, encompasses the transition from the synrift interval into the demonstrably passive margin megasequences. Here we divide it into a lower transition megasequence and an upper transition megasequence (Fig. 6). The lower transition megasequence only occurs west of the basement high and is

characterized by onlap onto either the uppermost reflection of the synrift or onto the basement/pre-rift ridge to the east. The megasequence (MSB2-3), which is the deepest penetrated by the Kudu wells (Bagguley 1997; Schmidt 2004), is of Hauterivian–Barremian age and consists of interbedded volcanic and fluvial strata that form the main clastic reservoir interval for the area. The upper part of this megasequence is characterized by high-amplitude reflections with a fanning, west-dipping geometry that corresponds to SDRs typical of volcanic passive margins (e.g. Franke 2013). In plan view the lower transition megasequence thickens noticeably to the west, where there is lateral transition into the SDRs (Fig. 5c). It is evident that seismically resolvable faulting had switched off by this stage and the accumulation of volcanics was controlled by accommodation space outboard of the main early continental rift system.

The upper transition megasequence (delineated by MSB3–4) overlies the SDRs, is best developed west of the basement high and is Barremian–early Aptian in age. It is dominated by high-amplitude and laterally continuous reflections that overstep the Kudu Basement Ridge and which are shallowly truncated by the bounding reflection at the top of the package. This interval contains the organic-rich source rock encountered in the Kudu well and is a consequence of the restricted marine environment and the major flooding event during the early Aptian in which the regionally important black shale was deposited (Séranne & Anka 2005).

There is a distinct pattern of variation in the thickness of this upper package (Fig. 5d), in contrast with the lower transition megasequence. For much of the margin the unit is noticeably thin and is often condensed to below seismic resolution. Towards the north there is a clear thickening into the basin, forming an elliptical geometry that is bound on the east by the Kudu Basement Ridge.

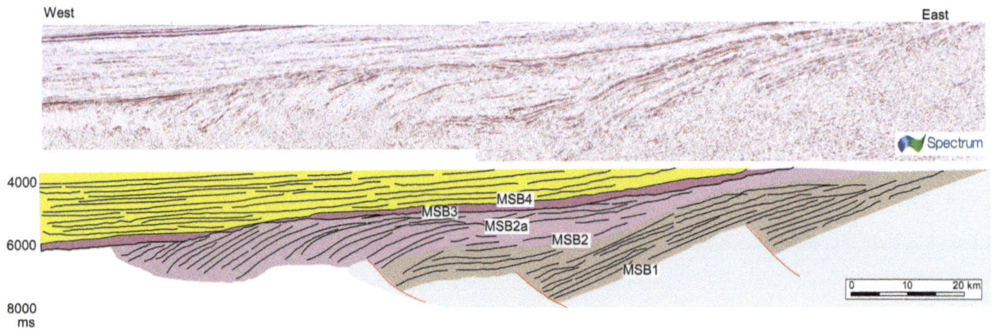

Fig. 6. This section (and its interpretation in the lower panel) is located to the west of the Kudu Basement Ridge. It shows the transition from continental crust through the volcanic synrift sequence and seaward-dipping reflections into oceanic crust (Fig. 1 for location).

Cretaceous megasequences (MSB4 to MSB6)

The two post-rift Cretaceous megasequences, which have a thickness of up to 4 km, account for most of the stratigraphy in both the inner and outer parts of the passive margin. The Cretaceous post-rift megasequences are bound by the MSB4 surface (mid-Aptian) at the base, showing some erosional truncation of the underlying transitional package, and by the MSB6 horizon at the top.

The lower of the two Cretaceous post-rift megasequences (MSB4–5) comprises the first broadly progradational unit of the margin. It progressively onlaps onto the top of the synrift and onto basement to the east (Fig. 3c). The westwards-prograding clinoform geometry typically has a 'height' (in two-way travel time) of approximately 100 ms (Fig. 3b), the steepest slope of which is inferred to represent a pro-delta (outer shelf) to slope mudstone facies transition. The package thins downdip to the west, where it becomes condensed and in places is not seismically resolvable. The MSB5 horizon, at the top of this lower post-rift package, is defined by a high-amplitude reflection correlatable across the entire basin that corresponds to the end-Cenomanian maximum flooding surface.

The progradation of the MSB4–5 megasequence towards the west marks the development of an established coastline or shelf break and the migration of that coastline basinwards; the Kudu wells indicate this is a mainly shallow-marine interval. The thickness of the megasequence is greater in the south (Fig. 3a) than in the middle or the north of the study area (Fig. 3b, c). This suggests that the dominant sediment input point for the basin was in a similar location to the present day Orange River to the south of the basin (Fig. 5e).

The overlying package (MSB5–6) is dominated by an aggradational to progradational architecture and the progressive development of a bathymetric break in slope of up to 800 ms. In the middle part of the margin it is noticeable that the package is dominated by continuous, high-amplitude and concordant reflections over a length scale of at least 80 km across the margin. The sedimentology of this megasequence is dominated by claystones, as documented in the Kudu wells. In contrast, the geometry of the outer margin is markedly different. In the southern part of the basin the upper part of the megasequence is dominated by gravity collapse structures (Fig. 3a); the updip extensional domain and the associated downdip compressional domain are clearly imaged and have been described in detail elsewhere (Butler & Paton 2010; De Vera *et al.* 2010; Dalton *et al.* 2015, 2016).

The timing of the upper horizon (MSB6) is diachronous across the margin, with a subcrop of Maastrichtian age (67 Ma) stratigraphy in the inner margin and an early Tertiary (53–65 Ma) stratigraphy in the outer margin (as constrained by the Kudu wells; Bagguley 1997; Schmidt 2004). This diachroneity is a consequence of a varying degree of erosion at the top of the Cretaceous megasequence, as shown by the erosional truncation of its uppermost reflections. In the outer margin there is no erosion, with concordant reflections throughout the Cretaceous and Tertiary megasequences. This is also the case at the break in slope of the Cretaceous margin. To the east of this, however, there is a progressive increase in erosion towards the inner shelf and the amount of erosion may be calculated by projecting the geometry of the underlying stratigraphy from the position at which erosion begins (Paton *et al.* 2008). This indicates a variable degree of erosion along 200 km of the Namibian margin, with a maximum of *c.* 800 m of section lost by erosion at the end of the Cretaceous.

A south to north reduction in interval thickness is also evident in the Turonian–Maastrichtian megasequence (Fig. 3a–c), but is less pronounced than in the preceding megasequence, with a much more uniform distribution of sediment. It is only in the most northern part of the basin that significant thinning is apparent and this corresponds to a northward progradation of the package. The thinning evident in the isochron map in the inner margin is a reflection of the latest Cretaceous to earliest Tertiary erosion rather than deposition (Fig. 5e).

Tertiary megasequence (MSB6 to seabed)

The seismic character and geometry of the Tertiary megasequence (MSB6 to seabed) varies significantly both across and along the margin. In the inner part of the margin, the megasequence is very thin and is characterized by medium- to high-amplitude, continuous, sub-parallel to parallel reflections that downlap onto the base Tertiary unconformity. The most eastern part of the megasequence has undergone tilting and erosional truncation at the seabed, where it is in places entirely absent.

For most of the southern area (Fig. 3a) the megasequence shallowly progrades and where the underlying base Tertiary unconformity forms a break of slope the Tertiary megasequence forms a shelf margin system. Within the megasequence there is a correlatable horizon characterized by erosional truncation below and onlap and downlap above, which divides the Tertiary into two packages. This horizon corresponds to the SCB-B surface of Weigelt & Uenzelmann-Neben (2004) and is inferred to be of Middle Miocene age. This horizon separates two wedges, both of which prograde from the underlying break in slope towards the west.

Stratigraphic data from the Kudu wells show that most of the Tertiary sediments are dominated by claystones and siltstones, with interbedded sandstones. On the shelf, the Tertiary sediments are very thin and are characterized by high-amplitude, parallel and continuous reflections interpreted to represent the presence of interbedded calcareous sandstones and shelly limestones (Gerrard & Smith 1982).

The internal architecture of the Tertiary megasequence is broadly similar along most of the margin, with a relatively thin package, probably representing significant sediment bypass of the shelf above the base Tertiary unconformity (Fig. 3a–c). A rapid thickening towards the west, beyond the shelf break, is also evident along the margin and the location of this is controlled by the geometry of the base Tertiary unconformity that mimics the Late Cretaceous break in slope.

The most outboard part of the basin is dominated by downlap onto the more distal portion of the base Tertiary unconformity. Despite the westwards switch in the location of sediment accumulation during the Early Tertiary (Fig. 5f), the margin is remarkably undeformed, with very little faulting and none of the significant gravity collapse structures that are so characteristic of the underlying Upper Cretaceous megasequence.

Development of the Namibian part of the Orange Basin

The Orange Basin of southern Namibia preserves an exceptional record of the development of a continental margin from rifting through to the present day. We now present a summary of the development of the margin and consider both its spatial and temporal development (Figs 7 & 8).

The Late Jurassic to Early Cretaceous early rift phase shows distributed extension with a number of relatively long north–south-trending normal fault sets (Fig. 8a). The onshore Gariep Belt structures converge with depth onto a west-dipping crustal deformation zone (Gray *et al.* 2008), which, when extrapolated offshore, can explain the

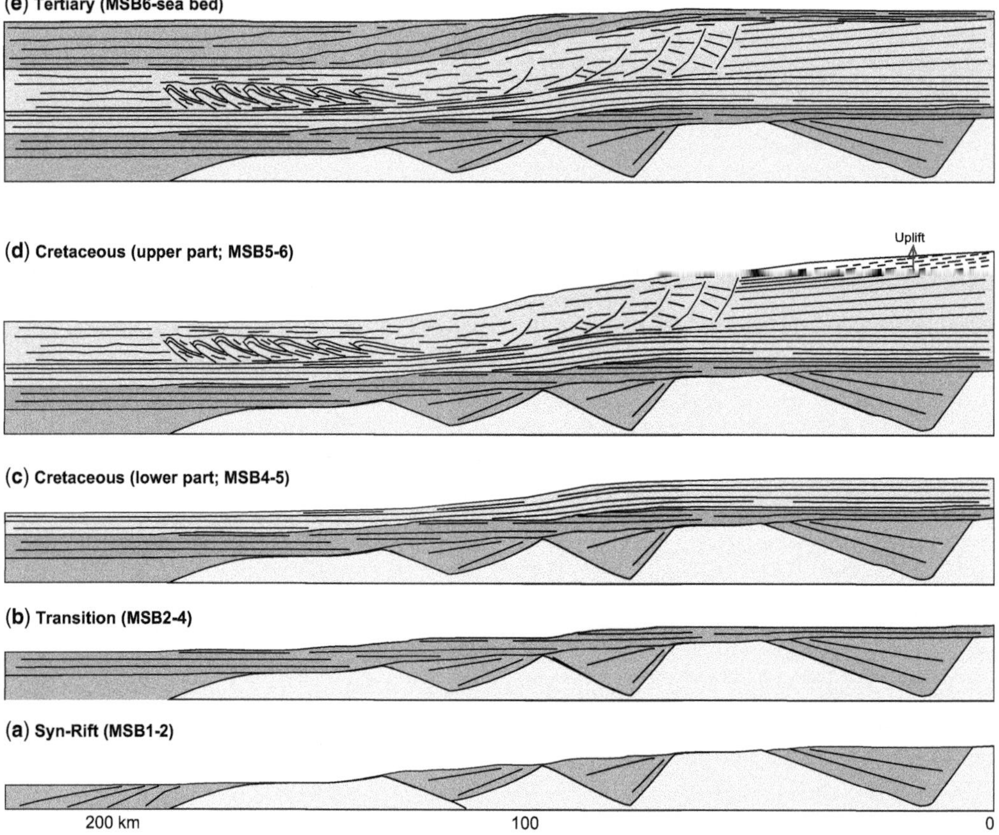

Fig. 7. Schematic diagram showing the evolution of the Namibian margin and the progressive evolution of the synrift, transition phase and the post-rift deposition.

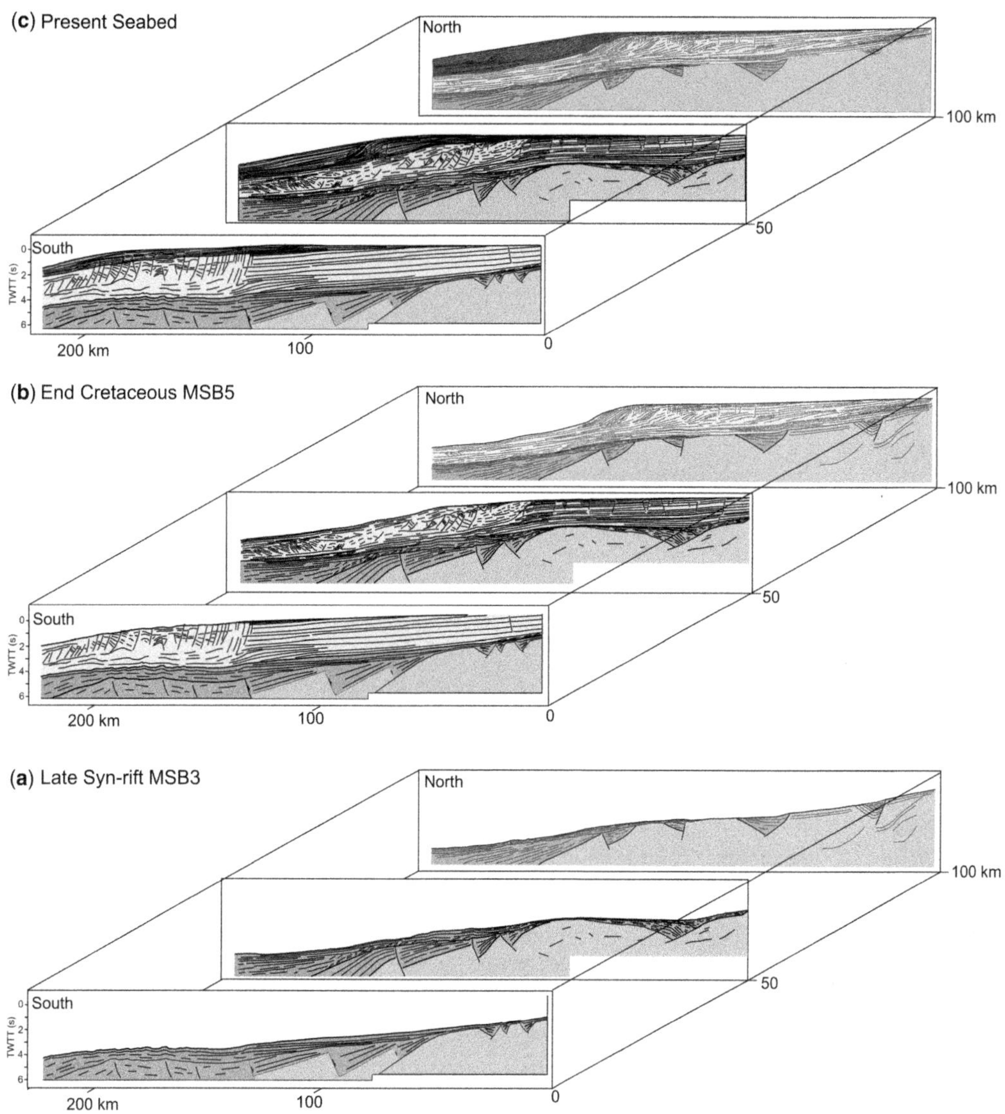

Fig. 8. Schematic 3D diagram with a view towards the NW, showing the variation in the margin geometry through time (end of transition phase (MSB4), end Cretaceous (MSB6) and the present day) based on the three sections in Figure 3.

observed west-dipping, mid-crustal reflections. The early rift faults, at least those east of the Kudu Basement Ridge, also appear to converge onto this zone, suggesting that this basement heterogeneity localized the deformation, at least during the early rift phase (Figs 7 & 8a). This early evolution of a rift system corresponds well with equivalent histories documented in many rift systems that were at least partly controlled by crustal heterogeneity (Clemson 1997). The Kudu Basement Ridge is an interesting feature as it separates two pre-rift basement zones

of more continuous reflections, but has little internal character itself. We interpret this to represent a Precambrian granitoid body around which the Pan-African deformation of the Gariep Belt is localized. The shape and scale of this feature is consistent with other similar bodies along the southern African margin, such as the Cape Granites. From the perspective of continental rifting, it appears to have acted to localize stress at its edges and so influenced where the rift system developed and its spatial geometry.

The exact configuration of the main rift system is relatively poorly constrained given the widely spaced 2D seismic dataset that is available over much of the study area, but geometrically it resembles many rift systems with normal faults that show segmentation along their length. Comparisons with fault growth models, such as those developed by Gupta *et al.* (1998) and Cowie *et al.* (2000) appear to be appropriate.

As the rift evolved, there was a marked migration of the system towards the incipient spreading centre in the west, while the system in the east progressively switched off (Fig. 6). The end of the rift phase is characterized by an erosion event in the east and by the widespread development of the two packages within the transition phase. The lower transitional megasequence includes the emplacement of SDRs as well as the extrusion of more extensive volcanics across the basin that cap the synrift half-graben fills. The distribution of volcanics is often associated with lower crustal bodies (Hirsch *et al.* 2010). As these bodies are considered to be emplaced at the time of rifting, it is likely that these lower crustal bodies represent the remnants of the magma chambers that fed the surface volcanics. The end of the SDR package is marked by the onset of the upper transition megasequence, which includes the main organic source rock unit, suggesting rapid basin deepening and anoxic conditions. As seen in Figure 3a, there is a significant variation in thickness of this upper transitional unit, with the greatest thickness occurring immediately to the west of the Kudu Basement Ridge. This is likely to be a consequence of where post-rift thermal subsidence was greatest at this earliest stage of break-up.

There is relatively little variation along the 200 km of the margin throughout the synrift and transitional phases of the Namibian margin (Fig. 8a). The Upper Cretaceous package, by contrast, exhibits significant variation (Fig. 8b). The lower portion is dominated throughout by the progradation of the shelf margin clinoforms and by the development of a clastics-dominated margin sequence. The sediment input at this stage was obviously determined by its provenance, but at this stage it was most likely to have been derived from the erosion of the remnant rift margin and proximal continental sequences.

To the south of the study area, in the South African portion of the Orange Basin, the Aptian clastic shallow-marine succession was sourced from a number of smaller river systems along the margin (Paton *et al.* 2008). This resulted in the relatively uniform deposition of this mainly deltaic package along much of the margin to the south of the Orange River. In contrast, on the southern Namibian margin, the isochron map for this interval (encompassed by Fig. 5e) conforms much more to a single entry point for deposition, with the location of the source being very close to the present day Orange River delta. This is in agreement with previous studies that proposed the evolution of the Oliphants drainage system into the Orange system during the Albian (e.g. Dingle & Hendry 1984). The sedimentology of the package in the Kudu wells confirms that it is a clastics-dominated shallow-marine sequence. The result is a radial distribution of the clastic shallow-marine deltaic sediments to the north and west. This is represented in the north by a very thin and, most likely, highly condensed sequence that can be interpreted as the transition from pro-delta to distal shelf deposits.

In contrast, the Upper Cretaceous megasequence is much more extensively developed both across and along the Namibian margin (Figs 7 & 8b). Only in the most distal, northwestern part of the margin are accumulations of limited thickness (equivalent to 200 ms two-way travel time or less). The progradational geometry of the MSB4–5 megasequence is replaced initially by a much more aggradational geometry, but then becomes more progradational again in the later stages of the Upper Cretaceous MSB5–6 megasequence (e.g. Fig. 3c). This change in depositional geometry is coincident with a dramatic change from a coarser clastics-dominated system to a claystone-dominated system. A similar relationship between geometry and sedimentology is also observed in the southern Orange Basin (Paton *et al.* 2008). The abrupt switch in sediment grade is considered to reflect the emergence of the Orange River catchment area capturing, eroding and draining the Karoo Basin over much of southern Africa. Despite significant sediment accumulation in the Late Cretaceous, the concordant reflections of the inner margin imply uniform subsidence across much of the margin, over an area at least 100 km in width and 300 km in length. The Late Cretaceous shelf was remarkably stable, apart from small normal faults with throws equivalent to reflector offsets of the order of 50–80 ms two-way travel time. The end of the Turonian–Maastrichtian megasequence is marked by an erosional event that is most evident in the south (Fig. 3a), but which also extends towards the north (Fig. 3b); this is equivalent to a similarly aged erosion event in the southern portion of the Orange Basin (Paton *et al.* 2008).

A comparison of the Cretaceous post-rift and Tertiary post-rift isochron maps (Fig. 5e, f) reveals the abrupt reorganization of the shelf depositional system. The relatively thin (200 ms two-way travel time) Tertiary megasequence of the middle margin probably reflects modest subsidence post-erosion, but also represents significant bypass of sediments across a shelf that was 100 km wide, leading to deposition in the available accommodation space that lay outboard of the inherited Cretaceous shelf break. The lateral variation along the Tertiary continental slope is a reflection of the switching of the Bengula current, which redistributes sediments from south to north along the margin (Weigelt & Uenzelmann-Neben 2004).

Discussion

Control of fault evolution from the continental rift to drift stages

A number of studies have documented the influence of crustal heterogeneity on the evolution of passive margins (Ring 1994; Schumacher 2002; Korme *et al.* 2004). In this part of the South Atlantic, the use of the Pan-African basement fabric in the Mesozoic extension has been discussed previously (Light *et al.* 1993; Clemson 1997) and, in particular, the inheritance of Gariep Belt fabrics as a control on the north–south structural trend (Hälbich & Alchin 1995; Frimmel *et al.* 1996). Although our study agrees with this for the nearshore portion of the margin, we propose that an alternative explanation of the Kudu Basement Ridge, especially given its transparent internal character, is that it is a Precambrian granitoid body. This is supported by the ridge being directly along the trend of the granite bodies contained within the onshore Gariep Belt (Fig. 1). This suggests that rifting to the west of this basement high may not have reactivated the pre-existing fabric on through-going structures. Caution must therefore be applied before extrapolating the basement fabric as continuous to the more distal parts of the basin margin.

In addition to affecting the location of the subsequent rift structure, our results suggest that the pre-existing crustal heterogeneity influenced the evolution of the rift faults. A considerable number of studies have documented the nature by which faults grow and interact with sedimentary basin-fill through field observations (e.g. Cartwright *et al.* 1995; Dawers & Anders 1995), analogue and numerical models (e.g. Gupta *et al.* 1998; Marchal *et al.* 1998; Cowie *et al.* 2000; McClay *et al.* 2002) and subsurface seismic studies at a variety of scales from regional 2D to high-resolution 3D data (e.g. Morley 1999; Contreras *et al.* 2000; Dawers & Underhill 2000; McLeod *et al.* 2000, 2002). Given the caveat that we cannot clearly map the earliest fault development, we present evidence of smaller faults growing and linking to form larger structures. These early faults do not show evidence of significant en echelon patterns as predicted by numerical and analogue models (e.g. Cowie *et al.* 2000; McClay *et al.* 2002). Even accounting for the line

spacing of the available data, we would expect such a spatial geometry to be evident in our data; therefore we propose that the absence of en echelon faults is because the inheritance of the basin fabric itself plays a role in modifying the nature of how the faults grow and interact (Walsh *et al.* 2002; Paton 2006; Mortimer *et al.* 2007). It has been documented within the Cape Fold Belt of South Africa that the basement fabric can result in the collinear development of normal faults and we invoke a similar process here (after Paton 2006). The faults that we do observe evidently have length scales of up to 30 km in the early synrift, but these had coalesced by the late synrift into linked segments of up to 100 km in length; again, these observations are consistent with normal fault developments in southern Africa (Paton *et al.* 2006) and play a significant role in both fairway distribution and reservoir compartmentalization (Mortimer *et al.* 2016). However, we propose that there are significant offsets in the fault locations on a wavelength of 100 km and that these steps are accommodated through accommodation zones or strike-slip faults, similar to continental transfer faults (Bosworth 1985).

Numerical models of rift evolution show that as lithospheric stretching evolves, the incipient break-up strain should progressively localize towards the spreading centre. Huismans & Beaumont (2013) predicted that, during the rift stage, the distributed strain and hence a wide rift basin progressively localize onto the rift axis. Although previous studies have demonstrated that this may happen during the continental rift phase (e.g. Skogseid 2001), this study suggests that this was the case throughout much of the rift history and into the oceanic spreading phase.

Segmentation and margin evolution

The segmentation of passive margin basins has been much discussed and clearly played a significant role in the tectonic evolution of the South Atlantic (Clemson *et al.* 1997; Franke *et al.* 2007). Of note is the correspondence between segment boundaries and onshore structural lineaments, specifically with the well-established trends of the Damara fold belt and Gariep structures, (e.g. Corner 1983; Clemson *et al.* 1997; Holzförster *et al.* 1999; Stollhofen *et al.* 2000; Corner *et al.* 2002). Koopmann *et al.* (2014) provided further insights into the role of segmentation in the evolution of the Orange Basin, including a discussion of how these formed propagation barriers to the northward progression of seafloor spreading, and they associated these features with onshore structures, e.g. Cape Cross. On the conjugate margin in Argentina and Uruguay, Stica *et al.* (2014) outlined how these segments played a fundamental role in ocean spreading and margin development, as well as influencing plate reconstruction (e.g. Nürnberg & Müller 1991; Heine *et al.* 2013; Moulin *et al.* 2013). Although there is evidence that not all onshore structures control offshore transfer zones – for example, on the northern Norwegian margin (Tsikalas *et al.* 2001) – there is broad agreement that pre-existing structures can influence the location of fracture zones (e.g. Lister *et al.* 1986; Stica *et al.* 2014). A point recognized by Stica *et al.* (2014) is that no study to date can relate the first evolution with break-up segmentation. Clemson *et al.* (1997) discussed the importance of segmentation for the development of the passive margin along the entire Namibian basin. Our results agree with their conclusion, but, in addition, demonstrate that some rift segments will become inactive as a function of how faults grow and interact; however, where continental transform faults have formed major structures, they remain significant features during the post-rift phase and into the oceanic domain (Franke *et al.* 2010).

Continuity of post-rift megasequences and thermal subsidence

Our mapping of the rift architecture, and hence the level of the top pre-rift surface, shows that is it very variable, both along and across the margin, with half-graben interspersed with basement highs (Fig. 8). When compared with the South African portion of the Orange Basin, the variability increases and is a reflection of the presence of the Cape Fold Belt reactivation that cross-cuts the north–south-trending Atlantic rift system (Paton *et al.* 2008, 2016). The results of this study, in conjunction with Paton *et al.* (2008, 2016), demonstrate significant variations in the extent to which the upper crust has been stretched along the margin. Yet there is remarkable continuity in the post-rift megasequences along the 1000 km length of the Orange Basin. The early post-rift fill (MSB4–5) is dominated by margin-derived clastic deposits prograding westwards into the ocean basin, having been sourced from a number of smaller fluvial systems. Sediment thicknesses are broadly comparable throughout the Aptian–Albian, suggesting that erosion rates of the hinterland were similar along its length. The Cenomanian saw the development of the Orange River as the dominant sediment entry point (Jungslager 1999) and, because there was no significant ocean current system at the time, there is a broad symmetry in deposits to the north (Namibia) and south (South Africa). The stratigraphic successions are almost identical and have remarkably similar thicknesses, except in the furthest north beyond the edge of the Orange River system.

The geometry and accommodation space created along a passive margin should be a function of crustal thinning (McKenzie 1978; Steckler & Watts 1978). It is expected that the areas with the greatest extension should correspond to the areas of greatest post-rift thermal subsidence, at least during the earlier phase of the post-rift period. However, the observations from the combined South African–Namibian Orange Basin (this study and Paton *et al.* 2008) demonstrate that the earliest post-rift phase shows little variation in thickness along the 1000 km margin, regardless of the rift architecture on which it sits. This is strong evidence that the magnitude of upper crustal extension does not always correspond to that of the lower crust, as is invoked by depth-dependent stretching models (e.g. Huismans & Beaumont 2011).

Conclusion

Using the southern Namibian margin as an example, we have demonstrated that rift evolution and the subsequent passive margin architecture are the products of a complex interplay of lithospheric processes and crustal heterogeneity. The latter plays a role in influencing the dimensions and style of fault evolution and in establishing the position of cross-cutting structures. We also highlighted the role of pre-rift granitic bodies in influencing the location of deformation.

MM acknowledges funding from the Iraqi Ministry of Higher Education and Scientific Research for this project, the release of data from Spectrum ASA and the provision of Petrel software from Schlumberger. We would like to acknowledge Namcor and Spectrum for release of the data for this project. This research was undertaken under the auspices of the Basin Structure Group and we acknowledge the support of Getech, BG and EON. We also thank two anonymous reviewers and Teresa Ceraldi for invaluable comments on the initial manuscript.

References

Bagguley, J.G. 1997. *The application of seismic and sequence stratigraphy to the post-rift megasequence offshore Namibia.* PhD thesis, Oxford Brookes University.

Beglinger, S.E., Van Wees, J.-D., Cloetingh, S. & Doust, H. 2012. Tectonic subsidence history and source maturation in the Campos Basin, Brazil. *Petroleum Geoscience*, **18**, 153–172, http://doi.org/10.1144/1354-079310-049

Blaich, O.A., Faleide, J.L. & Tsikalas, F.T. 2011. Crustal breakup and continent-ocean transition at South Atlantic conjugate margins. *Journal of Geophysical Research: Solid Earth*, **B116**, 1978–2012.

Bosworth, W. 1985. Geometry of propagating continental rifts. *Nature*, **316**, 625–627.

BROWN, L.F., JR, BENSON, J.M. *ET AL.* 1995. *Sequence Stratigraphy in Offshore South African Divergent Basins. An Atlas on Exploration for Cretaceous Lowstand Traps.* AAPG Studies in Geology, **41**.

BUTLER, R.W.H. & PATON, D.A. 2010. Evaluating lateral compaction in deepwater fold and thrust belts: how much are we missing from 'nature's sandbox'? *GSA Today*, **20**, 4–10.

CARTWRIGHT, J.A., TRUDGILL, B.D. & MANSFIELD, C.S. 1995. Fault growth by segment linkage: an explanation for scatter in maximum displacement and trace length data from the Canyonlands Grabens of SE Utah. *Journal of Structural Geology*, **17**, 1319–1326.

CLEMSON, J. 1997. *Segmentation of the Namibian passive margin*. PhD thesis, University of London.

CLEMSON, J., CARTWRIGHT, J. & BOOTH, J. 1997. Structural segmentation and the influence of basement structure on the Namibian passive margin. *Journal of the Geological Society, London*, **154**, 477–482, http://doi.org/10.1144/gsjgs.154.3.0477

CONTRERAS, J., ANDERS, M.H. & SCHOLZ, C.H. 2000. Growth of a normal fault system: observations from the Lake Malawi basin of the East African rift. *Journal of Structural Geology*, **22**, 159–168.

CORNER, B. 1983. An interpretation of the aeromagnetic data covering the western portion of the Damara Orogen in South West Africa/Namibia. *In*: MILLER, R.M. (ed.) *Evolution of the Damara Orogen of Southwest Africa/Namibia*. Geological Society of South Africa, Special Publications, **11**, 339–354.

CORNER, B., CARTWRIGHT, J. & SWART, R. 2002. Volcanic passive margin of Namibia: a potential fields perspective. *In*: MENZIES, M.A., KLEMPERER, S.L. & EBINGER, C.J. & BAKER, J. (eds) *Volcanic Rifted Margins*. Geological Society of America, Special Papers, **362**, 203–220.

COWARD, M.P. 1981. The junction between Pan African mobile belts in Namibia: its structural history. *Tectonophysics*, **76**, 59–73.

COWIE, P.A., GUPTA, S. & DAWERS, N.H. 2000. Implications of fault array evolution for synrift depocentre development: insights from a numerical fault growth model. *Basin Research*, **12**, 241–261.

DALTON, T., PATON, D.A., NEEDHAM, T. & HODGSON, N. 2015. Temporal and spatial evolution of deep water fold thrust belts; implications for quantifying strain imbalance. *Interpretation*, **3**, SAA59–SAA70, http://doi.org/10.1190/INT-2015-0034.1

DALTON, T.J.S., PATON, D.A. & NEEDHAM, D.T. 2016. Influence of mechanical stratigraphy on multi-layer gravity collapse structures: insights from the Orange Basin, South Africa. *In*: SABATO CERALDI, T., HODGKINSON, R.A. & BACKE, G. (eds) *Petroleum Geoscience of the West Africa Margin*. Geological Society, London, Special Publications, **438**. First published online January 19, 2016, http://doi.org/10.1144/SP438.4

DAWERS, N.H. & ANDERS, M.H. 1995. Displacement-length scaling and fault linkage. *Journal of Structural Geology*, **17**, 607–614.

DAWERS, N.H. & UNDERHILL, J.R. 2000. The role of fault interaction and linkage in controlling synrift stratigraphic sequences: Late Jurassic, Statfjord East Area, northern North Sea. *AAPG Bulletin*, **84**, 45–64.

DE VERA, J., GRANADO, P. & McCLAY, K. 2010. Structural evolution of the Orange Basin gravity-driven system, offshore Namibia. *Marine and Petroleum Geology*, **27**, 233–237, http://doi.org/10.1016/j.marpetgeo.2009.02.003

DINGLE, R.V. & HENDRY, Q.B. 1984. Late Mesozoic and Tertiary sediment supply to the east Cape Basin (SE Atlantic) and palaeo-drainage systems in southwestern Africa. *Marine Geology*, **56**, 13–26.

EBINGER, C.J., JACKSON, J.A., FOSTER, A.N. & HAYWARD, N.J. 1999. Extensional basin geometry and the elastic lithosphere. *Philosophical Transactions of the Royal Society, London, A*, **357**, 741–765.

EMERY, K.O., UCHUPI, E., BOWIN, C., PHILLIPS, J. & SIMPSON, E.S.W. 1975. Continental margin off western Africa: Cape St Francis (South Africa) to Walvis Ridge (South West Africa). *AAPG Bulletin*, **59**, 3–59.

FRANKE, D. 2013. Rifting, lithosphere break-up and volcanism: comparison of magma-poor and volcanic rifted margins. *Marine and Petroleum Geology*, **43**, 63–87.

FRANKE, D., NEBEN, S., LADAGE, S., SCHRECKENBERGER, B. & HINZ, K. 2007. Margin segmentation and volcano-tectonic architecture along the volcanic margin off Argentina/Uruguay, South Atlantic. *Marine Geology*, **244**, 46–67.

FRANKE, D., LADAGE, S. *ET AL.* 2010. Birth of a volcanic margin off Argentina, South Atlantic. *Geochemistry, Geophysics, Geosystems*, **11**, http://doi.org/10.1029/2009GC002715

FRIMMEL, H.E., HARTNADY, C.J.H. & KOLLER, F. 1996. Geochemistry and tectonic setting of magmatic units in the Pan-African Gariep Belt, Namibia. *Chemical Geology*, **130**, 101–121.

GAWTHORPE, R.L. & LEEDER, M.R. 2000. Tectono-sedimentary evolution of active extensional basins. *Basin Research*, **12**, 195–218.

GERRARD, I. & SMITH, G.C. 1982. Post-Paleozoic succession and structure of the southwestern African continental margin: studies in continental margin geology. *AAPG Bulletin*, **34**, 49–74.

GLADCZENKO, T.P., HINZ, K., ELDHOLM, O., MEYER, H., NEBEN, S. & SKOGSEID, J. 1997. South Atlantic volcanic margins. *Journal of the Geological Society, London*, **154**, 465–470, http://doi.org/10.1144/gsjgs.154.3.0465

GRAY, D.R., FOSTER, D.A., MEERT, J.G., GOSCOMBE, B.D., ARMSTRONG, R., TROUW, R.A.J. & PASSCHIER, C.W. 2008. A Damara orogen perspective on the assembly of southwestern Gondwana. *In*: PANKHURST, R.J., TROUW, R.A.J., DE BRITO NEVES, B.B. & DE WIT, M.J. (eds) *West Gondwana: Pre-Cenozoic Correlations Across the South Atlantic Region*. Geological Society, London, Special Publications, **294**, 257–278, http://doi.org/10.1144/SP294.14

GUPTA, S., COWIE, P.A., DAWERS, N.H. & UNDERHILL, J.R. 1998. A mechanism to explain rift-basin subsidence and stratigraphic patterns through fault-array evolution. *Geology*, **26**, 595–598.

HÄLBICH, I.W. & ALCHIN, D.J. 1995. The Gariep Belt: stratigraphic-structural evidence for obliquely transformed grabens and back-folded thrust stacks in a combined thick-skin thin-skin structural setting. *Journal of African Earth Sciences*, **21**, 9–33.

HEINE, C., ZOETHOUT, J. & MÜLLER, R.D. 2013. Kinematics of the South Atlantic rift. *Solid Earth.* **4**, 215–253, http://doi.org/10.5194/se-4-215-2013

HIRSCH, K.K., SCHECK-WENDEROTH, M., PATON, D.A. & BAUER, K. 2007. Crustal structure beneath the Orange Basin, South Africa. *South African Journal of Geology,* **110**, 249–260.

HIRSCH, K.K., SCHECK-WENDEROTH, M., VAN WEES, J.-D., KUHLMANN, G. & PATON, D.A. 2010. Tectonic subsidence history and thermal evolution of the Orange Basin. *Marine and Petroleum Geology,* **27**, 565–584.

HOLZFÖRSTER, F., STOLLHOFEN, H. & STANISTREET, I.G. 1999. Lithostratigraphy and depositional environments in the Waterberg-Erongo area, central Namibia, and correlation with the main Karoo Basin, South Africa. *Journal of African Earth Sciences,* **29**, 105–123.

HUISMANS, R. & BEAUMONT, C. 2011. Depth-dependent extension, two-stage breakup and cratonic underplating at rifted margins. *Nature,* **473**, 74–78.

HUISMANS, R.S. & BEAUMONT, C. 2013. Symmetric and asymmetric lithospheric extension: relative effects of frictional-plastic and viscous strain softening. *Journal of Geophysical Research: Solid Earth,* **108**, 2496.

JUNGSLAGER, E.H.A. 1999. Petroleum habitats of the Atlantic margin of South Africa. *In*: CAMERON, N.R., BATE, R.H. & CLURE, V.S. (eds) *The Oil and Gas Habitats of the South Atlantic.* Geological Society, London, Special Publications, **153**, 153–168, http://doi.org/10.1144/GSL.SP.1999.153.01.10

KOOPMANN, H., SCHRECKENBERGER, B., FRANKE, D., BECKER, K. & SCHNABEL, M. 2014. The late rifting phase and continental break-up of the southern South Atlantic: the mode and timing of volcanic rifting and formation of earliest oceanic crust. *In*: WRIGHT, T.J., AYELE, A., FERGUSON, D.J., KIDANE, T. & VYE-BROWN, C. (eds) *Magmatic Rifting and Active Volcanism.* Geological Society, London, Special Publications, **420**. First published online 18 December 2014, http://doi.org/10.1144/SP420.2

KORME, T., ACOCELLA, V. & ABEBE, B. 2004. The role of pre-existing structures in the origin, propagation and architecture of faults in the Main Ethiopian Rift. *Gondwana Research,* **7**, 467–479.

LIGHT, M.P.R., MASLANYJ, M.P., GREENWOOD, R.J. & BANKS, N.L. 1993. Seismic sequence stratigraphy and tectonics offshore Namibia. *In*: WILLIAMS, G.D. & DOBB, A. (eds) *Tectonics and Seismic Sequence Stratigraphy.* Geological Society, London, Special Publications, **71**, 163–191, http://doi.org/10.1144/GSL.SP.1993.071.01.08

LISTER, G.S., ETHERIDGE, M.A. & SYMONDS, P.A. 1986. Detachment faulting and the evolution of passive continental margins. *Geology,* **14**, 246–250.

MARCHAL, D., GUIRAUD, M., RIVES, T. & VAN DEN DRIESSCHE, J. 1998. Space and time propagation processes of normal faults. *In*: JONES, G., FISHER, Q.J. & KNIPE, R.J. (eds) *Faulting, Fault Sealing and Fluid Flow in Hydrocarbon Reservoirs.* Geological Society, London, Special Publications, **147**, 51–70, http://doi.org/10.1144/GSL.SP.1998.147.01.04

MASLANYJ, M.P., LIGHT, M.P.R., GREENWOOD, R.J. & BANKS, N.L. 1992. Extension tectonics offshore Namibia and evidence for passive rifting in the South Atlantic. *Marine and Petroleum Geology,* **9**, 590–601.

McCLAY, K.R., DOOLEY, T. & WHITEHOUSE, P. 2002. 4-D evolution of rift systems: insights from scaled physical models. *AAPG Bulletin,* **86**, 935–959.

McKENZIE, D. 1978. Some remarks on the development of sedimentary basins. *Earth and Planetary Science Letters,* **40**, 25–32.

McLEOD, A.E., DAWERS, N.H. & UNDERHILL, J.R. 2000. The propagation and linkage of normal faults: insights from the Strathspey-Brent-Statfjord fault array, northern North Sea. *Basin Research,* **12**, 263–284.

McLEOD, A.E., UNDERHILL, J.R., DAVIES, S.J. & DAWERS, N.H. 2002. The influence of fault array evolution on synrift sedimentation patterns: controls on deposition in the Strathspey–Brent–Statfjord half graben, northern North Sea. *AAPG Bulletin,* **86**, 1061–1093.

McMILLAN, I.K. 2003. Foraminiferally defined biostratigraphic episodes and sedimentation pattern of the Cretaceous drift succession (Early Barremian to Late Maastrichtian) in seven basins on the South African and southern Namibian continental margin. *South African Journal of Science,* **99**, 537–576.

MILLER, R.M. 1983. The Pan-African Damaran Orogen of Namibia. *In*: MILLER, R.M. (ed.) *Evolution of the Damara Orogen of Southwest Africa/Namibia.* Geological Society of South Africa, Special Publications, **11**, 431–515.

MORLEY, C.K. 1999. Patterns of displacement along large normal faults: implications for basin evolution and fault propagation, based on examples from East Africa. *AAPG Bulletin,* **83**, 613634.

MORLEY, C.K., NELSON, R.A., PATTON, T.L. & MUNN, S.G. 1990. Transfer zones in the East African Rift System and their relevance to hydrocarbon exploration in rifts (1). *AAPG Bulletin,* **74**, 1234–1253.

MORTIMER, E., PATON, D.A., SCHOLZ, C.A., STRECKER, M.R. & BLISNIUK, P. 2007. Orthogonal to oblique rifting: effect of rift basin orientation in the evolution of the North basin, Malawi Rift, East Africa. *Basin Research,* **19**, 393–407, http://doi.org/10.1111/j.1365-2117.2007.00332.x

MORTIMER, E.J., PATON, D.A., SCHOLZ, C.A. & STRECKER, M.R. 2016. Implications of structural inheritance in oblique rift zones for basin compartmentalization: Nkhata Basin, Malawi Rift (EARS). *Marine and Petroleum Geology,* **72**, 110–121, http://doi.org/10.1016/j.marpetgeo.2015.12.018

MOULIN, M., ASLANIAN, D., RABINEAU, M., PATRIAT, M., & MATIAS, L. 2013. Kinematic keys of the Santos–Namibe basins. *In*: MOHRIAK, W.U., DANFORTH, A., POST, P.J., BROWN, D.E., TARI, G.C., NEMČOK, M. & SINHA, S.T. (eds) *Conjugate Divergent Margins.* Geological Society, London, Special Publications, **369**, 91–107, http://doi.org/10.1144/SP369.3

MUNTINGH, A. & BROWN, L.F.J. 1993. Sequence stratigraphy of petroleum plays, post-rift Cretaceous rocks (lower Aptian to upper Maastrichtian), Orange Basin, South Africa. *In*: WEIMER, P. & POSAMENTIER, H.W. (eds) *Siliciclastic Sequence Stratigraphy: Recent Developments and Applications.* AAPG Memoir, **58**, 71–98.

NÜRNBERG, D. & MÜLLER, R.D. 1991. The tectonic evolution of the South Atlantic from Late Jurassic to present. *Tectonophysics,* **191**, 27–53.

PATON, D.A. 2006. Influence of crustal heterogeneity on normal fault dimensions and evolution: southern

South Africa extensional system. *Journal of Structural Geology*, **28**, 868–886.

PATON, D.A., MACDONALD, D.I., & UNDERHILL, J.R. 2006. Applicability of thin or thick skinned structural models in a region of multiple inversion episodes; southern South Africa. *Journal of Structural Geology*, **28**, 1933–1947.

PATON, D.A., DI PRIMIO, R., KUHLMANN, G., VAN DER SPUY, D. & HORSFIELD, B. 2007. Insights into the petroleum system evolution of the southern Orange Basin, South Africa. *South African Journal of Geology*, **110**, 261–274.

PATON, D.A., VAN DER SPUY, D., DI PRIMIO, R. & HORSFIELD, B. 2008. Tectonically induced adjustment of passive-margin accommodation space: influence on the hydrocarbon potential of the Orange Basin, South Africa. *AAPG Bulletin*, **92**, 589–609.

PATON, D.A., MORTIMER, E.J., HODGSON, N. & VAN DER SPUY, D. 2016. The missing piece of the South Atlantic jigsaw: when continental break-up ignores crustal heterogeneity. *In*: SABATO CERALDI, T., HODGKINSON, R.A. & BACKE, G. (eds) *Petroleum Geoscience of the West Africa Margin*. Geological Society, London, Special Publications, **438**. First published online March 9, 2016, http://doi.org/10.1144/SP438.8

PERON-PINVIDIC, G., MANATSCHAL, G. & OSMUNDSEN, P.T. 2013. Structural comparison of archetypal Atlantic rifted margins: a review of observations and concepts. *Marine and Petroleum Geology*, **43**, 21–47.

RING, U. 1994. The influence of preexisting structure of the evolution of the Cenozoic Malawi Rift (East African rift system). *Tectonics*, **13**, 313–326.

SCHMIDT, S. 2004. *The Petroleum Potential of the Passive Continental Margin of South Western Africa – a Basin Modelling Study*. Rheinisch-Westfälische Technische Hochschule, Aachen.

SCHUMACHER, M.E. 2002. Upper Rhine Graben: role of preexisting structures during rift evolution. *Tectonics*, **21**, 6-1–6-17.

SÉRANNE, M. & ANKA, Z. 2005. South Atlantic continental margins of Africa: a comparison of the tectonic vs climate interplay on the evolution of equatorial West Africa and SW Africa margins. *Journal of African Earth Sciences*, **43**, 283–300.

SIBUET, J.C., & TUCHOLKE, B.E. 2014. The geodynamic province of transitional lithosphere adjacent to magma-poor continental margins. *In*: MOHRIAK, W.U., ANFORTH, A., POST, P.J., BROWN, D.E., TARI, G.C., NEMČOK, M. & SINHA, S.T. (eds) *Conjugate Divergent Margins*. Geological Society, London, Special Publications, **369**, 429–452, http://doi.org/10.1144/SP369.15

SKOGSEID, J. 2001. Volcanic margins: geodynamic and exploration aspects. *Marine and Petroleum Geology*, **18**, 457–461.

STECKLER, M.S. & WATTS, A.B. 1978. Subsidence of the Atlantic type continental margin of New York. *Earth and Planetary Science Letters*, **42**, 1–13.

STICA, J.M., ZALAN, P.V. & FERRARI, A.L. 2014. The evolution of rifting on the volcanic margin of the Pelotas Basin and the contextualization of the Paraná–Etendeka LIP in the separation of Gondwana in the South Atlantic. *Marine and Petroleum Geology*, **50**, 1–21.

STOLLHOFEN, H., STANISTREET, I.G., ROHN, R., HOLZFÖRSTER, F. & WANKE, A. 2000. The Gaias lake system, northern Namibia and Brazil. *In*: GIERLOWSKI-KORDESCH, E.H. & KELTS, K.R. (eds) *Lake Basins through Space and Time*. AAPG Studies in Geology, **46**, 87–108.

TSIKALAS, F., FALEIDE, J.I. & ELDHOLM, O. 2001. Lateral variations in tectono-magmatic style along the Lofoten–Vesteralen volcanic margin off Norway. *Marine and Petroleum Geology*, **18**, 807–832.

UCHUPI, E. 1989. The tectonic style of the Atlantic Mesozoic rift system. *Journal of African Earth Sciences (and the Middle East)*, **8**, 143–164.

WALKER, J.D., GEISSMAN, J.W., BOWRING, S.A. & BABCOCK, L.E. 2012. *Geologic Time Scale v. 4.0*. Geological Society of America, Boulder.

WALSH, J.J., NICOL, A. & CHILDS, C. 2002. An alternative model for the growth of faults. *Journal of Structural Geology*, **24**, 1669–1675.

WEIGELT, E. & UENZELMANN-NEBEN, G. 2004. Sediment deposits in the Cape Basin: indications for shifting ocean currents? *AAPG Bulletin*, **88**, 765–780.

The missing piece of the South Atlantic jigsaw: when continental break-up ignores crustal heterogeneity

D. A. PATON[1]*, E. J. MORTIMER[1], N. HODGSON[2] & D. VAN DER SPUY[3]

[1]*Basin Structure Group, School of Earth and Environment, University of Leeds, Leeds LS2 9JT, UK*

[2]*Spectrum, Dukes Court, Woking GU21 5BH, UK*

[3]*Petroleum Agency South Africa, Bellville 7530, Cape Town, South Africa*

**Corresponding author (e-mail: d.a.paton@leeds.ac.uk)*

Abstract: Crustal heterogeneity is considered to play a critical role in the position of continental break-up, yet this can only be demonstrated when a fully constrained pre-break-up configuration of both conjugate margins is achievable. Limitations in our understanding of the pre-break-up crustal structure in the offshore region of many margins preclude this. In the southern South Atlantic, which is an archetypal conjugate margin, this can be achieved because of the high confidence in plate reconstruction. Prior to addressing the role of crustal heterogeneity, two questions have to be addressed: first, what is the location of the regionally extensive Gondwanan Orogeny that remains enigmatic in the Orange Basin, offshore South Africa; and, second, although it has been established that the Argentinian Colorado rift basin has an east–west trend perpendicular to the Orange Basin and Atlantic spreading, where is the western continuation of this east–west trend? We present here a revised structural model for the southern South Atlantic by identifying the South African fold belt offshore. The fold belt trend changes from north–south to east–west offshore and correlates directly with the restored Colorado Basin. The Colorado–Orange rifts form a tripartite system with the Namibian Gariep Belt, which we call the Garies Triple Junction. All three rift branches were active during the break-up of Gondwana, but during the Atlantic rift phase the Colorado Basin failed while the other two branches continued to rift, defining the present day location of the South Atlantic. In addressing these two outstanding questions, this study challenges the premise that crustal heterogeneity controls the position of continental break-up because seafloor spreading demonstrably cross-cuts the pre-existing crustal heterogeneity. Furthermore, we highlight the importance of differentiating between early rift evolution and subsequent rifting that occurs immediately prior to seafloor spreading.

Over the last few years there has been renewed interest in the evolution of both sides of the southern South Atlantic and in their influence on plate configuration and reconstruction (Franke *et al.* 2006, 2010; Paton *et al.* 2008; Moulin & Aslanian 2010; Blaich *et al.* 2011; Pángaro & Ramos 2012; Heine *et al.* 2013; Koopmann *et al.* 2014). The region is considered to be the quintessential passive continental margin and provides type localities for investigating the processes involved in lithospheric stretching and margin evolution, including the role of volcanism, rift margin development and passive margin petroleum systems (Hirsch *et al.* 2007; Heine *et al.* 2013; Sibuet & Tucholke 2013; Péron-Pinvidic *et al.* 2014).

Common to all these studies, and their precursors, is the premise that much of the rift geometry that controlled the subsequent Atlantic spreading was inherited from pre-Jurassic crustal heterogeneity and that these structures were reactivated during Mesozoic rifting (Light *et al.* 1993; Clemson *et al.*

1997). In the southern portion in particular, the onshore Cape Fold Belt (CFB) of South Africa, which was reactivated during the Mesozoic, is laterally continuous with the Ventana Fold Belt (VFB) of Argentina (Hälbich *et al.* 1983; De Wit & Ransome 1992; Macdonald *et al.* 1996; Paton *et al.* 2006; Stankiewicz *et al.* 2009; Tankard *et al.* 2009; Pángaro & Ramos 2012; Pángaro *et al.* 2015). Despite the substantial evidence for this continuity both onshore South Africa and onshore and offshore Argentina, the position offshore South Africa remains enigmatic.

By integrating deep seismic reflection data with gravity and magnetic data we have been able to constrain the basin configuration of the deeper portion of the South Africa Orange Basin (Fig. 1). This enabled us to differentiate between structures associated with Mesozoic rifting and the reactivation of the underlying basement. In doing so, we were able to identify the position of the CFB offshore South Africa and hence provide the

From: SABATO CERALDI, T., HODGKINSON, R. A. & BACKE, G. (eds) 2017. *Petroleum Geoscience of the West Africa Margin*. Geological Society, London, Special Publications, **438**, 195–210.
First published online March 9, 2016, http://doi.org/10.1144/SP438.8

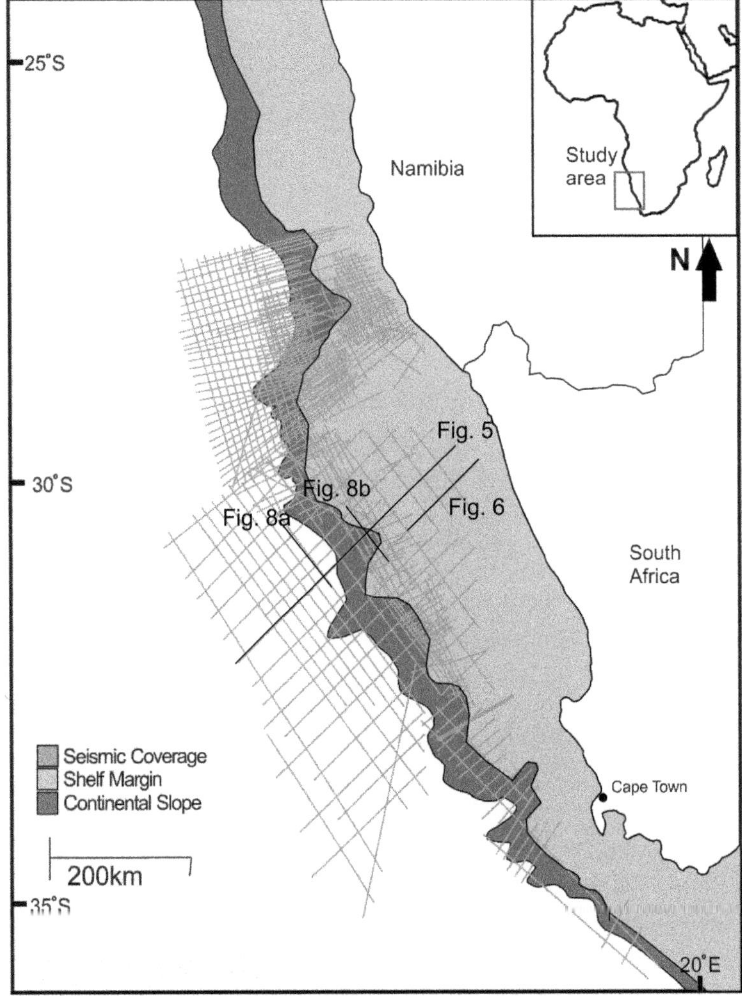

Fig. 1. Location of study area showing the distribution of the data and the location of key lines. The colours denote the sediment thickness in the Orange Basin.

missing piece of the jigsaw for South Atlantic reconstructions.

Our results provide insights into the processes of continental break-up. Although many established studies advocate that crustal heterogeneity plays an important role in crustal break-up (e.g. Ring 1994; Piqué & Laville 1996; Tommasi & Vauchez 2001), a number of recent studies have questioned the influence of inheritance, both with respect to crustal segmentation influencing the development of oceanic transform faults and also the localization of the break-up itself (e.g. Lundin & Doré 2011; Manatschal *et al.* 2015). We have used our reconstruction to test the influence of crustal heterogeneity on the position of break-up.

Regional setting

Throughout the last 1200 myr, southern South Africa has experienced repeated episodes of compressional and extensional deformation that have been demonstrably superimposed onto the same existing structural heterogeneity (Fig. 2; Tankard *et al.* 1982; Dingle *et al.* 1983; Hälbich 1993; Thomas *et al.* 1993). Of significance is the Gondwana suture, evident from a large positive magnetic anomaly (the Beattie anomaly) and an electrically conductive zone (the Southern Cape conductive belt) in the lower crust or upper mantle (de Beer 1983; Pitts *et al.* 1992). This suture developed between 950 and 900 Ma during the Namaqua-Natal

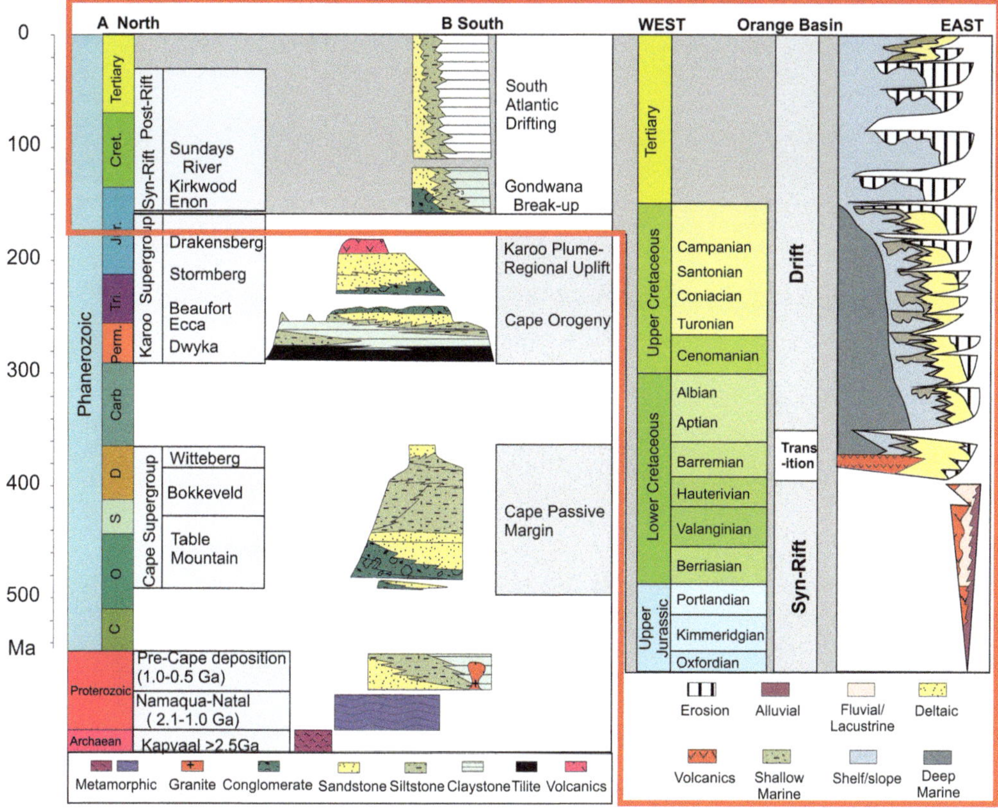

Fig. 2. Chronostratigraphy of southern South Africa and the Orange Basin (after Hälbich 1993; Veevers *et al.* 1994; Brown *et al.* 1995; Paton *et al.* 2006, 2008; Tankard *et al.* 2009).

Orogeny (Thomas *et al.* 1993) when compression resulted in the inversion of the passive margin that existed along the southern margin of the Kapvaal craton. Following this compression, between 900 and 600 Ma, a series of east–west-trending extensional basins evolved into which the Pre-Cape Group sediments were deposited. A subsequent period of compression (the Pan-African Orogeny; 600–450 Ma) led to the inversion of these extensional basins (Tankard *et al.* 1982; Gresse 1983; Krynauw 1983; Shone *et al.* 1990). A proposed south-dipping mega-décollement and northwards-verging thrust system may also be attributed to this episode of compression, although this remains speculative (Hälbich 1993).

Following the Pan-African Orogeny, clastic sediments of the Ordovician to Early Carboniferous Cape Supergroup were deposited along either an intra-continental or passive margin. This deposition ceased due to renewed compression during the Gondwanan Orogeny, the onset of which is identified by the regionally correlatable glacigenic deposits of the Permian Dwyka Formation. This

Gondwanan compression led to the deformation of the Cape Supergroup which, in South Africa, is manifested within the CFB and the Karoo foreland basin to the north (Fig. 3; Hälbich *et al.* 1983; Hälbich 1993; Veevers *et al.* 1994). This Gondwanan Fold Belt is traceable from the Sierra de la Ventana in eastern Argentina to the Pensacola Mountains of the Trans-Antarctic Mountains (Du Toit 1937; Pángaro & Ramos 2012).

The final stage in the tectonic evolution of the southern South Atlantic margins occurred with the break-up of the Gondwanan continent. Continental rifting, which initiated in the Middle Jurassic with the deposition of terrestrial and shallow-marine sediments (Dingle *et al.* 1983; McMillan *et al.* 1997), resulted in the superimposition of Mesozoic extensional structures onto the CFB (De Wit & Ransome 1992; Hälbich 1993). Rifting continued with the deposition of shallow or non-marine sediments onshore, while in the offshore portions the rate of extension rapidly increased, resulting in an abrupt transition to a deep-water setting (McLachlan & McMillan 1976; Shone 1978; Dingle *et al.* 1983;

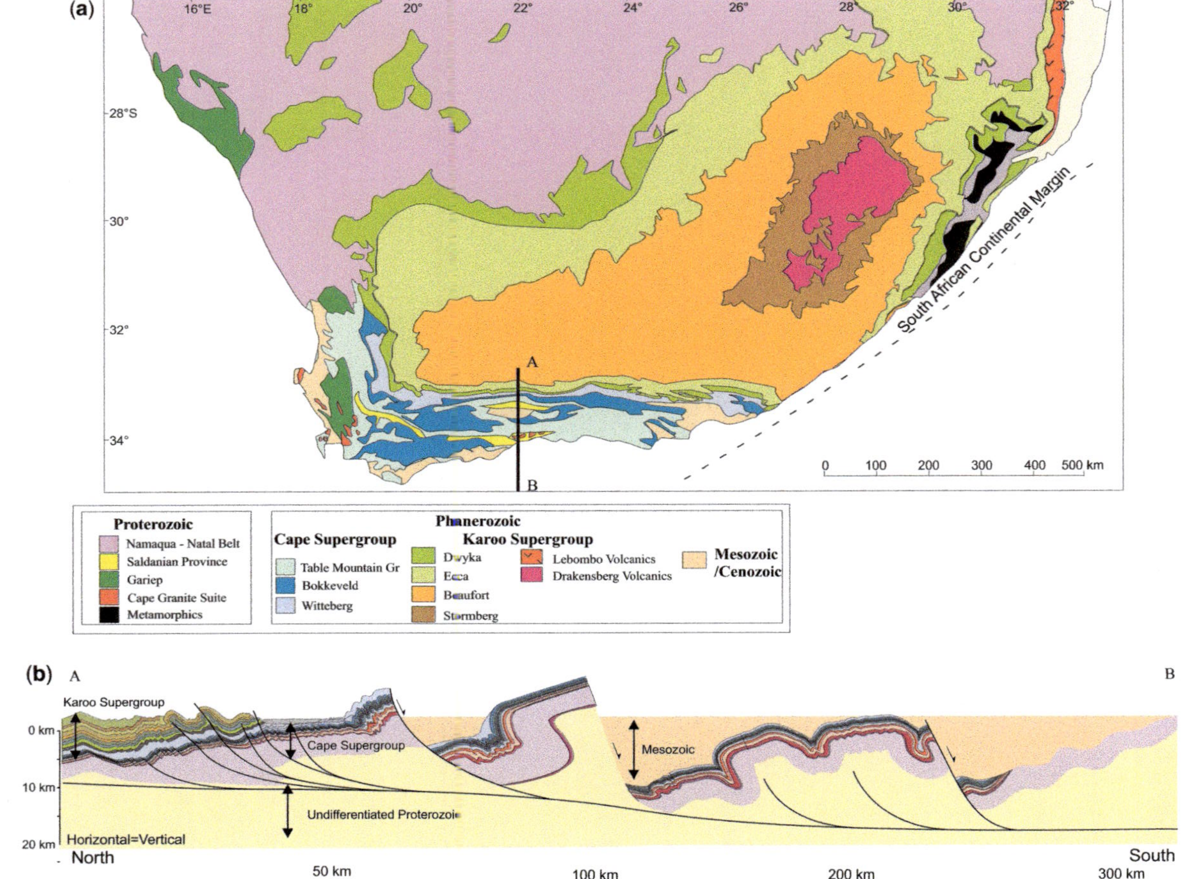

Fig. 3. Regional geology of southern South Africa. (**a**) Summary geological map of South Africa showing the location of the main stratigraphic units in the area, including the Precambrian inliers, the Cape Supergroup, the Karoo foreland basin and the Mesozoic rift basins. The location of the broadly east–west-trending Cape Fold Belt is also shown. (**b**) North–south cross-section through the fold belt illustrating the south-dipping décollement on to which the Mesozoic faults décolle, the thin-skinned nature of the deformation in the north of the fold belt and the internal deformation of the Cape Supergroup (after Hälbich 1993; Veevers *et al.* 1994; Paton *et al.* 2006; Tankard *et al.* 2009).

Paton & Underhill 2004). The rift–drift transition is considered to have occurred during the Valanginian and is marked by the onset of deposition of shallow-marine post-rift sediments (McMillan *et al.* 1997) across the majority of the basin.

Central Cape regional section

The geometry of the pre-rift sequence (pre-Jurassic) is generally poorly imaged in seismic reflection data, even in more recent studies. Therefore, to understand the geometry of the CFB, we used the onshore exposures of the central Cape (Figs 3–5). These exposures allowed us to characterize the style of CFB deformation and, through the integration of the onshore geometry with offshore seismic data, to consider how the structural configuration of the fold belt influenced the subsequent Mesozoic extensional structures.

The CFB is controlled by a regional south-dipping crustal detachment onto which both the Permo-Triassic compression and the Mesozoic

extensional structures décolle (Stankiewicz *et al.* 2009). Structural sections and restorations of the onshore portion demonstrate that, despite this common detachment horizon, there is a significant variation in deformation style from north to south (Fig. 4; Paton *et al.* 2006). In the north, deformation is dominated by thin-skinned, north-verging shallow-angle thrusts (Spikings *et al.* 2015). Structural modelling of these suggests a common detachment at *c.* 10 km depth, in agreement with the detachment derived from deep reflection and refraction studies through the central Cape (Stankiewicz *et al.* 2009). This strongly contrasts with the style of deformation further south (Fig. 3), which consists of a series of box folds with wavelengths of *c.* 8 km and amplitudes of 5 km. Internally, these folds have steeply dipping northern and southern limbs with significant shortening accommodated in both the fold limbs and the intervening flat by multi-layered chevron folding (Fig. 4). As the regional cross-section reveals (Fig. 3b), the CFB is dominated by these box folds, in particular to the south of the Karoo deformation front.

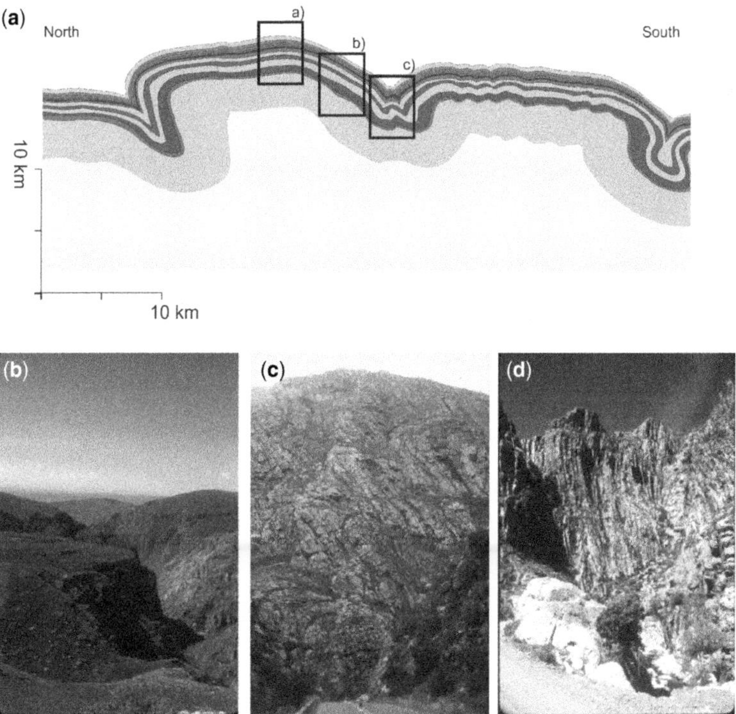

Fig. 4. (**a**) Detailed structural section through a box fold within the central Cape Fold Belt showing that the broad geometry is dominated by consistent, very steeply dipping panels forming the box fold limbs and a relatively flat fold top. (**b–d**) Photographs showing the internal geometry, including: (**b**) the relatively undeformed upper limb of the fold; (**c**) internal chevron folding of the quartzite-dominated Cape Supergroup; and (**d**) the relatively undeformed southern limb of a box fold. This style of deformation is representative across the fold belt (Paton *et al.* 2006).

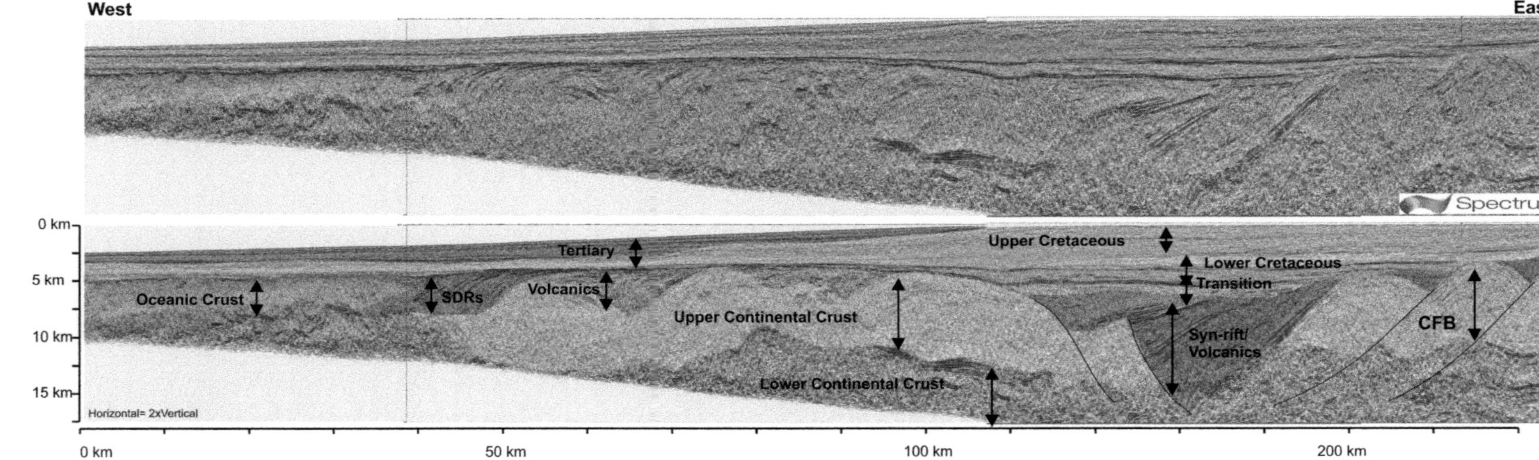

Fig. 5. Regional seismic section across the South African Orange Basin (see Fig. 1 for location). This section shows the entire margin geometry from the relatively unstretched continental crust, across the rift basins and break-up volcanics into oceanic crust. The main megasequences used in this study are identified. CFB, Cape Fold Belt; SDRs, seawards-dipping reflectors.

The position of subsequent Mesozoic extensional faulting is controlled by this pre-Mesozoic fold belt architecture (Fig. 3). The extensional faults are consistently positioned immediately to the south of the northern limb of the box folds, which Paton *et al.* (2006) suggested is a consequence of two stages of structural inversion. They postulated that much earlier normal faults influenced the deposition of the Cape Supergroup during the Cambro-Ordovician intra-continental margin; these faults were subsequently inverted in compression (positive inversion) during the Permian Gondwanan Orogeny and finally reactivated once more in extension (negative inversion) during Mesozoic rifting.

Orange Basin regional transect

The seismic reflection data provide imaging of the entire continental margin from the relatively unthinned present day continental crust, across the attenuated continental crust/volcanic terranes and into the oceanic crust (Fig. 5). The ages of seismic packages are constrained in the nearshore areas by a number of wells penetrating the post-rift, synrift and basement units, thereby providing a framework for the interpretation of the megasequence (Muntingh 1993; Paton *et al.* 2008).

A megasequence approach to the seismic interpretation was taken and, although it is based on the existing seismic stratigraphic framework, some modification was required to account for the additional imaging at depth (Fig. 5; Muntingh 1993; Brown *et al.* 1995). In total, nine packages were identified (from youngest to oldest): Tertiary; Late Cretaceous; Early Cretaceous; oceanic crust; seaward-dipping reflections; synrift; pre-rift; and mid- to lower continental crust.

The Tertiary package was very similar to that observed elsewhere along the Orange Basin in both the Republic of South Africa and Namibia (Paton *et al.* 2008; Mohammed *et al.* 2016) and consists of a very thin, or absent, package in the inner part of the margin, but it rapidly thickens across the break in slope to form a break-of-slope fan system in the deep water portion of the basin. In the section presented here (Fig. 5), the stratigraphy is relatively undeformed, although it should be noted that localized gravity collapse features are evident throughout the Orange Basin (Butler & Paton 2010; De Vera *et al.* 2010; Dalton *et al.* 2015, 2016).

In contrast with the areally restricted Tertiary package, the late Cretaceous (Cenomanian–Maastrichtian) package is widespread along the margin and dominates the post-rift stratigraphy. The upper sequence was deposited across the majority of the margin and has concordant reflections in the inner part and an aggradational geometry at the break in slope. Significant erosional truncation is also evident at the top of the sequence immediately below the Tertiary package. The middle Cretaceous sequence (mid-Aptian to Cenomanian), which consists of fluvio-deltaic sediments, forms a westwards-prograding shelf margin sequence associated with distributed deposition prior to the establishment of the Orange River Basin as a single point source of sediment entry.

The Lower Cretaceous sequence (Barremian to mid-Aptian) exhibits significant thickness variation across the margin, forming a broad, shallow basin with onlap onto both the present day margin in the east and also onto the pre-rift stratigraphy in the outboard portion.

The horizon that defines the base of the Barremian sequence can be correlated across the entire margin; however, the subcrop to that horizon is highly variable and it is difficult to determine whether the packages are time-equivalent. At the westernmost extent of the data coverage the subcrop to the Barremian sequence is characterized by chaotic/transparent reflectivity with a high-amplitude, rubbly top, characteristic of oceanic crust. The oceanic crust transitions eastwards into continuous, high-amplitude reflections that show a consistent oceanwards dip; these are the seaward-dipping reflectors (SDRs) that are common along volcanic passive margins (Franke 2013). Towards the east, the package beneath the Barremian sequence is dominated by continuous, high-amplitude reflections, which are penetrated in the nearshore position by well A-J1 and consist of interbedded fluvial and volcanic sequences (Fig. 6). Geometrically, the sequence in the inner part of the margin is defined by diverging reflections, indicating a synrift sequence. This is likely to be of Upper Jurassic to lowest Cretaceous age (Jungslager 1999).

The basement/pre-rift megasequence has a highly variable seismic character that changes significantly with depth. At relatively shallow levels, immediately beneath the Barremian sequence, there is significant concordant reflectivity. In places this has a consistent dip, although occasionally short-wavelength folding is present. At a mid-crust level (*c.* 10 km) the seismic character is dominated by a series of highly reflective, continuous, shallow-dipping reflections that can be mapped across the dataset in three dimensions. Given their depth and character, these are interpreted to be mid-crustal shear zones and are likely to be structurally below the regional décollement surface (Fig. 4). Immediately above the high-angle reflectivity is a relatively transparent signature at a depth of *c.* 8 km, which is traceable across much of the section up to the oceanic crust. Increasing reflectivity is observed within this transparent package at shallower levels (*c.* 6 km) and up to the top basement unconformity.

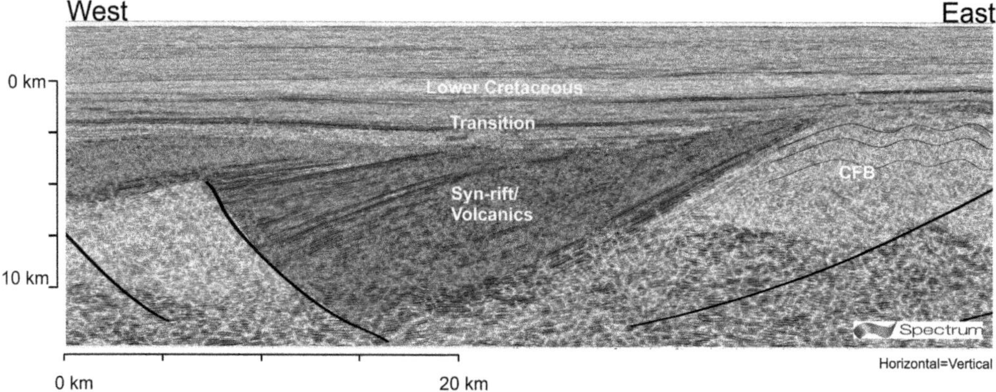

Fig. 6. East–west-trending seismic section showing the geometry of the main rift sequence within the Orange Basin. This sequence is penetrated by wells (see Fig. 1 for location) and the basin consists of interbedded fluvial sandstones and volcanics. CFB, Cape Fold Belt.

Variations in structure along the margin

There is considerable variation in the fault geometry and pre-rift seismic character both along and across the margin. In the northern part of the Orange Basin (northern South Africa and southern Namibia) is a series of north–south-trending half-graben separated by the Kudu Basement High. The faults to the east of the basement high are west-dipping and appear to have reactivated the west-dipping pre-rift reflectivity that is most likely to be the Proterozoic Gariep Belt (Gray *et al.* 2008), while the faults to the west of the high are associated with Atlantic rifting.

In contrast, in the southern Orange Basin, the pre-rift is divided into two very different packages. The package dominated by concordant reflectivity is penetrated in nearshore wells and is either Cambrian quartzites, consistent with the Cape Supergroup, or Devonian metasediments of the Bokkeveld Group (Fig. 2). There are two important points to note. First, the continuity of the seismic character across the section (Fig. 5) suggests the presence of continental lithosphere significantly further to the west than has previously been considered. Second, this upper pre-rift package broadly defines high-amplitude box folds with significant internal folding in the anticlines, relatively undeformed southern limbs and normal-faulted northern limbs. These structures are considered to be the equivalent of the box folds that characterize the onshore CFB (Fig. 7).

Mapping of the structures revealed a significant variation in trend and geometry. In the SE the basin is dominated by a small population of very large faults (throws of 10 km) with little evidence of small, distributed faulting. These are very similar to the faults in a comparable position along-trend in the CFB, e.g. the Gamtoos and Plettenberg faults

(Paton & Underhill 2004). This structural grain trends east–west in the central Cape (underpinning the Gamtoos and Plettenburg faults) and then abruptly becomes north–south-trending at the Cape Syntaxis (see Fig. 9). This north–south trend continues offshore and is mapped in the subsurface (Fig. 7). Further to the north, the north–south-trending faults do not continue along the margin and instead their orientation changes to east–west-trending. North–south-orientated sections reveal that both the CFB structures and subsequent Mesozoic rift faults have this east–west orientation (Fig. 8a, b).

Gravity and magnetic interpretation

The gravity high along the margin has long been used to define the continent–ocean boundary, with continental crust being taken to be on the landward side and oceanic crust on the outboard side (Rabinowitz & LaBrecque 1979). This has often been used as a constraint for the extensive nature of the distribution of break-up volcanics and as a pinpoint for plate reconstructions (Franke *et al.* 2010; Moulin & Aslanian 2010; Blaich *et al.* 2011; Heine *et al.* 2013).

When our interpretation of the deep seismic reflection data was compared with this gravity signature, it revealed that the gravity high corresponds to a position significantly inboard of not only the position of the principal SDRs, but also much of the late-stage rift basin (Fig. 9). The diffuse outboard gravity anomaly corresponds to the transition between attenuated continental crust and oceanic crust, as identified by the seismic reflection data. The variability of the crustal architecture observed illustrates that a sharp transition between oceanic and continental crust is not appropriate for this

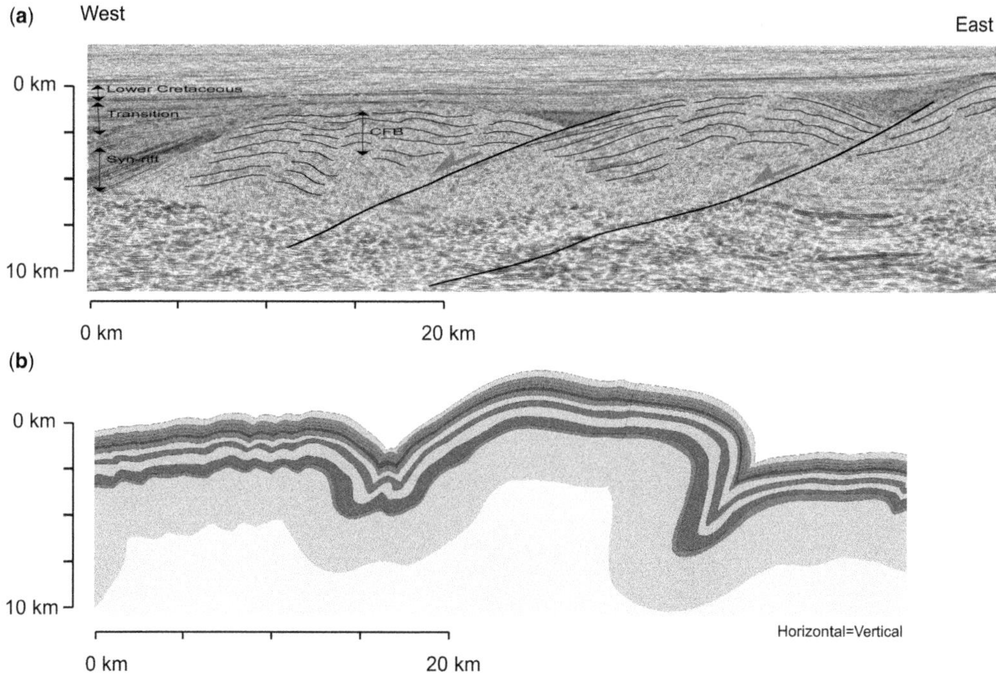

Fig. 7. Comparison of the pre-rift megasequence in the regional seismic sections and the onshore Cape Fold Belt (CFB) structures. (**a**) Folding of concordant reflectivity is imaged in the seismic data and overall forms an anticlinal geometry with a wavelength of approximately 10 km. This is very similar in geometry to (**b**), the Cape Supergroup box folds observed in the central Cape. This section is orientated east–west and provides evidence of the continuity of the Cape Fold Belt to the NW of the Cape Syntaxis.

margin. However, we do note that the transition between the volcanic province and definitive oceanic crust is not only a very narrow zone identifiable on seismic reflection data, but that it corresponds to a discrete step in the intensity of the magnetic anomaly and that this step can be used with significant certainty to trace this boundary.

Based on these observations, we superimposed our structural modelling onto both the gravity and magnetic intensity maps (Fig. 10a, b). The magnetic maps reflect the north–south orientation of the eastern oceanic crust and also illustrate the variation in magmatic bodies. The inboard magnetic highs do not correspond to identifiable magmatic bodies (based on the observed seismic reflection character), but rather reflect basement highs and record the high degree of magnetic susceptibility of the Precambrian basement.

The gravity maps exhibit more variability, although both the gravity high anomaly and the diffuse outer high coincident with the oceanic crust are north–south-trending. Further gravity modelling is required to constrain this signature better; however, existing modelling along the margin suggests a lower crustal origin rather than a shallow crustal

cause (Hirsch *et al.* 2007). We propose that the variation in gravity signature, especially the shorter wavelength (therefore shallower) component is a reflection of the upper crustal architecture. When the fault map is superimposed on the gravity anomaly map (Sandwell *et al.* 2014), the earlier north–south-trending faults that become east–west-trending are evident, as are the later north–south-trending structures.

Continuity of the structures within Western Gondwana

The Phanerozoic structures and stratigraphy of eastern Argentina and its adjacent offshore shelf are dominated by Upper Palaeozoic and Mesozoic sequences. The Palaeozoic consists of Patagonian foreland basin deposits (Rapalini 2005; Milani & De Wit 2008; Ramos 2008; López de Luchi *et al.* 2010; see also references cited in Pángaro & Ramos 2012) and the VFB (López Gamundi *et al.* 1995; Tomezzoli & Vilas 1999; Ramos 2008; Pángaro & Ramos 2012). It is the latter that is the focus of this study. The VFB is a NW-trending

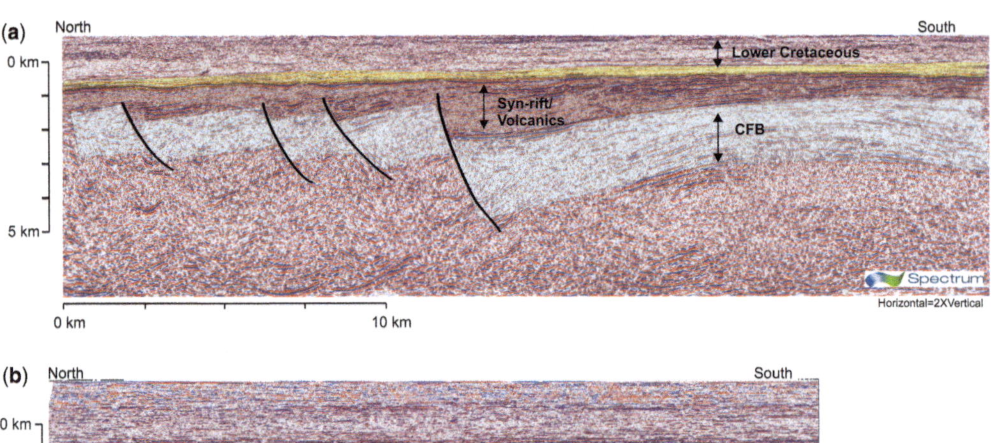

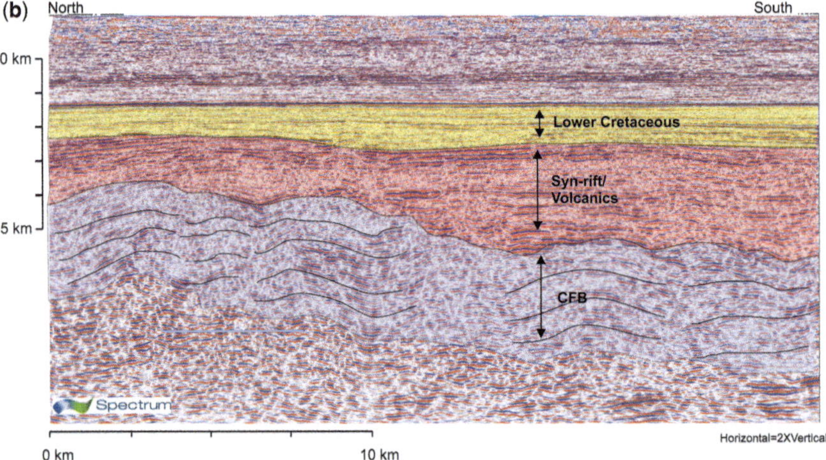

Fig. 8. North–south-trending sections in the central area of the Orange Basin showing the change in trend of the fold belt (see Fig. 1 for location). (**a**) This section shows the presence of east–west-trending Mesozoic normal faults. These faults are the lateral equivalent of the north–south-trending faults in Figure 6. (**b**) This section corresponds to the northern end of the Cape Fold Belt (CFB) and shows that the folding present within the pre-rift stratigraphy is geometrically identical to that in Figure 7.

structural feature that extends at least 800 km across both onshore and offshore areas.

The onshore stratigraphy consists of Cambrian to Devonian quartz-rich platform deposits that overlie the Rio de la Plata Craton and are therefore both the lateral and time-equivalent depositional sequence of the Cape Supergroup, as first postulated by Keidel (1916) and Du Toit (1937). The late Carboniferous to Permian sequence, which overlies the Devonian, is an up to 7 km thick sequence of sediments deposited in a wide range of depositional environments, including glaciogenic, low-energy marine and shallow-marine to continental settings (Pángaro & Ramos 2012; Pángaro *et al.* 2015). These are the direct equivalent of the Permian sediments of South Africa that represent the onset of the Gondwanan Orogeny.

The offshore continuity of the VFB has been proposed previously (Fryklund *et al.* 1996), although this required further substantiation. The oldest sequence penetrated offshore was a syn-Gondwanan Orogeny, Upper Carboniferous sequence below the base of the rift half-graben. This is correlated to the Sauce Grande Formation, which is the equivalent of the South African Dwyka Formation. The overlying half-graben have been recognized in the subsurface, but an absence of clearly imaged seismic reflection data has prevented the correlation of structural trends into the offshore region. Franke *et al.* (2006) proposed that VFB equivalent structures were present offshore, but it is only recently that definitive evidence has been produced. Pángaro & Ramos (2012) integrated seismic reflection and gravity data and proposed that the nature of folding within the seismic data was correlated with offshore structures. They demonstrated a north-verging geometry consistent with the east–west-trending VFB. Furthermore, they proposed that a seismic stratigraphic package characterized by small-scale folding was the Lower Palaeozoic sequence.

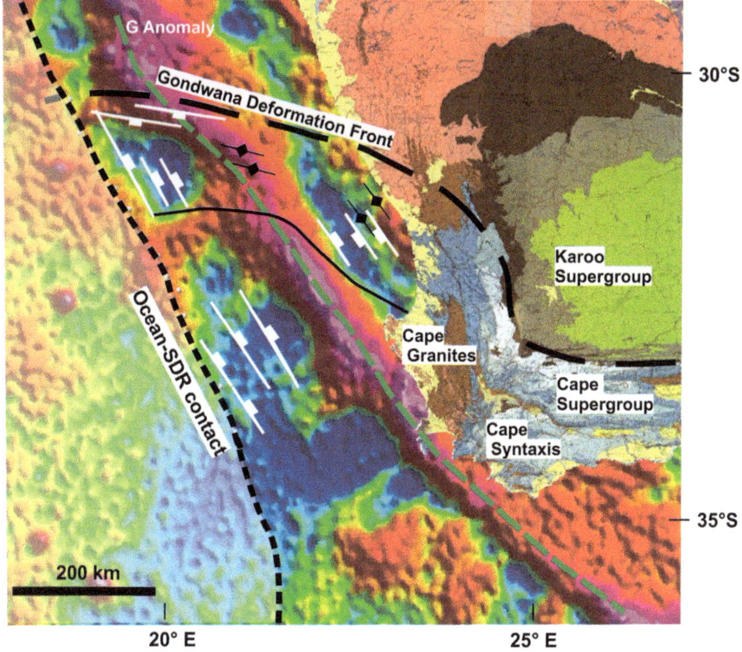

Fig. 9. Structural map of the study area superimposed on offshore Bouger gravity data. The structural map shows the continuity of the Cape Fold Belt (CFB) to the north of the Cape Syntaxis with a north–south orientation. There is an abrupt change to an east–west orientation that defines a second syntaxis. The east–west trend in the west, which also corresponds to a gravity low, is overprinted by later north–south-trending faults. As discussed in the text, although the gravity anomaly is dominated by the G Anomaly, this is more likely to be associated with lower crustal bodies rather than a shallower crustal structure.

Missing link of the CFB in South Africa

There is a significant dichotomy when current plate reconstructions of western Gondwana are considered. The CFB is east–west-trending in southern South Africa (Hälbich *et al.* 1983; De Wit & Ransome 1992; Paton *et al.* 2006; Tankard *et al.* 2009) and abruptly changes to a north–south orientation at the Cape Syntaxis; however, because of limited exposure its continuity towards the north and offshore remains unconstrained. Despite the extensive analysis of offshore data (Muntingh 1993; Brown *et al.* 1995; Paton *et al.* 2008; Koopmann *et al.* 2014), the relatively limited imaging of these reflection data at depth has previously prevented the mapping of basement-involved structures with any degree of certainty (Fig. 11). This has resulted in the existing structural models being dominated by the shallower north–south-trending normal faults of Early Cretaceous rifting. Existing studies of the Argentinian crustal architecture have also interpreted the geometry of the fold belt from the VFB into the CFB and the presence of both Cape and Colorado syntaxes (Pángaro & Ramos 2012).

Our interpretation of these new seismic reflection profiles provides evidence of the continuity of the western Gondwanan Fold Belt. Our new structural model identifies the fact that the CFB does indeed continue to trend northwards for *c.* 100 km, but as it continues into the Orange Basin a further syntaxis is identifiable within the seismic reflection and gravity data, which results in an abrupt change in the orientation of the fold belt to an east–west trend. This trend then continues towards the west and is directly equivalent to the Colorado Basin of Argentina and the onshore Sierra de le Ventana.

Further evidence for this correlation of the western Gondwanan Fold Belt is provided when the structural styles of the Permian compression are compared. Not only are the structures similarly manifested as small-scale north-verging asymmetrical folding, but the style and location of the subsequent negative inversion, identified by normal faults, are comparable.

A tale of two rifts

Although our new structural model provides evidence for the location of the CFB and, therefore,

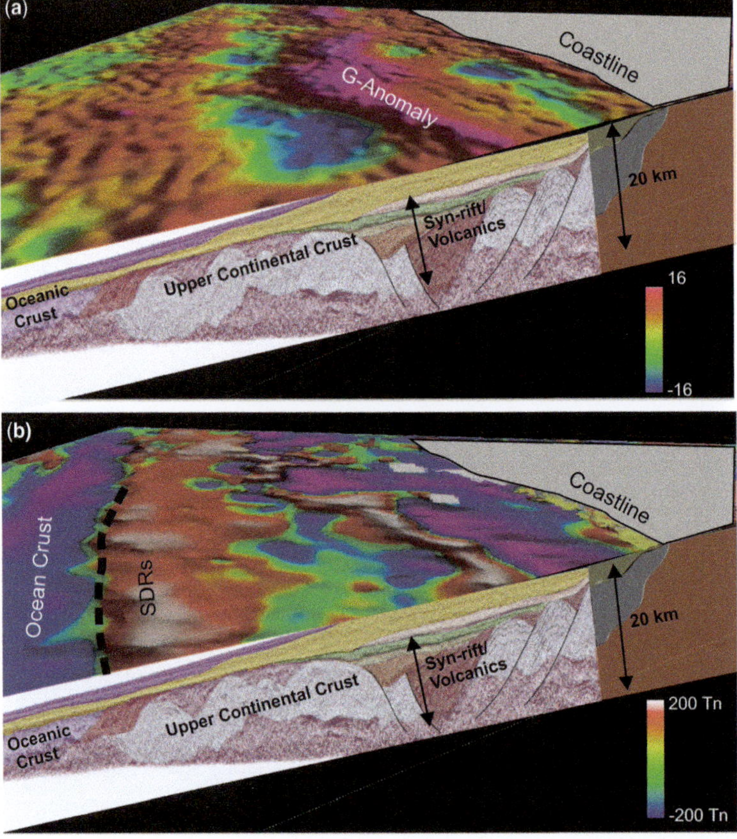

Fig. 10. This 3D perspective from the south of the basin shows the relationship between the basin architecture and (**a**) the gravity anomalies (Sandwell *et al.* 2014) and (**b**) the magnetic intensity. It shows that there is no correspondence between the G Anomaly and the position of the continent–ocean boundary. In contrast, the largest gradient magnetic intensity reflects the transition from seawards-dipping reflectors (SDRs) to oceanic crust.

constrains the missing link for all of the plate reconstructions of the South Atlantic, in doing so it creates a conundrum fundamental to our understanding of continental break-up.

The present day depocentres on both margins of the southern South Atlantic are north–south-trending, parallel to the present day rift orientation. However, this is at odds with the common consensus that the location of continental rifting is determined by heterogeneities within the crust, which subsequently control the position of continental break-up. Our revised structural model of the southern South Atlantic reveals a more complex development.

The initial rifting along western Gondwana was a consequence of the reactivation of the western Gondwanan Fold Belt, with broadly SW-orientated extension and hence east–west-trending faults. This led to the development of the main rift basins of the Orange Basin and the Colorado Basin in Argentina (Muntingh 1993; Brown *et al.* 1995; Franke *et al.*

2010; Pángaro & Ramos 2012). As these rift basins formed through the negative inversion of the fold belts, the rift basin geometry was controlled by the underlying fold belt geometry, including the Cape and Colorado syntaxes, the northern extent of which was likely to have been controlled by the presence of the Kalahari and Rio del Plata cratons (de Beer 1995; Pángaro & Ramos 2012). The timing of this rifting was likely to have been mid-Jurassic to early Cretaceous.

During the mid-Cretaceous, the rift configuration changed significantly and was much more dominated by South Atlantic rifting, which superimposed a north–south trend. As Mohammed *et al.* (2016) have demonstrated, the Namibian rift system utilized the north–south-trending Gariep Belt. At some locations this north–south trend must have intersected with the east–west-trending Gondwanan Fold Belt. Given the resultant geometry, we speculate that this formed a triple junction with

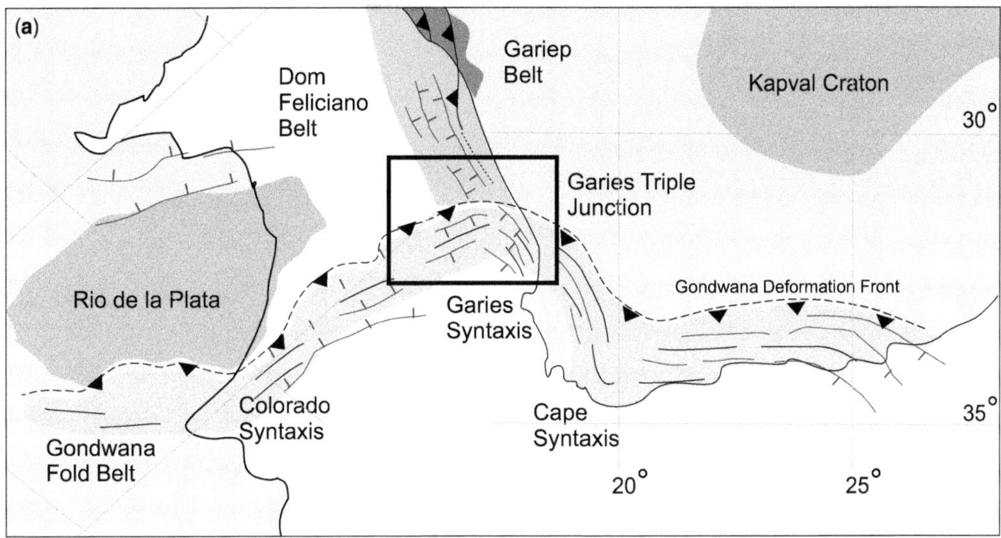

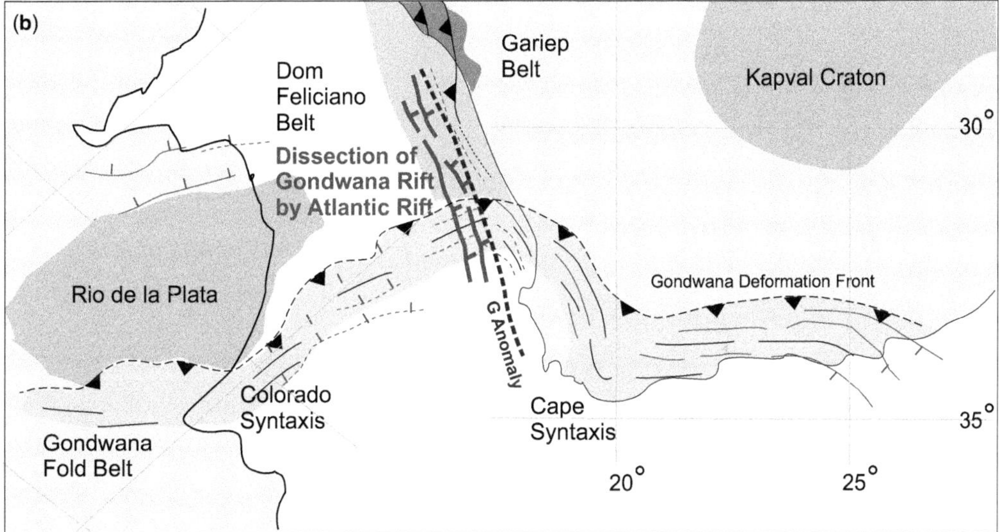

Fig. 11. Reconstruction of the southern South Atlantic at 140 Ma. (**a**) The Cape Fold Belt structures and Mesozoic rift faults, as interpreted in this study, have been plotted, as have the structures from Namibia (Mohammed *et al.* 2016) and the Colorado Basin (Pángaro & Ramos 2012). This configuration implies that there was a rift triple junction at the intersection of the three rift systems. (**b**) During the principal phase of South Atlantic rifting, the Colorado Basin became inactive and rifting was focused on the Orange and Gariep rift systems.

three active rift arms; we call this the Garies Triple Junction. As South Atlantic rifting became established, the Gariep Belt and the north–south-trending portion of the western Gondwanan Fold Belt ultimately controlled the position of the Atlantic rift system.

This example of the South Atlantic emphasizes that the pre-break-up configuration can be complex and is likely to be overprinted by the latter stages of rifting, making it difficult to determine the early structural configuration. The influence of crustal heterogeneity on the location of subsequent continental break-up and seafloor spreading has been much debated, with some studies suggesting that there is a strong relationship (e.g. Piqué & Laville 1996; Vauchez *et al.* 1997; Tommasi & Vauchez 2001; Gernigon *et al.* 2015), while others advocate very limited control (Lundin & Doré 2011; Manatschal *et al.* 2015). Manatschal *et al.* (2015) relate observations from a number of hyperextended margins with numerical stretching models and suggest that there is little influence on the final lithospheric necking position. There is clearly a variation in the role that crustal heterogeneity will have, but our results demonstrate an example where the pre-existing structure has no control on the development of a volcanic margin. Furthermore, the cross-cutting Atlantic rift system occurs within the Hauterivian, suggesting that there was a relatively rapid switch from a rift system dominated by crustal structures and east–west-trending faults to the north–south-orientated Atlantic rift system.

Conclusions

The use of newly acquired deep seismic reflection data has allowed us to re-evaluate the structural configuration of the Orange Basin, South Africa. This has enabled determination of the location of the CFB in the offshore area for the first time and correlation of its continuity across into the VFB of Argentina. This new structural model therefore provides the missing link in the South Atlantic for the location of the western Gondwanan Fold Belt.

In addition, our new structural model demonstrates that prior to break-up there was a tripartite rift system in the Orange Basin consisting of the rifts of South Africa, Gariep and Colorado. During South Atlantic rifting the Colorado system become inactive and the other two became the ultimate location of continental break-up.

This new model provides evidence that, in this part of the South Atlantic, the final position of break-up was not controlled by the pre-existing crustal heterogeneity in the Gondwanan Fold Belt that it demonstrably cross-cuts. These results therefore provide a demonstrable example of crustal heterogeneity not controlling the position of continental break-up.

We thank the sponsors of the Basin Structure Group (Getech, EON, BG) for providing funding for this research, Spectrum and Petroleum Agency South Africa for providing the data and Schlumberger for Petrel licences. We are especially grateful to two anonymous reviewers for their insightful comments and suggestions.

References

BLAICH, O.A., FALEIDE, J.I. & TSIKALAS, F. 2011. Crustal breakup and continent–ocean transitions at South Atlantic conjugate margins. *Journal of Geophysical Research: Solid Earth*, **116**, http://doi.org/10.1029/2010JB007686

BROWN, L.F., Jr., BENSON, J.M. ET AL. 1995. *Sequence Stratigraphy in Offshore South African Divergent Basins. An Atlas on Exploration for Cretaceous Lowstand Traps.* AAPG Studies in Geology, **41**.

BUTLER, R.W.H. & PATON, D.A. 2010. Evaluating lateral compaction in deepwater fold and thrust belts: how much are we missing from 'nature's sandbox'? *GSA Today*, **20**, 4–10. http://doi.org/10.1130/GSAT G77A.1

CLEMSON, J., CARTWRIGHT, J. & BOOTH, J. 1997. Structural segmentation and the influence of basement structure on the Namibian passive margin. *Journal of the Geological Society, London*, **154**, 477–482, http://doi.org/10.1144/gsjgs.154.3.0477

DALTON, T.J.S., PATON, D.A., NEEDHAM, D.T. & HODGSON, N. 2015. Temporal and spatial evolution of deep water fold thrust belts; implications for quantifying strain imbalance. *Interpretation*, **3**, SAA59–SAA70, http://doi.org/10.1190/INT-2015-0034.1

DALTON, T.J.S., PATON, D.A., NEEDHAM, D.T. 2016. Influence of mechanical stratigraphy on multi-layer gravity collapse structures: insights from the Orange Basin, South Africa. *In*: SABATO, CERALDI, T., HODGKINSON, R.A. & BACKE, G. (eds) *Petroleum Geoscience of the West Africa Margin*. Geological Society, London, Special Publications, **438**. First published online January 19, 2016, http://doi.org/10.1144/SP438.4

DE BEER, C.H. 1995. Fold interference from simultaneous shortening in different directions: the Cape Fold Belt syntaxis. *Journal of African Earth Sciences*, **21**, 157–169.

DE BEER, J.H. 1983. Geophysical studies in the southern Cape Province and models of the lithosphere in the Cape Fold Belt. *In:* SÖHNGE, A.P.G. & HÄLBICH, I.W. (eds) *Geodynamics of the Cape Fold Belt*. Geological Society of South Africa, Special Publications, **12**, 57–64.

DE VERA, J., GRANADO, P. & McCLAY, K. 2010. Structural evolution of the Orange Basin gravity-driven system, offshore Namibia. *Marine and Petroleum Geology*, **27**, 223–237.

DE WIT, M.J. & RANSOME, I.G. 1992. Regional inversion tectonics along the southern margin of Gondwana. *In*: DE WIT, M.J. & RANSOME, I.G.D. (eds) *Inversion Tectonics of the Cape Fold Belt, Karoo and Cretaceous Basins of Southern Africa*. Balkema, Rotterdam, 15–22.

DINGLE, R.V., SIESSER, W.G. & NEWTON, A.R. 1983. *Mesozoic and Tertiary Geology of Southern Africa*. Balkema, Rotterdam.

DU TOIT, A. 1937. *Our Wandering Continents: an Hypothesis of Continental Drifting*. Oliver and Boyd, London.

FRANKE, D. 2013. Rifting, lithosphere breakup and volcanism: comparison of magma-poor and volcanic rifted margins. *Marine and Petroleum Geology*, **43**, 63–87.

FRANKE, D., NEBEN, S., SCHRECKENBERBER, B., SCHULZE, A., STILLER, M. & KRAWCZYK, C.M. 2006. Crustal structure across the Colorado Basin, offshore Argentina. *Geophysical Journal International*, **165**, 850–864.

FRANKE, D., LADAGE, S. ET AL. 2010. Birth of a volcanic margin off Argentina, South Atlantic. *Geochemistry, Geophysics, Geosystems*, **11**, http://doi.org/10.1029/2009GC002715

FRYKLUND, B., MARSHALL, A. & STEVENS, J. 1996. La Cuenca del Colorado. *In*: RAMOS, V.A. & TURIC, M.A. (eds) *Geologia y recursos naturales de la plataforma continental Argentina*. Relatario XIII Congreso Geologico Argentino y III Congreso de Exploración de Hidrocarburos, **8**. Asociación Geologica Argentina & Instituto Argentino del Petróleo, Buenos Aires, 135–158.

GERNIGON, L., BLISCHKE, A., NASUTI, A. & SAND, M. 2015. Conjugate volcanic rifted margins, seafloor spreading, and microcontinent: Insights from new high-resolution aeromagnetic surveys in the Norway Basin. *Tectonics*, **34**, 907–933, http://doi.org/10.1002/2014TC003717

GRAY, D.R., FOSTER, D.A., MEERT, J.G., GOSCOMBE, B.D., ARMSTRONG, R., TROUW, R.A.J. & PASSCHIER, C.W. 2008. *A Damara orogen perspective on the assembly of southwestern Gondwana. In*: PANKHURST, R.J., TROUW, R.A.J., DE BRITO NEVEES, B.B., DE WIT, M.J. (eds) *West Gondwana: Pre-Cenozoic Correlations Across the South Atlantic Region.* Geological Society, London, Special Publications, **294**, 257–278, http://doi.org/10.1144/SP294.14

GRESSE, P. 1983. Lithostratigraphy and structure of the Kaaimans Group. *In*: SOHNGE, A.P.G. & HÄLBICH, I.W. (eds) *Geodynamics of the Cape Fold Belt.* Geological Society of South Africa, Special Publications, **12**, 7–19.

HÄLBICH, I.W. (ed.) 1993. *Cape Fold Belt–Agulhas Bank Transect Across Gondwana Suture, Southern Africa.* Global Geoscience Transects, **9**. American Geophysical Union, Boulder, 1–16.

HÄLBICH, I.W., FITCH, F.J. & MILLER, J.A. 1983. Dating the Cape Orogeny. *In*: SOEHNGE, A.P.G. & HÄLBICH, I.W. (eds). *Geodynamics of the Cape Fold Belt.* Geological Society of South Africa, Special Publications, **12**, 131–148.

HEINE, C., ZOETHOUT, J. & MULLER, R.D. 2013. Kinematics of the South Atlantic rift. *Solid Earth Discussions*, **5**, 41–115.

HIRSCH, K.K., SCHECK-WENDEROTH, M., PATON, D.A. & BAUER, K. 2007. Crustal structure beneath the Orange Basin, South Africa. *South African Journal of Geology*, **110**, 249–260, http://doi.org/10.2113/gssajg.110.2-3.249

JUNGSLAGER, E.H.A. 1999. Petroleum habitats of the Atlantic margin of South Africa. *In*: CAMERON, N.R., BATE, R.H. & CLURE, V.S. (eds) *The Oil and Gas Habitats of the South Atlantic.* Geological Society, London, Special Publications, **153**, 153–168, http://doi.org/10.1144/GSL.SP.1999.153.01.10

KEIDEL, J. 1916. La geología de las sierras de la Provincia de Buenos Aires y sus relaciones con las montañas de Sud África y Los Andes. *Anales del Ministerio de Agricultura de la Nación, Sección Geología, Mineralogía y Minería*, **11**, 1–78.

KOOPMANN, H., BRUNE, S., FRANKE, D. & BREUER, S. 2014. Linking rift propagation barriers to excess magmatism at volcanic rifted margins. *Geology*, **42**, 1071–1074.

KRYNAUW, J.R. 1983. Granite intrusion and metamorphism in the Kaaimans Group. *In*: SÖHNGE, A.P.G. & HÄLBICH, I.W. (eds) *Geodynamics of the Cape Fold Belt.* Geological Society of South Africa, Special Publications, **12**, 21–32.

LIGHT, M.P.R., KEELEY, M.L., MASLANYJ, M.P. & URIEN, C.M. 1993. The tectono-stratigraphic development of Patagonia, and its relevance to hydrocarbon exploration. *Journal of Petroleum Geology*, **16**, 465–482.

LÓPEZ DE LUCHI, M.G., RAPALINI, A.E. & TOMEZZOLI, R.N. 2010. Magnetic fabric and microstructures of Late Paleozoic granitoids from the North Patagonian Massif: evidence of a collision between Patagonia and Gondwana? *Tectonophysics*, **494**, 118–137, http://doi.org/10.1016/j.tecto.2010.09.003

LÓPEZ GAMUNDI, O.R., CONAGHAN, P., ROSSELLO, E.A. & COBBOLD, P.R. 1995. The Tunas Formation (Permian) in the Sierras Australes Fold Belt, East-Central Argentina: evidence of syntectonic sedimentation in a Varis-ican foreland basin. *Journal of South American Earth Sciences*, **8**, 129–142.

LUNDIN, E.R. & DORÉ, A.G. 2011. Hyperextension, ser-pentinization, and weakening: a new paradigm for rifted margin compressional deformation. *Geology*, **39**, 347–350.

MACDONALD, D., GOMEZ-PEREZ, I., FRANZESE, J., SPAL-LETTI, L., LAWVER, L., GAHAGAN, L. & PATON, D. 1996. Mesozoic break-up of SW Gondwana: implica-tions for regional hydrocarbon potential of the southern South Atlantic. *Marine and Petroleum Geology*, **20**, 287–308.

MANATSCHAL, G., LAVIER, L. & CHENIN, P. 2015. The role of inheritance in structuring hyperextended rift systems: some considerations based on observations and numerical modeling. *Gondwana Research*, **27**, 140–164, http://doi.org/10.1016/j.gr.2014.08.006

MCLACHLAN, I.R. & MCMILLAN, I.K. 1976. Review and stratigraphic significance of Southern Cape Mesozoic paleontology. *Transactions of the Geological Society of South Africa*, **79**, 197–212.

MCMILLAN, I.K., BRINK, G.J., BROAD, D.S. & MAIER, J.J. 1997. Late Mesozoic sedimentary basins of the south coast of South Africa. *In*: SELLY, R.C. (ed.) *African Basins.* Sedimentary Basins of the World, **3**. Elsevier, Amsterdam, 319–376.

MILANI, E.J. & DE WIT, M.J. 2008. Correlations between the classic Paraná and Cape Karoo sequences of South America and southern Africa and their basin infills flanking the Gondwanides: Du Toit revisited. *In*: PANKHURST, R.J., TROUW, R.A.J. ET AL. (eds) *West Gondwana: Pre-Cenozoic Correlations Across the South Atlantic Region.* Geological Society, Special Publications, **294**, 319–342, http://doi.org/10.1144/SP294.17

MOHAMMED, M., PATON, D., COLLIER, R.E.L., HODGSON, N., NEGONGA, M. 2016. Interaction of crustal heteroge-neity and lithospheric processes in determining passive margin architecture on the southern Namibian margin. *In*: SABATO CERALDI, T., HODGKINSON, R.A. & BACKE, G. (eds) *Petroleum Geoscience of the West Africa Margin.* Geological Society, London, Special Publi-cations, **438**. First published online March 16, 2016, http://doi.org/10.1144/SP438.9

MOULIN, M. & ASLANIAN, D. 2010. A new starting point for the South and Equatorial Atlantic Ocean. *Earth Science Reviews*, **98**, 1–37.

MUNTINGH, A. 1993. Geology, prospects in Orange Basin offshore western South Africa. *Oil and Gas Journal*, **91**, 105–109.

PÁNGARO, F. & RAMOS, V.A. 2012. Paleozoic crustal blocks of onshore and offshore central Argentina: new pieces of the southwestern Gondwana collage and their role in the accretion of Patagonia and the evolution of Mesozoic south Atlantic sedimentary basins. *Marine and Petroleum Geology*, **37**, 162–183.

PÁNGARO, F., RAMOS, V.A. & PAZOS, P.J. 2015. The Hesperides basin: a continental-scale upper Palaeozoic to Triassic basin in southern Gondwana. *Basin Research*, http://doi.org/10.1111/bre.12126

PATON, D.A. & UNDERHILL, J.R. 2004. Role of crustal anisotropy in modifying the structural and sedimentological evolution of extensional basins: the Gamtoos Basin, South Africa. *Basin Research*, **16**, 339–359, http://doi.org/10.1111/j.1365-2117.2004.00237.x

PATON, D.A., MACDONALD, D. & UNDERHILL, J.R. 2006. Applicability of thin or thick skinned structural models in a region of multiple inversion episodes: southern South Africa. *Journal of Structural Geology*, **28**, 1933–1947, http://doi.org/10.1016/j.jsg.2006.07.002

PATON, D.A., VAN DER SPUY, D., DI PRIMIO, R. & HORSFIELD, B. 2008. Tectonically induced adjustment of passive-margin accommodation space; influence on the hydrocarbon potential of the Orange Basin, South Africa. *AAPG Bulletin*, **92**, 589–609, http://doi.org/10.1306/12280707023

PÉRON-PINVIDIC, G., MANATSCHAL, G. & OSMUNDSEN, P. 2014. Structural comparison of archetypal Atlantic rifted margins: a review of observations and concepts. *Marine and Petroleum Geology*, **43**, 21–47.

PIQUÉ, A. & LAVILLE, E. 1996. The central Atlantic rifting: reactivation of Palaeozoic structures? *Journal of Geodynamics*, **21**, 235–255.

PITTS, B., MAHER, M.J., DE BEER, J.H. & GOUGH, D.I. 1992. Interpretation of magnetic, gravity and magnetotelluric data across the Cape Fold Belt and Karoo Basin. *In*: DE WIT, M.J. & RANSOME, I.G.D. (eds) *Inversion Tectonics of the Cape Fold Belt, Karoo and Cretaceous Basins of Southern Africa*. Balkema, Rotterdam, 33–45.

RABINOWITZ, P.D. & LABRECQUE, J. 1979. The Mesozoic South Atlantic Ocean and evolution of its continental margins. *Journal of Geophysical Research: Solid Earth*, **84**, 5973–6002.

RAMOS, V. 2008. Patagonia: a Paleozoic continent adrift? *Journal of South American Earth Sciences*, **26**, 235–251.

RAPALINI, A.E. 2005. The accretionary history of southern South America from the latest Proterozoic to the Late Paleozoic: some palaeomagnetic constraints. *In*: VAUGHAN, A.P., LEAT, P.T. & PANKHURST, R.J. (eds) *Terrane Processes at the Margins of Gondwana*. Geological Society, London, Special Publications, **246**, 305–328, http://doi.org/10.1144/GSL.SP.2005.246.01.12

RING, U. 1994. The influence of preexisting structure on the evolution of the Cenozoic Malawi rift (East African rift system). *Tectonics*, **13**, 313–326.

SANDWELL, D.T., MÜLLER, R.D., SMITH, W.H.F., GARCIA, E. & FRANCIS, R. 2014. New global marine gravity model from CryoSat-2 and Jason-1 reveals buried tectonic structure. *Science*, **346**, 65–67, http://doi.org/10.1126/science.1258213

SHONE, R.W. 1978. A case for lateral gradation between the Kirkwood and Sundays River formations, Algoa Basin. *Transactions of the Geological Society of South Africa*, **81**, 319–326.

SHONE, R.W., NOLTE, C.C. & BOOTH, P.W.K. 1990. Pre-Cape rocks of the Gamtoos area – a complex tectonostratigraphic package preserved as a horst block. *South African Journal of Geology*, **93**, 616–621.

SIBUET, J.C. & TUCHOLKE, B.E. 2013. The geodynamic province of transitional lithosphere adjacent to magma-poor continental margins. *In*: MOHRIAK, W.U., DANFORTH, A., POST, P.J., BROWN, D.E., TARI, G.C., NEMČOK, M. & SINHA, S.T. (eds) *Conjugate Divergent Margins*. Geological Society, London, Special Publications, **369**, 429–452, http://doi.org/10.1144/SP369.15

SPIKINGS, A.L., HODGSON, D.M., PATON, D.A. & SPYCHALA, Y.T. 2015. Palinspastic restoration of an exhumed deep-water system: a workflow to improve paleogeographic reconstructions. *Interpretation*, **3**, 71–87.

STANKIEWICZ, J., PARSIEGLA, N., RYBERG, T., GOHL, K., WECKMANN, U., TRUMBULL, R. & WEBER, M. 2009. Crustal structure of the southern margin of the African continent: results from geophysical experiments. *Journal of Geophysical Research: Solid Earth*, **113**, http://doi.org/10.1029/2008JB005612

TANKARD, A., WELSINK, H., AUKES, P., NEWTON, R. & STETTLER, E. 2009. Tectonic evolution of the Cape and Karoo basins of South Africa. *Marine and Petroleum Geology*, **26**, 1379–1412.

TANKARD, A.J., JACKSON, M.P.A., ERIKSSON, K.A., HOBDAY, D.K., HUNTER, D.R. & MINTER, W.E.L. 1982. *Crustal Evolution of Southern Africa*. Springer, New York.

THOMAS, R.J., VON VEH, M.W. & MCCOURT, S. 1993. The tectonic evolution of southern Africa: an overview. *Journal of African Earth Sciences*, **16**, 5–24.

TOMEZZOLI, R.N. & VILAS, J.F. 1999. Paleomagnetic constraints on the age of deformation of the Sierras Australes thrust and fold belt, Argentina. *Geophysical Journal International*, **138**, 857–870.

TOMMASI, A. & VAUCHEZ, A. 2001. Continental rifting parallel to ancient collisional belts: an effect of the mechanical anisotropy of the lithospheric mantle. *Earth and Planetary Science Letters*, **185**, 199–210.

VAUCHEZ, A., BARRUOL, G. & TOMMASI, A. 1997. Why do continents break-up parallel to ancient orogenic belts? *Terra Nova*, **9**, 62–66.

VEEVERS, J.J., COLE, D.I. & COWAN, E.J. 1994. Southern Africa: Karoo Basin and Cape Fold Belt. *In*: VEEVERS, J.J. & POWELL, C.McA. (eds) *Permian–Triassic Pangean Basins and Fold Belts along the Panthalassan Margin of Gondwanaland*. Geological Society of America, Memoirs, **184**, 223–279.

Influence of mechanical stratigraphy on multi-layer gravity collapse structures: insights from the Orange Basin, South Africa

T. J. S. DALTON[1]*, D. A. PATON[1] & D. T. NEEDHAM[2]

[1]*School of Earth and Environment, University of Leeds, Leeds LS2 9JT, UK*

[2]*Needham Geoscience Ltd, 10 Ghyll Wood, Ilkley LS29 9NR, UK*

Corresponding author (e-mail: T.J.Dalton@leeds.ac.uk)

Abstract: Gravity collapse structures are common features on passive margins and typically have a tripartite configuration including an updip extensional domain, a transitional domain and a downdip compressional domain with a common detachment underlying the system. A number of studies have classified these systems, yet few document the wide variations in geometry. This study documents the gravity collapse structures of the Namibian and South African Orange Basin; these structures represent some of the best imaged examples of this important process. We first demonstrate the geometry and kinematic evolution of these systems, focusing on examples of the tripartite configuration from a typical collapse. We then highlight the significant variability in the structures of the system and describe features such as cross-cutting in margin-parallel sections, portions of the system with multiple detachments, systems with stacked synchronous detachments and the temporal evolution of faults within the system. By integrating our observations from a number of sections, we present a model explaining the spatial and temporal evolution of the system. This enables us to discuss likely causes of collapse structures and also, by placing the system into a well-constrained stratigraphic context, how the presence of both maximum flooding surfaces and early margin deltaic sequences have a fundamental control on the resulting collapse geometry.

Deep water fold–thrust belts (DWFTBs) and their associated extensional systems occur in many passive margin systems throughout the world and provide an excellent opportunity to study the formation and development of both extensional and compressional faults. A considerable variation in the structural style of collapse systems is seen across different margins and this is generally accepted to result from differences in both the driving mechanisms for collapse and the geometry and nature of the detachment surface (Rowan *et al.* 2004; Krueger & Gilbert 2009; Morley *et al.* 2011).

Morley *et al.* (2011) classified DWFTBs into two broad categories: those controlled by near-field stress systems created by sediment loading and differential uplift/subsidence at passive margins (Type I) and those controlled by far-field stress regimes associated with active margins (Type II). Type I DWFTBs are further divided into Type Ia (shale detachment), such as those in the Orange Basin, and Type Ib (salt detachment), such as in Angola. Krueger & Gilbert (2009) proposed that DWFTBs should be divided into those found on active margins (caused by subduction) and those found on passive margins in a similar manner to the far- and near-field stress systems. Krueger & Gilbert (2009) then subdivided the passive margin systems into three categories based on the nature of their décollement: regional salt, regional shale and local non-discrete, where the local detachments are discontinuous and lead to a regional décollement crossing stratigraphic levels (Fig. 1).

In this study we focused on shale detachment systems that, regardless of the driving mechanism, commonly consist of three domains (Fig. 1): an updip extensional domain dominated by normal faulting; a downdip compressional domain composed of imbricate thrusts and folds; and a transitional domain. The transitional domain (sometimes referred to as the translational domain) is not referred to by all researchers, but is defined as an area between the extensional and compressional domains that is either a package of largely undeformed sediments (Corredor *et al.* 2005; Krueger & Gilbert 2009) or an area in which both compressional and extensional features overprint (Butler & Paton 2010; de Vera *et al.* 2010). This overprint arises from a shift in the location of the point of contact between the compressional and extensional domains. It is often difficult to resolve the internal geometry of the transitional domain because of limited seismic imaging. The basic premise of area balancing during deformation is expected to apply to these coupled systems; however, the work of de Vera *et al.* (2010) and Butler & Paton (2010) in the Orange Basin established an imbalance between the extension and compression domains of up to 25% in favour of extension, leaving a considerable

From: SABATO CERALDI, T., HODGKINSON, R. A. & BACKE, G. (eds) 2017. *Petroleum Geoscience of the West Africa Margin*. Geological Society, London, Special Publications, **438**, 211–228.
First published online January 19, 2016, updated October 3, 2016, http://doi.org/10.1144/SP438.4

(a)

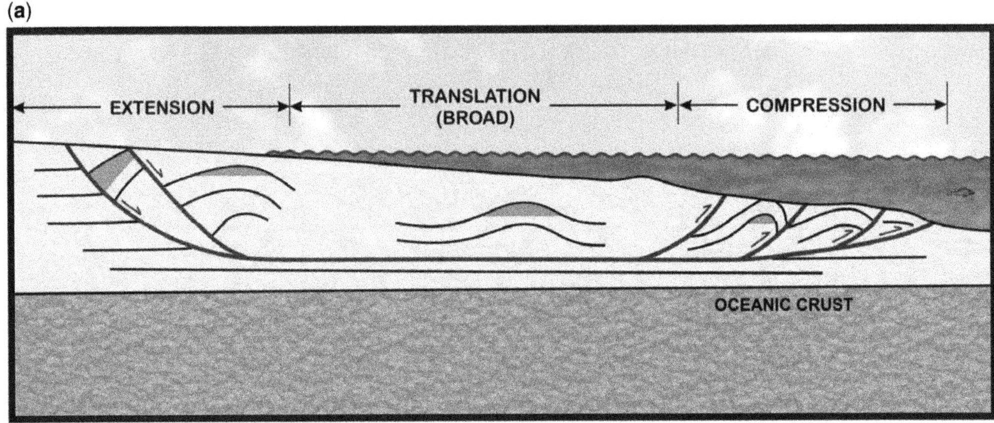

(b)

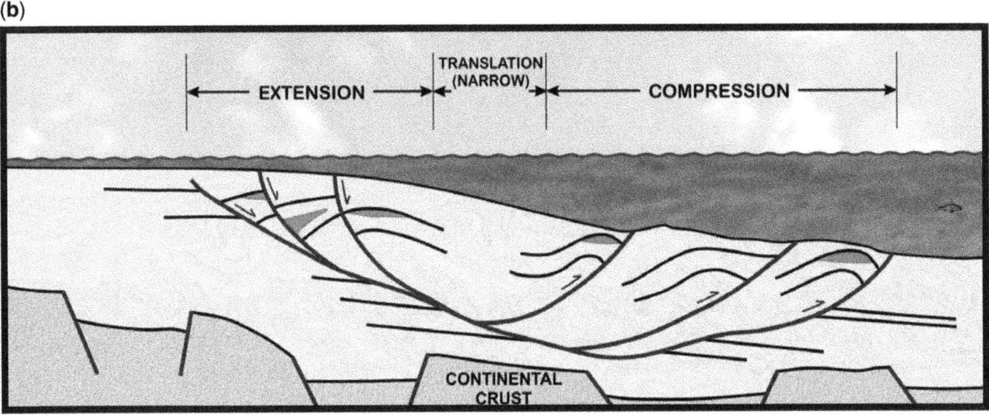

Fig. 1. Model of gravitational collapse (Krueger & Gilbert 2009). (**a**) Typical features and geometry of gravity system controlled by a regional detachment. (**b**) Geometry where no regional décollement is present.

missing component of contractional strain to be explained.

In this study, we looked in detail at the three domains along a typical section from the Orange Basin system and compared them with other portions of the same collapse structure to observe variations along-strike. From these observations we constructed a model to explain the temporal evolution of this important margin process. Finally, we considered how the margin stratigraphy played a critical role in the nature of the deformation and propose that this had a significant and, until now, unrecognized control on this process, which occurs on many passive margins.

Regional setting

The Orange Basin is the southernmost basin on the West African passive margin. It formed during the break-up of Gondwana and subsequent spreading of the South Atlantic Ocean (Muntingh & Brown 1993; Brown *et al.* 1995; Paton *et al.* 2008; Koopmann *et al.* 2014).

It underwent significant rifting during the Late Jurassic to Early Cretaceous, forming graben and half-graben infilled with synrift siliciclastic and lacustrine sediments (Jungslager 1999; Mohammed *et al.* 2015, this volume, in press). This was followed by continental break-up in the Barremian and the establishment of a passive continental margin, onto which a thick post-rift sedimentary sequence was deposited (Gerrard & Smith 1982). The thickness of the post-rift sediments ranges from 3 km in the south and north to up to 5.6 km in the centre of the basin. This sediment was largely sourced from the Orange River (Paton *et al.* 2008) and is broadly separated into two phases: black shales and claystones were deposited during an early drift phase and then a later drift phase deposited a thick succession of interbedded heterolithic sediments composed of shales and claystones

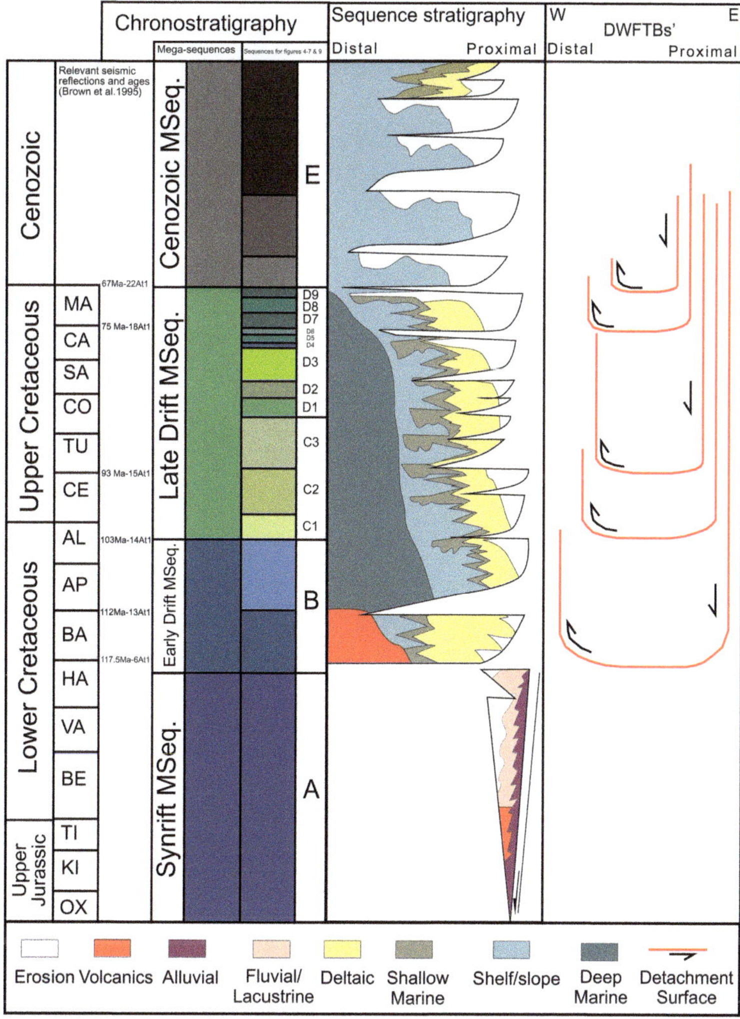

Fig. 2. Chronostratigraphy of Orange Basin adapted from Paton *et al.* (2008) showing the depths to which the deep water fold–thrust belts penetrate across the entire basin.

(Fig. 2). It is within the latter phase that we observed the greatest number of gravity collapse structures. Although much of the margin stratigraphy is claystone, we defined the system as dominated by a shale detachment because the décollement surfaces are shale intervals. These correspond to maximum flooding surfaces, with some identified as proved source rocks (van der Spuy *et al.* 2003). This Cretaceous succession underwent considerable tilting, up to 750 m in the inner margin (Paton *et al.* 2008), at the end of the Maastrichtian to produce a considerable proximal unconformity with the overlying Cenozoic sequence. Most of the Cenozoic sequence was deposited outboard into the basin and varies in thickness from 250 to 450 m on the

margin and from 500 to 1400 m on the continental slope and beyond. The hydrocarbon system sequence stratigraphy and deeper structures of the Orange Basin are well established (Light *et al.* 1993; Muntingh & Brown 1993; Paton *et al.* 2007, 2008; Hirsch *et al.* 2010).

Data and methods

This study combined 38 480 km of vintage two-dimensional pre-stack time migration seismic data released by the Petroleum Agency of South Africa with 45 386 km of two-dimensional seismic data from Spectrum (Fig. 3), of which 24 042 km was

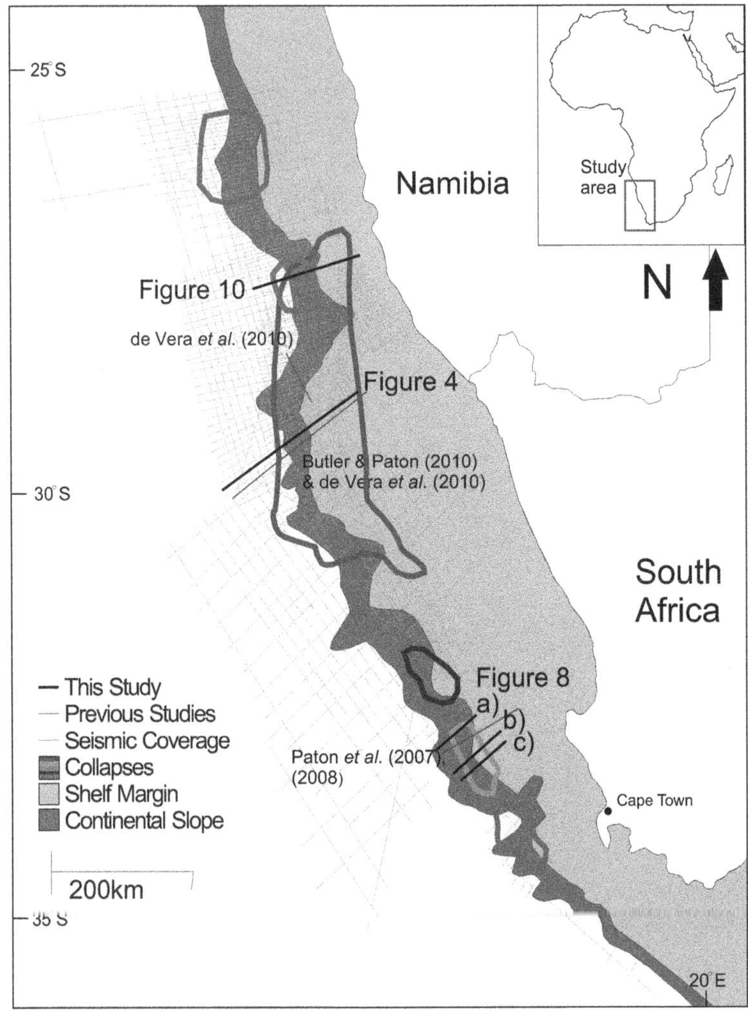

Fig. 3. Map of Orange Basin indicating the location of lines used in this study, the location of lines used in previous studies (Paton *et al.* 2007, 2008; Butler & Paton 2010; de Vera *et al.* 2010) and an outline representing the total data coverage used in this study.

pre-stack depth migration (PSDM) data obtained using fast-track Kirchoff migration. The maximum recording lengths were in the range 7–10 s two-way travel time; the vintage data were acquired between 1976 and 2012 and the Spectrum data were acquired in 2012. The line spacing of this deep seismic coverage was between 8 and 15 km and it provided unprecedented data coverage of the basin and allowed us to review the entire margin (Paton *et al.*, this volume, in press). We present here our interpretations of the Spectrum PSDM seismic lines because these provided better images. Well data for the basin are extensive, although limited to the shelf margin, and have been used in previous

studies (e.g. Muntingh & Brown 1993; Paton *et al.* 2008) to define the margin stratigraphy and to define stratigraphic ages for our seismic intervals (Fig. 2). Stratigraphic megasequences and associated regional and local unconformities have been identified using reflection termination, cut-offs and onlap relationships with the sequence stratigraphic system based on the work of Muntingh & Brown (1993).

As we focused on the detailed architecture of the syn-kinematic packages, we subdivided the megasequences into sequences based on variations in the seismic character that indicated changes over time in the depositional or structural environment and rates of deposition. We used a unified stratigraphic system to define these sequences across the margin (Fig. 2). We defined the packages regionally across the section as megasequences A–E and further subdivided megasequences C and D numerically in a temporal succession. However, because the packages were often isolated, any one section may not show a complete sequence.

Regional sections

We present here a 160 km long east–west-oriented regional seismic profile (Fig. 4) that illustrates both the main structural elements of the margin and one of the simpler collapse structures in the centre of the basin (Fig. 3). This is shown down to a depth of 6.4 km. Based on the previous regional interpretations of Paton *et al.* (2008), we divided the stratigraphy into four megasequences: Synrift, Late Jurassic to Hauterivian (A); Early Drift, Barremian to Aptian (B); Late Drift, Aptian to Maastrichtian (C and D); and Cenozoic (E). To aid the interpretation across the margin, including the structural features and local unconformities, we divided the megasequences into seismic sequences based on the character of internal seismic reflections. Although the synrift packages were imaged in the dataset (Paton *et al.*, this volume, in press), we focused on the sequences stratigraphically above the top-synrift reflection. The top of the synrift package was delineated by a package of high-amplitude parallel reflections, present across the basin, onto which Barremian-age stratigraphy was deposited. In this section, the nature of the contact was represented by an aggradational sequence of reflections conformable with the top of the synrift package. It is important to note that elsewhere in the basin this boundary has a progradational relationship marked by downlapping reflections onto the synrift sediments prior to an aggradational phase. The Late Drift megasequence is deposited conformably on the Early Drift package and is defined by its higher reflectivity.

Evident within this Late Drift megasequence are numerous unconformities represented by the truncation and onlapping of reflections. These unconformities only occur off the palaeoslope margin and are often restricted to fault blocks; they are therefore not regional in extent. Reflections within the centre of this package are both folded and faulted. In the proximal portion of the basin, westwards-dipping normal faults are identified by dislocated packages shifting downdip of one another (eastern end of Fig. 4); this is the extensional portion of the gravity collapse structure. Continuing westwards and downdip, the seismic character becomes increasingly chaotic and complex and we define this as the transition domain.

The most distal part of the system, the compressional domain, is characterized by a series of east-dipping thrust faults identifiable by high-amplitude, steeply dipping reflections that appear to be stacking packages on top of one another.

These structural features will be discussed in more detail later; however, it is important to note at this point that although the main décollement for this collapse (red reflection in Fig. 4) is broadly coincident with the top of the Early Drift megasequence, it does not have a constant slope and shows significant changes in the direction and angle of dip. The décollements are picked based on where the faults terminate, as identified through cut-offs and changes in the dip of the reflections. The Late Drift megasequence is capped by a regional truncation at the top of the palaeoshelf picked out by the top of the uppermost green package (Fig. 4), whereas these sequences are conformable at the base of the palaeoslope. The unconformity is, however, still interpretable by a change from a low- to a high-amplitude reflection along this boundary. The Cenozoic package is defined by a change in the location of deposition from proximal to a more distal position on the continental slope. The package is considerably thinner than the Cretaceous sequence on the central margin and thickens significantly to the west. Each of the structural elements is considered in more detail in the following sections.

Extensional domain

To define the structures in more detail, we subdivided the megasequence into a number of sequences (A–E, Fig. 2). Figure 5 shows a typical interpreted section from the upper portions of the extensional system, where the latest faults formed prior to detaching onto a regional décollement. This interpretation focuses on the upper part of the extensional portion of the structure and shows a more detailed breakdown of the Late Drift megasequence (sequences C and D). These packages are defined based on the internal seismic facies and reflection termination and reveal relative changes in sediment supply and fault-controlled accommodation space (Brown *et al.* 1995).

Internally, sequence C3 has an absence of seismic impedance contrast, resulting in only limited internal geometry being imaged, but it is conformable with the high-amplitude reflections at the base of sequence D1 and is truncated by sequence D3. This suggests that D1 was being deposited as the upper portions of C3 were being eroded and that the discontinuity was a direct result of faulting. D3 has a thick package of high-amplitude reflections that allow several horizons to be tracked internally. When restored, sequence D3 forms a westwards-thickening wedge. Changes in spacing between

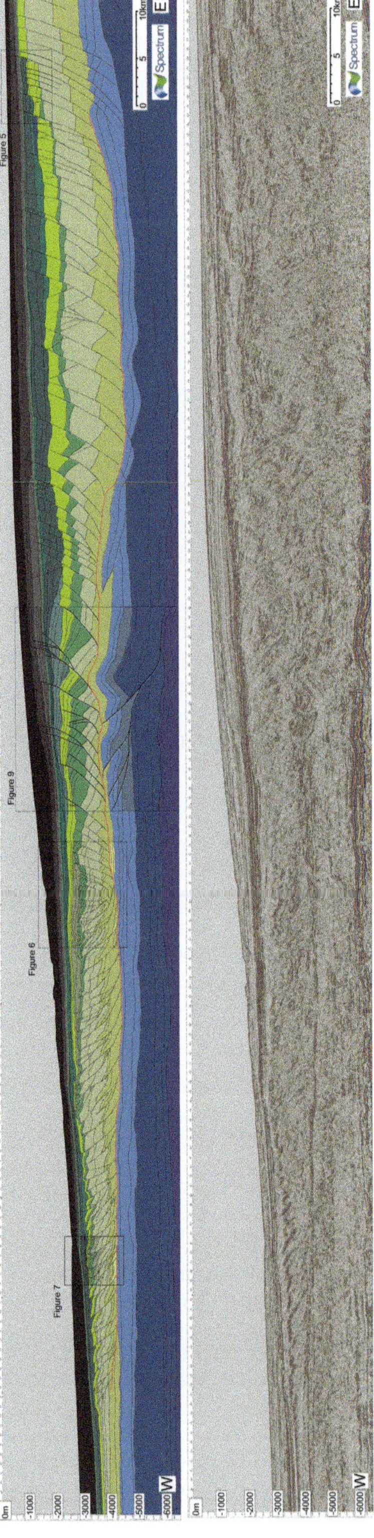

Fig. 4. Pre-stack depth migrated uninterpreted and interpreted sections of Line 1 (see Fig. 3 for location) shown with a vertical exaggeration of 3:1. The colours correspond to each megasequence (Fig. 2). The Synrift megasequence is purple, the Early Drift megasequence is blue, the Late Drift megasequence is green and the Cenozoic megasequence is grey; different shades correspond to discrete packages within each megasequence. The detachments are picked out in red.

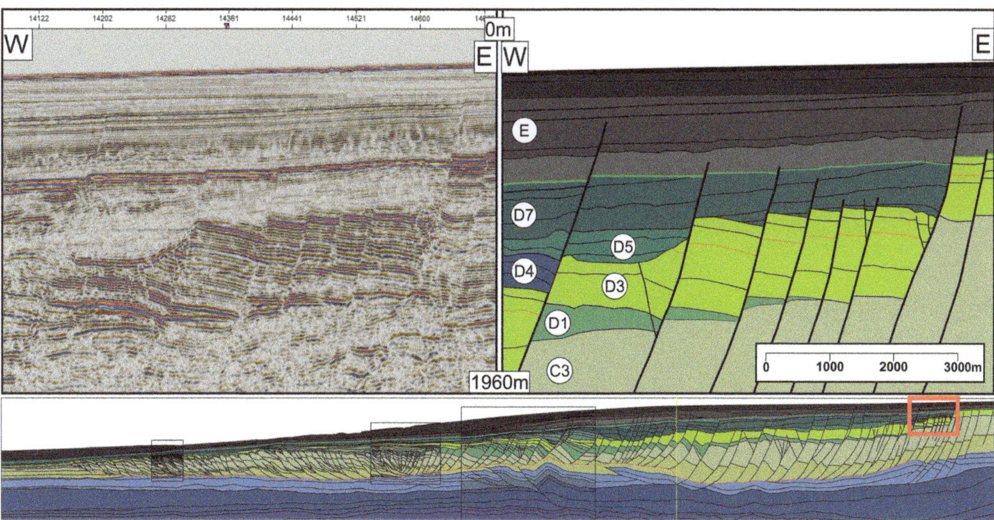

Fig. 5. Detailed interpreted and uninterpreted sections of the extensional domain from Figure 4. Section is vertically exaggerated 3:1.

traceable horizons in D3 show slight changes in thickness along its length, indicating that different faults were active at different points. Of particular note are the small changes in thickness in the packages above and below the orange horizon (Fig. 5), which indicate the movement of small faults in the package between two faults with far larger throws. As the largest thickness changes occur on the faults on which D4 and D7 truncate, this implies that the deformation tends to concentrate onto a few larger, more widely spaced faults – that is, large faults become large and stay large, thus stopping smaller faults from growing. Sequence D4, defined by a package of low-amplitude reflections, reinforces this point. Its presence in only the west of the section abutting a large fault plane implies that it grew more rapidly at this point than faults to the east. This created a larger accommodation space that was rapidly infilled, as indicated by folding of the reflections into the fault. D3 is clearly truncated by the base of sequence D5 with a rugose contact that appears to represent the collapse of the top of the fault block. D5 has chaotic and poorly imaged reflectance that infills the eroded section truncated at the top of D4. As shown in Figure 4, D5 extends for 16 km west of Figure 5 and continues to erode earlier fault blocks. Its chaotic seismic character and erosive base suggest that it is a mass transport complex (MTC). Several similar MTCs can be seen throughout the extensional portion of the collapse features (e.g. Posamentier & Kolla 2003; McGilvery & Cook 2004).

Sequence D7 is defined by a series of reflections that onlap onto the top of D4 and D5 and are clearly imaged on the tops of the fault blocks. This implies that fault movement outstripped sediment supply at this point. It also appears that several of the faults had switched off by this time. As the sediment supply increased and the faults switched off, the sediments began to bury the fault blocks and deformation was again concentrated into the larger faults, where we observed some limited sediment growth into the fault plane. The truncation of D7 by megasequence E – that is, the boundary between the Late Drift and Cenozoic megasequences – can be seen throughout the palaeo-continental margin and the upper palaeoslope. It is unclear how much sediment has been eroded, although Paton *et al.* (2008) suggested it may have been as much as 750 m. Only the two largest faults were active after this unconformity formed, although they offset it with only small throws.

Transitional domain

The section in Figure 6 shows the point at which the extensional domain changed into the compressional domain and indicates that this occurred predominantly within megasequences C and D.

Megasequence B contains the high-amplitude parallel reflections that denote the top of the Early Drift megasequence, although these reflections become less distinct immediately beneath the most deformed section of the transition zone, probably due to signal attenuation. These reflections are composed of coarsening upwards silt–medium sandstone packages (Paton *et al.* 2007) that represent the progradation of the Aptian deltaic margin and

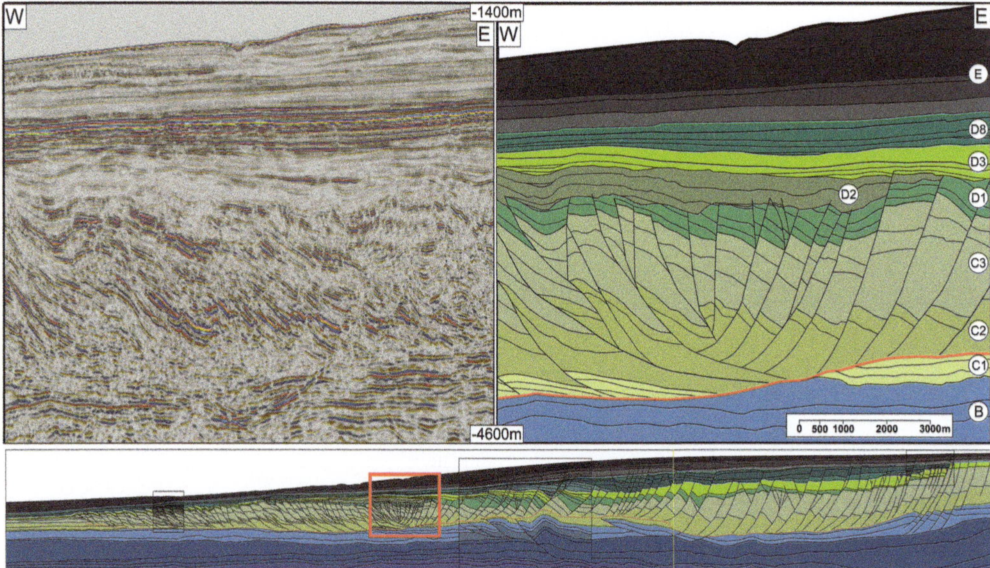

Fig. 6. Detailed interpreted and uninterpreted sections of the transitional domain from Figure 4. Section is vertically exaggerated 3:1.

are capped by a maximum flooding surface that forms the detachment. Sequence C1 is defined as a set of lower amplitude parallel reflections that have a variable relationship across the section with the surrounding sequences. In the east of the section they are conformable with megasequence B; the reflections become truncated towards the west by the base of sequence C2. They reappear in the west as a set of reflections conformable with C2 and downlap onto megasequence B. We interpret this as a shift in the depth of the main décollement that is immediately above B in the east and cuts down to an inter-megasequence B layer with the consequence of translating the C1 package downslope by *c.* 2300 m towards the west.

C2 is defined by low-amplitude, largely discontinuous reflections and is conformable with C3, which consists of higher amplitude, more continuous reflections. The lowest reflections in C2 to the east and centre downlap onto B and C1 and are directly above the detachment at this point; they are conformable with C1 in the west as the detachment cuts down-sequence, as described earlier. The division between the C2 and C3 intervals is identifiable by an easily correlated, high-amplitude reflection package. This allows us to define fault cut-offs with confidence. In the east, these faults dip steeply landwards with normal offsets and detach onto the main basal décollement. Progressing west they become more closely spaced and detach onto a shallow basinwards-dipping thrust fault

located above the regional décollement. A shift from extensional to compressional tectonics occurs on top of this thrust fault. The low amplitudes at the base of C2 are probably a result of the coalescing of multiple faults at this level, causing increased stress at this depth. The thickness of C2/C3 is largely maintained throughout the margin, including the area of intense faulting, which suggests that it is largely a pre-kinematic sequence deposited prior to collapse, although some reflections in the top of C3 show limited thickening into fault planes, suggesting some degree of syn-kinesis.

Sequence D1 is defined here by a set of low-amplitude reflections that are largely conformable with C3; as in the previous sequence, cut-offs are used to define the location of faults. Many packages have wedge-like geometries that thicken into fault planes, implying fault growth during deposition and making this a syn-kinematic succession. From the position of cut-offs in the region in which normal faulting gives way to thrusting, it can be seen that several faults stopped moving. The upper boundary is truncated by sequence D2. This sequence is defined upwards by a series of low-amplitude continuous reflections that onlap onto the erosional truncation defining the top of D1. They form a tapering wedge to the east, where the formation onlaps onto significant faults with throws of *c.* 120 and *c.* 250 m. There appear to be numerous minor truncations of horizons against one another within the formation, possibly due to limited deposition in what are effectively minibasins. Some faults do persist into the base of D2, but most are truncated by it, implying an erosional episode followed by progressive infill during which limited reactivation occurred, causing minor folding as opposed to faulting in the overlying sequence. At this point of the section, sequence D3 is defined by low-amplitude continuous reflections that downlap onto D2, infilling its uneven upper surface, before latterly adopting a more aggradational geometry. Minor folding of some reflections at the western end of the section imply limited localized reactivation on some thrusts. Sequence D8 is composed of high-amplitude reflections conformable with D3. The contact between these two horizons can be traced into the compressional domain downdip. The top of D8 appears conformable with megasequence E (Cenozoic).

The transitional domain in the centre of Figure 6 generally picks out a large fold structure detaching onto a thrust above the regional décollement. The back-limb of the fold is cut by normal faults, which progressively become thrust faults towards the crest, some of which are likely to be inverted normal faults. The precise contact between the compressional and extensional domains (e.g. the transitional zone) is narrow, similar to the modes proposed by Corredor *et al.* (2005) and Krueger & Gilbert (2009); however, the possible inversion may imply a more complex structural style, as suggested by de Vera *et al.* (2010) and Butler & Paton (2010).

Compressional domain

Figure 7 is a typical section from the distal end of the compressional domain and we have divided it into nine packages. Megasequence B, which is correlated from the transitional domain, is defined by several near-horizontal reflections that show a consistent increase in amplitude from east to west, probably reflecting a progressive change in its petrophysical properties. Sequences C1–C3, despite being of varying amplitudes, have the same geometry of stacked, steeply east-dipping reflections (*c.* 35° using PSDM data) that shallow and flatten with depth to become parallel with the top reflection of megasequence B. Definable packages of reflections stack with discrete cut-offs that pick out a set

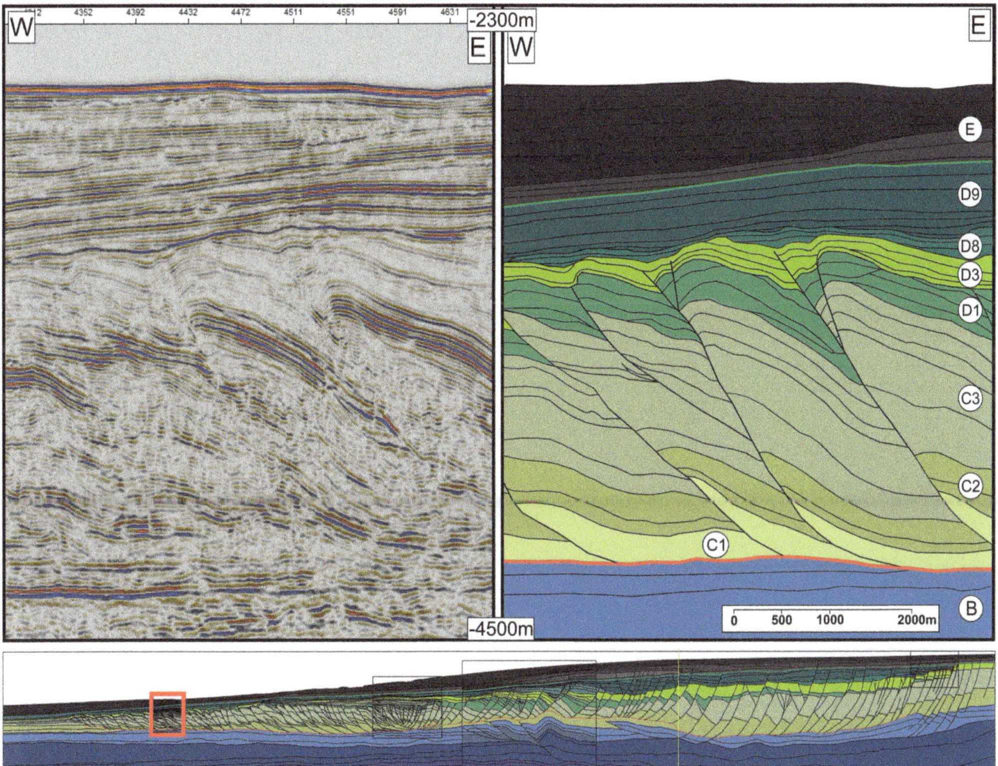

Fig. 7. Detailed interpreted and uninterpreted sections of the compressional domain from Figure 4, image is vertically exaggerated 3:1.

of imbricate thrusts. Sequence D1 is defined by a set of discontinuous low-amplitude reflections that onlap the thrust planes and downlap onto C3. The reflections are folded and have been truncated by both sequence D3 and by one another. The variation in thickness and the associated onlap onto anticlines of D1 suggest that thrusting was active during the deposition of D3 and was frequently emergent, leading to folding and erosion of the depositing sediments in a syn-kinematic fashion. This onlap implies that the deformation rates were greater than the sedimentation rates during this interval. D3 truncates D1 with a set of low-amplitude, but continuous, reflections. These reflections are folded above the underlying thrust planes, but are only cut by two of the thrusts with far smaller throws. Although it is clear that the faults remained active during this period, the rate of deformation relative to sedimentation had slowed significantly.

Sequence D8 truncates the crests of the folded reflections in D3 and onlaps in the synclines formed by the dipping reflections on the back-limb of the thrust faults. This suggests the end of deformation in this part of the compressional domain, with sediment infilling the remnant topography, although the sediment supply is insufficient to entirely fill the bathymetric lows. The last of these lows were filled by small onlapping packages at the base of sequence D9, which is otherwise conformable with D8. As with the transitional domain, the contact between the Late Drift (D) and the Cenozoic megasequences (E) is conformable. In a broader context, when viewing the compressional domain in Figure 4, the imbricates have a relatively equal spacing and become progressively less deformed away from the transition zone, while also deforming ever-younger sequences. The dips of the faults shallow from 40–50° at the transitional domain to 15–25° at the frontal thrusts. They progressively deform younger sequences, implying that once the dip of the thrusts becomes too high, it is preferential to deform more distal sediments.

Variations in DWFTB geometry

Having summarized the structural elements that comprise a typical section for gravity collapse, we now outline how the styles of deformation deviate from this typical section by looking at variations along the margin, as illustrated by a number of additional sections.

Lateral variation

The three sections in Figure 8 are modified from Dalton *et al.* (2015) and show three slip-parallel 35 km long sections running north–south (see Fig. 3 for locations) through a DWFTB in the southern portion of the Orange Basin. Growth strata indicate that collapse initiated during the deposition of the Cenozoic megasequence, which detaches onto a maximum flooding surface at the top of the Campanian in the Late Drift megasequence (Paton *et al.* 2008). Section (a) consists of an extensional domain with no corresponding compressional domain. Section (b) has a more classical geometry with both extensional and compressional domains detached onto the Campanian décollement; however, an additional set of thrusts detach onto the contact between the Cenozoic and Late Drift megasequences. Section (c) indicates that this upper detachment is far more developed with a separate set of normal faults detaching onto the base of the Cenozoic megasequence. The geometries of reflections in the extensional domain indicate slip occurring along both detachments synchronously, suggesting that gravitationally driven strain is distributed between both systems. The Campanian detachment has larger throws, suggesting that it has taken more of the strain, although the folding of the Cenozoic reflections in the far west appears to restrict its westerly development. The imbalance between the two detachments in section (c) suggests that the Cenozoic detachment is more efficient; however, as the system grows northwards the Campanian detachment becomes more important. This may relate to local variations in the slip potential of the detachment surfaces, such as changes in thickness, overpressure or lithology.

Multiple detachments

The presence of multiple detachment horizons in gravity collapse systems has been recognized previously (e.g. Totterdell & Krassay 2003; Rowan *et al.* 2004; Corredor *et al.* 2005; Briggs *et al.* 2006), but few studies have documented how the position and interaction of different slip horizons creates a range of complex geometries indicating changes in the timing and location of deformation.

Sub-Aptian failure. The focus of previous studies of the Orange Basin collapse structures (Butler & Paton 2010; de Vera *et al.* 2010) has been on the system contained within the Late Drift megasequence. Figure 9 presents a more detailed interpretation of a portion of the extensional domain of the collapse (see Fig. 4 for location). We see here that the main detachment in the Late Drift megasequence is underlain by a set of thrusts detaching onto the top of the Synrift megasequence. This lower detachment is formed along a maximum flooding surface between the top of the Synrift megasequence (Hauterivian) and the base of the Early Drift megasequence (Barremian) identified by Brown *et al.*

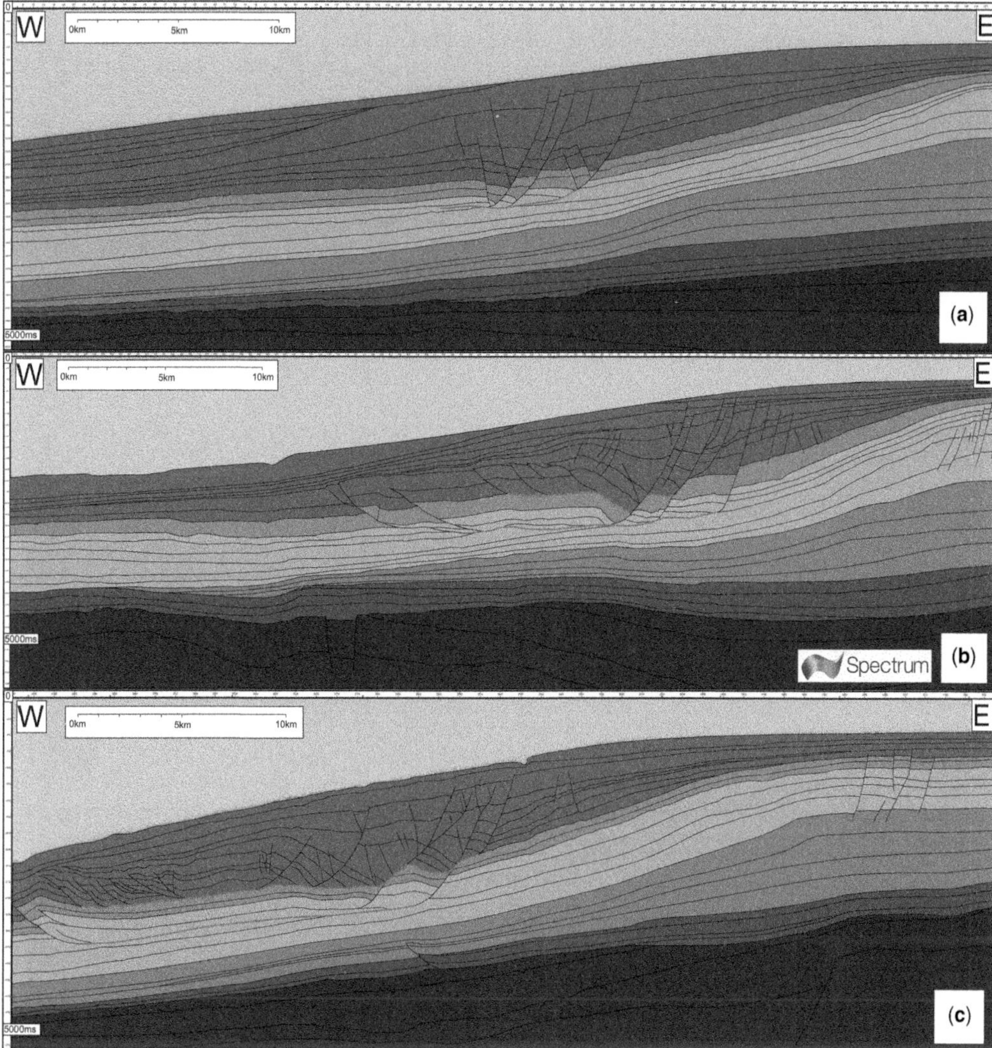

Fig. 8. Three interpreted sections (**a–c**) from the south of the Orange Basin (Fig. 3) adapted from Dalton *et al.* (2015). All sections are 35 km long and are presented as pre-stack time-migrated with vertical exaggerations of 3:1.

(1995). Folding of the upper detachment and overlying horizons by the developing underlying thrusts imply that they formed later. In addition, thickening in sequence D6 into the fault immediately above the fold suggests that its inception led to reactivation of this fault. The lack of significant thickening of sequence D7 suggests that thrusting had largely ceased by the time of its deposition. This suggests that although the upper system was initiated first, both systems existed coevally. The vergence of these lower thrusts is consistent with the same basinwards translation as the upper system. The most proximal normal faults present in the east of the section clearly penetrate into the Early Drift megasequence and are likely to link directly to these thrusts, although the seismic resolution prevents clear confirmation. D6 is not present above these faults (Fig. 5) and may have either been eroded out or not been deposited; however, the infilling of the subsequent D7 into the fault planes suggests that these faults were most active prior to its deposition.

Stacked detachments. Sections through the far north of the largest collapse structure provide further insights into the multi-layer detachment systems (Figs 3 & 10). In Figure 10, an 83 km long section shows several different detachment surfaces at a

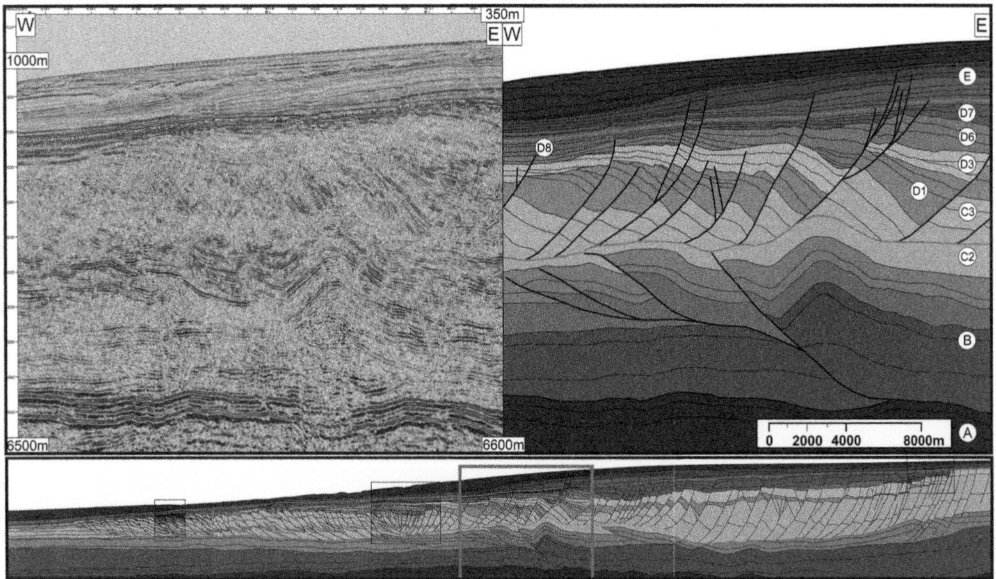

Fig. 9. Detailed interpreted and uninterpreted section from Figure 4 showing folding of the upper detachment by and lower detachment system. Section is vertically exaggerated 3:1.

number of stratigraphic intervals, picked by the identification of mutual fault terminations. In the east of the section, a 30 km long package of normal faults extends up to 2.5 km from a detachment layer within the Early Drift megasequence to the Cenozoic horizon. A second smaller extensional domain, 12 km long with faults extending vertically 600 m up from the detachment, is seen downdip and is contained entirely within the Early Drift megasequence, representative of an early phase of collapse. Two compressional domains also exist: a lower detachment in the Early Drift megasequence, along the Aptian–Barremian maximum flooding surface (Muntingh & Brown 1993), contains thrusts penetrating up to 1.1 km into the overlying Late Drift megasequence, whereas an upper 45 km long detachment, along the Cenomanian–Turonian maximum flooding surface (Paton *et al.* 2007), consists of widely spaced thrusts extending 700 m up from the detachment and is entirely within the Late Drift megasequence. Reflections in the lower compressional domain demonstrate two periods of activity; one period is synchronous with the lower extensional system, with which it shares a detachment and a later phase of reactivation leading to thrusting and folding of the Late Drift megasequence. The upper detachment may have been active synchronously with the lower compressional domain, but remained active for longer, as indicated by thrusting and folding of the uppermost Late Cretaceous package. It is interesting to note that the upper system

terminates at the location at which the first thrust of the lower system emerges. Altering the slope angle of the upper detachment at this point may have made further slip along it non-viable. The upper extensional system remained active throughout the Late Drift megasequence and clearly transferred considerable strain downdip. However, the upper compressional domain remained active during this period, although this domain does not appear to be genetically linked at this point, so the process of transmission of strain between the upper and lower compressive domains is not clear. No genetic link emerged in reviewing parallel sections; in fact, the lower system disappeared relatively rapidly. The transition from extensional to compressional domains along the upper detachment in this section was of a very different character to that seen in Figure 6 and appeared as a zone of largely deformed sediments, as described by Corredor *et al.* (2005) and Krueger & Gilbert (2009).

Discussion

The presence of gravity collapse structures has been documented on many margins and some of the inherent variability has been discussed in detail elsewhere (Morley & Guerin 1996; Rowan *et al.* 2004; Krueger & Gilbert 2009; Morley *et al.* 2011). Studies that have focused on systems driven by thin shale detachments generally propose that

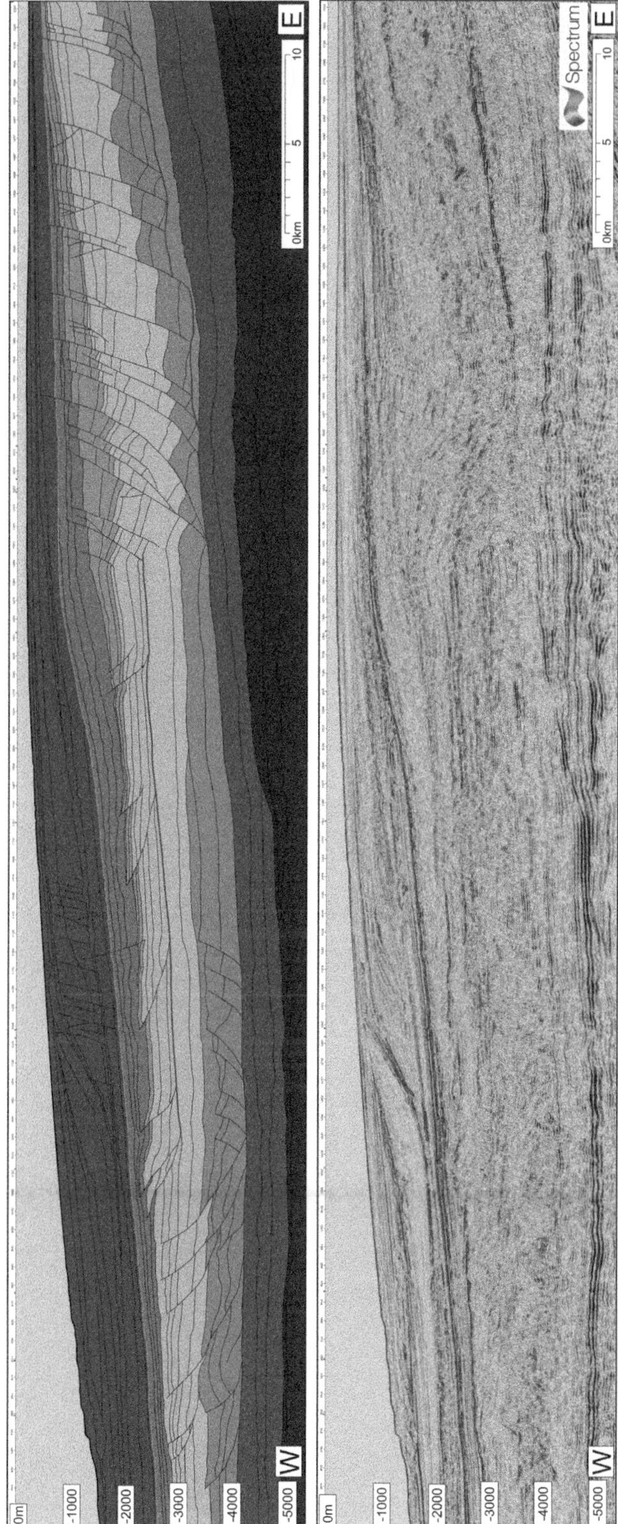

Fig. 10. Interpreted section taken from northern portion of the same collapse structure as featured in Figure 4. Section is vertically exaggerated 3:1

they are relatively coherent bodies presenting little variation within a single system. We discuss here how the observed lateral variability in geometries observed in this study has influenced our understanding of thin shale detachment systems. We also consider the greater complexity observed in these features to synthesize a new temporal model of collapse development in the Orange Basin.

Model for the temporal evolution of a collapse structure

Variations in the style and character of deformation appear consistently across the width of the Orange Basin, including the spacing between thrusts, the depth and location of slip detachment surfaces, and the nature of the transition zone. Although there is also considerable variation in the thickness of the Upper Cretaceous sediments across the basin, the same regional detachment is present throughout. This means that the changes in the styles of deformation observed are present within a single DWFTB so that any single end-member model is not applicable. Dalton *et al.* (2015) have demonstrated that the extensional domain was initiated prior to the formation of a later compressional phase. In this study, we have shown through growth packages that the earliest phase of collapse was located in the centre around the transition zone. For example, D8 in Figure 6 is a post-kinematic horizon, but in Figure 7 it is clearly a syn-kinematic package and is entirely eroded out to the west, where the overlying D9 package, here post-kinematic, becomes syn-kinematic, showing that later phases of movement occur progressively more distally than earlier phases. Few sequences can be tracked throughout the entire structure as they are either truncated by later sequences or are only locally present. However, analyses on the megasequence scale and of larger traceable sequences reinforce this finding. New faults formed and grew at the outer extents of the collapse, although older faults were still active with a reduction in offset. Successive younger faults formed out from the transition zone, downdip to the west in the case of the compressional domain and updip to the east in the case of the extensional domain. The high fidelity of the seismic imaging of our data showed that the transition domain represents a short-wavelength change from extensional to compressional tectonics as opposed to being a zone of overprinted regimes and, more importantly, that it appears to remain fixed. In general, the position of maximum strain migrates away from the transition domain, although we do observe fault reactivation occurring (Fig. 9).

It is similarly clear that if we can relate later, more proximal, movements to ever more distal thrusts, then this would reinforce the concept that these regimes preserve the original contact between them as a block of material that ceases to deform, allowing the translation of strain downslope. Observations of the underlying thrust systems, and the timing of structures above and beneath in Figure 9, indicate a synchronous relationship between the systems – that is, the overlying detachment was folded by the underlying system, which remained active throughout. This suggests that they are both part of a single system as opposed to being two stacked systems of different ages.

Although many studies make reference to multiple detachment horizons (Rowan *et al.* 2004; Krueger & Gilbert 2006; Morley *et al.* 2011; Peel 2014), their presence is generally not included in models of gravity collapse systems. Growth strata indicate that these alternative detachments are often not merely spatially and temporally separate collapse events, but are linked integral portions of the same system. They thus have an important role in terms of strain distribution. They preferentially appear on more mature systems and link to the youngest, most proximal, normal faults. This implies that they form after a point at which continued deformation along the extant distal compressional regime is no longer as efficient as linking to a lower detachment. Sequence-scale observations show that these structures take a long time to form and experience multiple reactivations, which control deposition and erosion along the margin. With this in mind, we have produced a model for the formation and growth of these systems in thin shale detachment systems (Fig. 11).

Our model assumes that continued lateral compaction and deformation of the sediments above and ahead of the original detachment reaches a point at which it is no longer the most efficient way of accommodating the gravitationally induced stress. Assuming that the underlying sediments are comparatively under-compacted and in the presence of an appropriate alternative slip horizon, strain is now accommodated along a lower décollement. However, it is not clear how the strain is transferred from normal faults connected to a lower system. The strain recorded in the upper compressional domain is shown in Figure 10, where both the upper compressional regime and the most proximal normal faults deform age-equivalent sediments and thus must be linked. The extensional domain in Figure 8c shows two slip surfaces that have been exploited by the same faults at different times and it is possible that the same relationships exist in the more mature system in Figure 10, although continued deformation has made this relationships difficult to ascertain.

Brown *et al.* (1995) indicated that our detachment horizons are maximum flooding surfaces,

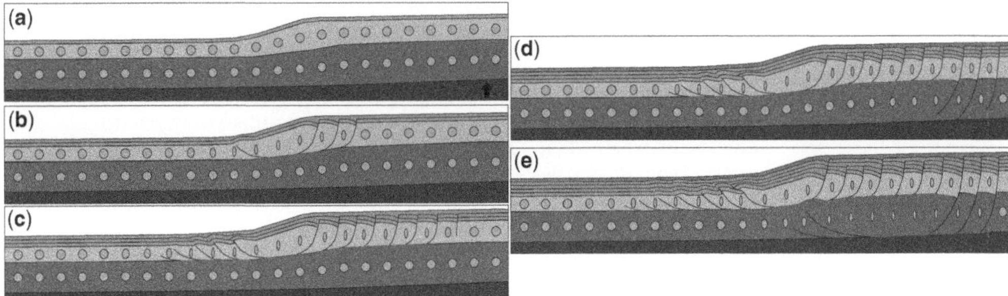

Fig. 11. Multistage model of gravity collapse culminating in the formation of a lower detachment. Orange ellipses represent the distribution of strain within the DWFTB and reflect the findings of Dalton *et al.* (2015) that considerable compaction of the margin is required prior to the formation of the compressional domain. (**a**) Undeformed margin; (**b**) initiation of failure and formation of primary thrust; (**c**) new normal faults form proximal to basin with thrust faults forming in response increasingly more distal; (**d**) upper system becomes increasingly more difficult to slide along, normal faults penetrate down to alternative lower slip surface and compact lower sediments; (**e**) once compacted sufficiently, a lower compressional domain develops.

presumably composed of low basal friction shales, which, as long as they are sufficiently thick and continuous, will continue to allow slip (Rowan *et al.* 2004). If the shale thins or is absent from a section, then the system will lock up. This locking up of the system while the overburden builds up sufficiently to lead to the re-initiation of failure by overcoming frictional cohesion results in the development of isolated sediment imbalances at the head of fault scarps (de Vera *et al.* 2010). This, in turn, leads to the formation of MTCs, which rework the sediments of the upper portion of the extensional domain. This explains why we tend to see the large-scale development of MTCs only on mature systems prone to more lock ups. They become more prevalent stepping back towards the coast, where fewer shale intervals were deposited, and provide potential slip surfaces on what were palaeo-continental margins.

The initial geometries are controlled by the original local accumulations of sediments. The most amenable slip horizons – for example, the shale with the lowest frictional cohesion – will be used in preference to other slip horizons. This cohesion may, however, vary across the basin depending on the original depositional conditions and better slip horizons may be used elsewhere (Dalton *et al.* 2015). The collapse systems in this study were commonly associated with maximum flooding surfaces or the base of slope systems.

Stratigraphic controls on margin collapse

Although the majority of the passive margin stratigraphy on the Orange Basin is claystone, our observations imply that there is a strong control on the location and evolution of the collapse structures from variations in stratigraphy. The principle slip surfaces have been well documented as being relatively thin (*c.* 100 m), organic-rich shale horizons (e.g. Muntingh & Brown 1993; Paton *et al.* 2008) that act as low friction surfaces. This depositionally controlled variation in the basin can be related to the two end-member model for gravity collapse structures on shale detachments (Krueger & Gilbert 2009). One end-member suggests slip along a single detachment horizon, whereas in the second end-member the detachment switches between local over-pressured shale horizons as variations in depositional occurrence and slip potential allow. In the Orange Basin, examples of both end-members are observed with the upper compressional domain in Figure 10. There is clearly slipping along a single regional plane, while the easterly extensional domain has a highly undulating character suggestive of smaller localized slip horizons.

The model presented by Morley *et al.* (2011) characterizes the collapse systems within the Niger Delta and Orange Basin as being of equivalent types (Type 1a). Both are detached on shale and, although there is much discussion as to the existence of shale diapirism, there appears to be distinct differences in the style of deformation between the two basins. The implications of a thick shale interval versus a thin horizon, as commented on by Rowan *et al.* (2004), alters the nature of the failure. Critical wedge concepts (Bilotti & Shaw 2005; Briggs *et al.* 2006) assume the oceanwards propagation of the system. As long as there is low basal friction, then the system will continue to propagate. If there is a thick detachment layer, then it will localize all of the deformation on to the basal system. For example, where the Akata shale in the Niger delta is thick, it internally deforms and the whole overburden can behave as a mechanically strong unit

(Corredor *et al.* 2005). This could cause long wavelength folding with some localized faulting (Costa & Vendeville 2002) and would not require significant intra-stratigraphic deformation. In the Orange Basin, in contrast, and in other basins dominated by interbedded heterolithic sediments with thin detachments, the mechanically strong unit above the detachment will need to undergo considerable intra-stratigraphic deformation, such as folding and intra-layer thrusting, to allow it to transfer strain downdip (Dalton *et al.* 2015).

Our observations also show that the collapse is controlled not just by the detachment thickness, but also by variations in the margin stratigraphy. Existing stratigraphic studies of the Orange Basin (Brown *et al.* 1995; Paton *et al.* 2007, 2008) show that there are two key stratigraphic variations in the evolution of the basin. During the Aptian (megasequence B in this study), the stratigraphy facies consists of a landwards-stepping clastic front. This results in the landward migration of the delta foreset to marine shale transition. Overlying the delta system is the main shelf margin sequence with interbedded organic-rich shale horizons. This results in a complex distribution of the décollement horizons and corresponding multiphase development, which is described in the following parts of Figure 12:

(a) Stratigraphic distribution of the stable passive margin.
(b) Extensional faulting initiates on the continental slope, detaching onto an advantageous shale horizon and subsequently leading to thrusting downdip on the abyssal plain.
(c) Continued gravitational imbalance on the margin leads to additional faults forming proximal and distal to the original collapse, which continues itself to deform.

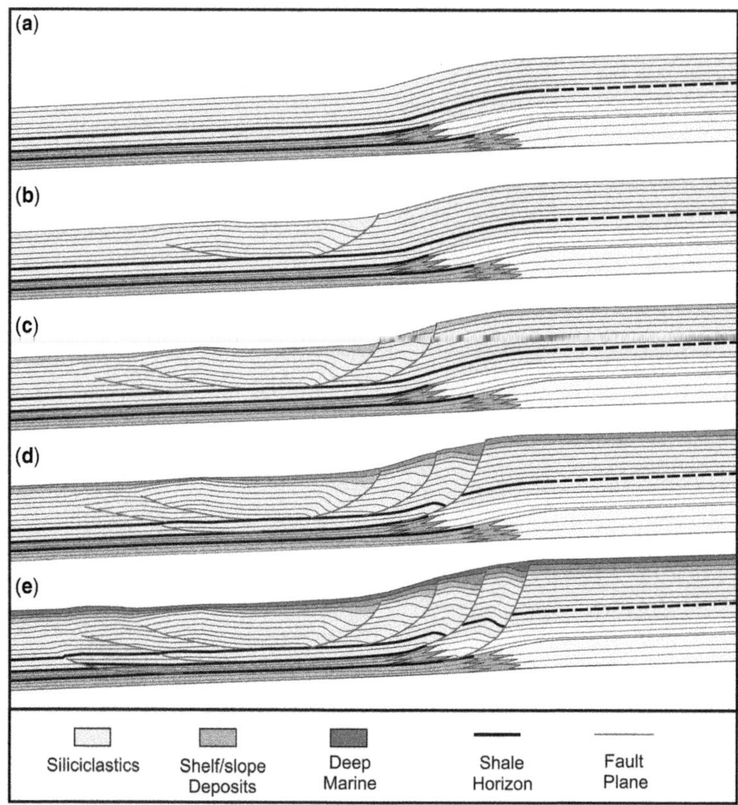

Fig. 12. Model explaining the role of deposition on the location and development of detachment horizons. (**a**) Section through a typical margin showing three stacked sequences, two with shale horizons at the base of slope and the upper shale horizon representing a maximum flooding surface. (**b**) Development of a simple single detachment system slipping along the maximum flooding surface. (**c**) System matures with the development of additional faults and eventually locks up. (**d**) In response to system locking up, alternative slip horizons along the lower base of slope shale are used instead. (**e**) Even lower detachment horizons are sought as the shale in (d) is restricted depositionally.

(d) The ability of the upper detachment to redistribute strain downdip becomes less efficient and so new extensional faults penetrate down to a lower shale horizon to compact lower, relatively under-compacted, sediments.

(e) This process continues to exploit lower shale horizons to redistribute strain; the original systems may also continue to deform, although lower systems may alter the structural development of the overlying systems. The propagation of the faults to the lower packages is, in part, controlled by the stratigraphy of the margin and the location of the delta front.

Conclusions

Using very well imaged examples of gravity collapse structures from the Namibian and South African Atlantic passive margin, we have illustrated the significant variation in the structures present in these tripartite systems. This variation includes the typical updip extensional faults and downdip thrust faults, but also multi-detachment faulting and folding, stacked detachments, cross-cutting and the complex progressive evolution of the system.

As this system is dominated by a series of relatively thin detachments, we suggest that the role of stratigraphy, especially the distribution of maximum flooding organic-rich units, plays a fundamental role in both the style and spatial distribution of the deformation. We propose that such a model helps to explain the differences that occur in thick shale systems, salt systems and thinly bedded heterolithic systems.

Correction notice: The original version was incorrect. Figure 9 was repeated in place of Figure 7.

References

BILOTTI, F. & SHAW, J.H. 2005. Deep-water Niger Delta fold and thrust belt modelled as a critical-taper wedge: the influence of elevated basal fluid pressure on structural styles. *AAPG Bulletin*, **89**, 1475–1491.

BRIGGS, S.E., DAVIES, R.J., CARTWRIGHT, J.A. & MORGAN, R. 2006. Multiple detachment levels and their control on fold styles in the compressional domain of the deepwater west Niger Delta. *Basin Research*, **18**, 435–450.

BROWN, L.F., JR, BENSON, J.M. *ET AL.* 1995. *Sequence Stratigraphy in Offshore South African Divergent Basins. An Atlas on Exploration for Cretaceous Lowstand Traps by Soekor (Pty) Ltd.* AAPG Studies in Geology, **41**.

BUTLER, R.W.H. & PATON, D.A. 2010. Evaluating lateral compaction in deepwater fold and thrust belts: how much are we missing from 'nature's sandbox'? *GSA Today*, **20**, 4–10.

CORREDOR, F., SHAW, J.H. & BILOTTI, F. 2005. Structural styles in the deep-water fold and thrust belts of the Niger Delta. *AAPG Bulletin*, **89**, 735–780.

COSTA, E. & VENDEVILLE, B.C. 2002. Experimental insights on the geometry and kinematics of fold-and-thrust belts above weak, viscous evaporitic décollement. *Journal of Structural Geology*, **24**, 1729–1739.

DE VERA, J., GRANADO, P. & MCCLAY, K. 2010. Structural evolution of the Orange basin gravity-driven system, offshore Namibia. *Marine and Petroleum Geology*, **27**, 233–237.

DALTON, T.J.S., PATON, D.A., NEEDHAM, D.T. & HODGSON, N. 2015. Temporal and spatial evolution of deepwater fold thrust belts: implications for quantifying strain imbalance. *Interpretation*, **3**, SAA71–SAA87.

GERRARD, T. & SMITH, G.C. 1982. Post-Palaeozoic succession and structure of the south-western African continental margin. *In*: WATKINS, J.S. & DRAKE, C.L. (eds) *Studies in Continental Margin Geology*. AAPG Memoir, **34**, 49–74.

HIRSCH, K.K., SCHECK-WENDEROTH, M., VAN WEES, J.-D., KUHLMANN, G. & PATON, D.A. 2010. Tectonic subsidence history and thermal evolution of the Orange Basin. *Marine and Petroleum Geology*, **27**, 565–584.

JUNGSLAGER, E.H.A. 1999. Petroleum habitats of the Atlantic margin of South Africa. *In*: CAMERON, N.R., BATE, R.H. & CLURE, V.S. (eds) *The Oil and Gas Habitats of the South Atlantic*. Geological Society, London, Special Publications, **153**, 153–168, http://doi.org/10.1144/GSL.SP.1999.153.01.10

KOOPMANN, H., SCHRECKENBERGER, B., FRANKE, D., BECKER, K. & SCHNABEL, M. 2014. The late rifting phase and continental break-up of southern South Atlantic: the mode and timing of volcanic rifting and formation of earliest oceanic crust. *In*: WRIGHT, T.J., AYELE, A., FERGUSON, D.J., KIDANE, T. & VYE-BROWN, C. (eds) *Magmatic Rifting and Active Volcanism*. Geological Society, London, Special Publications, **420**. First published online December 18, 2014, http://doi.org/10.1144/SP420.2

KRUEGER, A.C.V.A. & GILBERT, E. 2006. Mechanics and kinematics of back thrusting in deep water fold-thrust belts – observations from the Niger Delta (abstract). *2006 AAPG Annual Meeting*, Houston, TX, USA, Abstracts Volume, **15**, 58.

KRUEGER, A. & GILBERT, E. 2009. Deepwater fold-thrust belts: not all the beasts are equal. *AAPG Datapages/Search and Discovery*, Article #30085. http://www.searchanddiscovery.com/documents/2009/30085krueger/index.html

LIGHT, M.P.R., MASLANYJ, M.P., GREENWOOD, R.J. & BANKS, N.L. 1993. Seismic sequence stratigraphy and tectonics offshore Namibia. *In*: WILLIAMS, G.D. & DOBB, A. (eds) *Tectonics and Seismic Sequence Stratigraphy*. Geological Society, London, Special Publications, **71**, 163–191, http://doi.org/10.1144/GSL.SP.1993.071.01.08

MCGILVERY, T.A. & COOK, D.L. 2004. Depositional elements of the slope/basin depositional system offshore Brunei. *Proceedings of the Indonesia Petroleum Association Meeting*, 7–8 December, Jakarta, Indonesia, 407–420.

MOHAMMED, M., PATON, D.A., COLLIER, R.E.L.I., HODGSON, N. & NEGONGA, M. In press. Interaction of crustal heterogeneity and lithospheric process in determining passive margin architecture on the Namibian margin. *In*: SABATO CERALDI, T., HODGKINSON, R.A. & BACKE, G. (eds) *Petroleum Geoscience of the West Africa Margin*. Geological Society, London, Special Publications, **438**, http://doi.org/10.1144/SP438.9

MORLEY, C.K. & GUERIN, G. 1996. Comparison of gravity-deformation styles and behaviour associated with mobiles shales and salt. *Tectonics*, **15**, 1154–1170.

MORLEY, C.K., KING, R., HILLIS, R., TINGAY, M. & BACKE, G. 2011. Deepwater fold and thrust belt classification, tectonics, structure and hydrocarbon prospectivity: a review. *Earth Science Reviews*, **104**, 41–91.

MUNTINGH, A. & BROWN, L.F.J. 1993. Sequence stratigraphy of petroleum plays, post-rift Cretaceous rocks (Lower Aptian to Upper Maastrichtian), Orange Basin, South Africa. *In*: MASTERSON, A.R. (ed.) *Siliciclastic Sequence Stratigraphy. Recent Developments and Applications*. AAPG Memoir, **58**, 71–98.

PATON, D.A., DI PRIMIO, R., KUHLMANN, G., VAN DER SPUY, D. & HORSFIELD, B. 2007. Insights into the petroleum system evolution of the southern Orange Basin, South Africa. *South African Journal of Geology*, **110**, 261–274.

PATON, D.A., VAN DER SPUY, D., DI PRIMIO, R. & HORSFIELD, B. 2008. Tectonically induced adjustment of passive-margin accommodation space; influence on hydrocarbon potential of the Orange Basin, South Africa. *AAPG Bulletin*, **92**, 589–609.

PATON, D.A., MORTIMER, E.J. & HODGSON, N. In press. The missing piece of the South Atlantic jigsaw: when continental break-up ignores crustal heterogeneity. *In*: SABATO CERALDI, T., HODGKINSON, R.A. & BACKE, G. (eds) *Petroleum Geoscience of the West Africa Margin*. Geological Society, London, Special Publications, **438**, http://doi.org/10.1144/SP438.8

PEEL, F.J. 2014. The engines of gravity-driven movement on passive margins: quantifying the relative contributions of spreading v. gravity sliding mechanisms. *Tectonophysics*, **633**, 126–142, http://doi.org/10.1016/j.tecto.2014.06.023

POSAMENTIER, H.W. & KOLLA, V. 2003. Seismic geomorphology and stratigraphy of depositional elements in deep-water settings. *Journal of Sedimentary Research*, **73**, 367–388.

ROWAN, M.G., PEEL, F.J., VENDEVILLE, B.C. 2004. Gravity-driven fold belts on passive margins. *In*: MCCLAY, K.R. (ed.) *Thrust Tectonics and Hydrocarbon Systems*. AAPG Memoir, **82**, 157–182.

TOTTERDELL, J.M. & KRASSAY, A.A. 2003. The role of shale deformation and growth faulting in the Late Cretaceous evolution of the Bight Basin, offshore southern Australia. *In*: VAN RENSBERGEN, P., HILLIS, R.R., MALTMAN, A.J. & MORLEY, C.K. (eds) *Subsurface Sediment Mobilization*. Geological Society, London, Special Publications, **216**, 429–442, http://doi.org/10.1144/GSL.SP.2003.216.01.28

VAN DER SPUY, D., ZIEGLER, T. & BOWYER, M. 2003. Deepwater 2D data reveal Orange Basin objectives off western South Africa. *Oil & Gas Journal*, **101**, 44–49.

Index

Page numbers in *italics* refer to Figures. Page numbers in **bold** refer to Tables